Online Assessment and Measurement:
Foundations and Challenges

Mary Hricko
Kent State University, USA

Scott L. Howell
Brigham Young University, USA

 Information Science Publishing

Hers̄· • Melbourne • Singapore

Acquisitions Editor: Renée Davies
Development Editor: Kristin Roth
Senior Managing Editor: Amanda Appicello
Managing Editor: Jennifer Neidig
Copy Editor: Becky Shore
Typesetter: Cindy Consonery
Cover Design: Lisa Tosheff
Printed at: Yurchak Printing Inc.

Published in the United States of America by
     Information Science Publishing (an imprint of Idea Group Inc.)
     701 E. Chocolate Avenue, Suite 200
     Hershey PA 17033
     Tel: 717-533-8845
     Fax: 717-533-8661
     E-mail: cust@idea-group.com
     Web site: http://www.idea-group.com

and in the United Kingdom by
     Information Science Publishing (an imprint of Idea Group Inc.)
     3 Henrietta Street
     Covent Garden
     London WC2E 8LU
     Tel: 44 20 7240 0856
     Fax: 44 20 7379 3313
     Web site: http://www.eurospan.co.uk

        Library of Congress Cataloging-in-Publication Data

Online assessments and measurement : foundations and challenges / Mary Hricko and Scott L. Howell, editors.
        p. cm.
    Summary: "This book explores the development of online assessment and the way practitioners of online learning can modify their methodologies in the design, development, and delivery of their instruction to best accommodate their participants"--Provided by publisher.
    Includes bibliographical references and index.
    ISBN 1-59140-497-5 (hardcover) -- ISBN 1-59140-498-3 (soft cover) -- ISBN 1-59140-499-1 (ebook)
    1. Computer-assisted instruction--Evaluation. 2. Educational tests and measurements--Data processing. 3. Distance education--Evaluation. I. Hricko, Mary. II. Howell, Scott L.
    LB1028.3.O553 2005
    371.26'0285--dc22
                        2005004528

British Cataloguing in Publication Data
A Cataloguing in Publication record for this book is available from the British Library.

All work contributed to this book is new, previously-unpublished material. Each chapter is assigned to at least 2-3 expert reviewers and is subject to a blind, peer review by these reviewers. The views expressed in this book are those of the authors, but not necessarily of the publisher.

Online Assessment and Measurement:
Foundations and Challenges

Table of Contents

Preface vi
Mary Hricko, Kent State University, USA

Section I. Understanding Online Assessment and Measurement

Chapter I. Assessment is as Assessment Does:
A Conceptual Framework for Understanding Online Assessment
and Measurement 1
Jeanette M. Bartley, University of Technology, Jamaica

Chapter II. Defining Online Assessment for the Adult Learning
Market 46
Betty Bergstrom, Promissor, USA
Jim Fryer, Promissor, USA
Joel Norris, Promissor, USA

Chapter III. Definitions, Uses, and Benefits of Standards 67
Eric Shepherd, Questionmark Corporation, USA

Section II. Best Practices in Designing Online Assessment

Chapter IV. Ten Key Qualities of Assessment Online 86
Chris Morgan, Southern Cross University, Australia
Meg O'Reilly, Southern Cross University, Australia

Chapter V. Factors to Consider in the Design of Inclusive Online Assessments 102

    Sandra J. Thompson, University of Minnesota, USA
    Rachel F. Quenemoen, University of Minnesota, USA
    Martha L. Thurlow, University of Minnesota, USA

Chapter VI. Best Practices in the Assessment of Online Discussions 118

    Katrina A. Meyer, University of Memphis, USA

Section III. Challenges in Online Assessment and Measurement

Chapter VII. Challenges in the Design, Development, and Delivery of Online Assessment and Evaluation 132

    Clark J. Hickman, University of Missouri–St. Louis, USA
    Cheryl Bielema, University of Missouri–St. Louis, USA
    Margaret Gunderson, University of Missouri–Columbia, USA

Chapter VIII. Creating a Unified System of Assessment 165

    Richard Schuttler, University of Phoenix, USA
    Jake Burdick, University of Phoenix, USA

Chapter IX. Legal Implications of Online Assessment: Issues for Educators 182

    Bryan D. Bradley, Brigham Young University, USA

Chapter X. Legal Implications of Online Assessment: Issues for Test and Assessment Owners 200

    Robert R. Hunt, Caveon Test Security, USA

Chapter XI. Accessibility of Computer-Based Testing for Individuals with Disabilities and English Language Learners within a Validity Framework 214

    Eric G. Hansen, Educational Testing Services (ETS), Princeton, USA
    Robert J. Mislevy, University of Maryland, College Park, USA

Section IV. Security, Authentication, and Support

Chapter XII. Delivering Computerized Assessments Safely and Securely 263

    Eric Shepherd, Questionmark Corporation, USA
    John Kleeman, Questionmark Corporation, USA
    Joan Phaup, Questionmark Corporation, USA

Chapter XIII. Securing and Proctoring Online Assessments 280

    Jamie R. Mulkey, Caveon Test Security, USA
    John Fremer, Caveon Test Security, USA

Chapter XIV. Securing and Proctoring Online Tests 300

    Bernadette Howlett, Idaho State University, USA
    Beverly Hewett, Idaho State University, USA

Chapter XV. Supporting and Facilitating Academic Integrity in Distance Education through Student Services 330

    Brian F. Fox, Santa Fe Community College, USA

Chapter XVI. User Authentication and Academic Integrity in Online Assessment 341

    Boris Vilic, Duquesne University, USA
    Marie A. Cini, City University, USA

About the Authors 360

Index 369

Preface

Introduction

Assessment not only measures learning, but it also contributes to learning. To improve the quality of instruction, assessment is often used as a building block. However, applications of traditional-based assessment do not always meet the needs of the online environment. In turn, most research on assessment in distance education recommends a multilevel approach for evaluating instruction. Many educators do not know how to develop such assessment strategies or, if they can, indeed accept the validity of such measurement. Issues associated with developing valid and reliable assessment tools, challenges regarding the accessibility and legality of such tests, and matters of securing and proctoring exams all arise when one discusses topics on online assessment and measurement. For this reason, it is valuable to begin this series with a book that not only addresses the foundations of developing effective online assessment and measurement, but also identifies the potential challenges that educators may face in using assessment tools.

Existing Research and Information

Although there is a great deal of journal literature on online assessment and measurement, most books on the topic are somewhat limited in scope and offer only a general overview of the topic. Most of the information on assessment and measurement of student learning outcomes is focused on traditional-based educational practices. In some instances, there are attempts to create parallels between the traditional-based and online environments, but the applications are often generalized and do not adequately meet the needs of both students and educators. At present, in terms of book literature, most information about subjects pertinent to online assessment and measurement is found in chapters on computer based testing (CBT). Although this information provides important resources in terms of design and delivery of such testing, it does not often address other matters of concern that educators believe are equally important.

Of the titles currently available, there are only a few that provide a comprehensive overview of the subject. Two of the better resources on this topic include *Assessment Strategies for the Online Class: From Theory to Practice* (Anderson, Bauer, & Speck, 2002) and *Computer-Assisted Assessment in Higher Education* (Brown, Race, & Bull, 1999). Anderson's text focuses on the types of assessment strategies that educators can employ in their online courses and offers suggestions on how to use these assessments. Whereas the individual chapters of the book may mention some of the challenges in using these specific assessment strategies, they do not provide solutions for such issues.

Brown's text provides more of an overview of the issues associated with online assessment and measurement but only in the context of computer-assisted assessment (CAA). Other texts, such as *Developing and Implementing Assessment of Student Learning Outcomes* (Serban & Friedlander, 2004) examines assessment in both traditional and online formats. Although this book is useful for gaining a general understanding of the issues often associated with assessment practices, its discussion on the online format and the examples used are not very detailed. Finally, other books are simply too dated, such as *Technology Assessment in Education and Training* (Baker & O'Neil, 1994). Although this book offers a general overview of the subject, more up-to-date information is needed. In fact, many of the books available for review offer dated information about assessment tools educators no longer use.

Topics of Discussion

Online Assessment and Measurement: Foundations and Challenges is divided into four sections that focus on and outline important topics in online assessment and measurement. The objective of this book is to provide readers with a clear understanding of the foundations and challenges of designing, developing, and delivering online assessment and measurement. The book also provides readers with explanations and solutions for issues associated with the accessibility, security, and legal ramifications of online assessment.

Section I focuses discussion on the complexity of defining online assessment. Each of the authors offers perspectives of interpreting assessment and defining standards for effective practices in developing models. This section begins with Jeannette M. Bartley's chapter, a literature review of the most current research articles written about online assessment and measurement. Through her analysis, Bartley not only provides a thorough introduction to the key concepts and issues of online assessment and measurement, but is also able to develop a conceptual framework for understanding the dominant philosophies, trends, and perspectives of online assessment and measurement.

In Chapter II, Betty Bergstrom, Jim Fryer, and Joel Norris discuss the evolution of online assessment in adult education and why these practices require clear definitions. This chapter defines the types of assessments most commonly used in adult testing and explains the strengths and weaknesses of these types of assessments and their applications. This chapter gives a good overview of the process of establishing testing formats.

Eric Shepherd of the Questionmark Corporation presents a detailed overview of the response from industry in Chapter III. In his discussion, Shepherd describes the process by which standards are formulated and why standards must be developed in terms of designing reliable tools for assessment. Shepherd also provides an overview of the primary developers and leading standards. Understanding how assessment is defined by its industry proves informative and useful to educators interested in the rationale behind development.

Section II focuses on the best practices in designing online assessment. This section addresses strategies to ensure consistency and accessibility in assessment; it also examines specific tools that are used. In Chapter IV, Chris Morgan and Meg O'Reilly outline the 10 key qualities of assessment online, and offer a detailed overview of the best methods for educators to employ when designing, developing, and delivering online assessment. Morgan and O'Reilly base their discussion on an examination of the research of proven models and practices. Their research articulates the most important values and practices educators should consider when developing online assessments. Their analysis also identifies ways in which educators can modify traditional testing practices for an online context.

Chapter V, by Sandra J. Thompson, Rachel F. Quenemoen, and Martha L. Thurlow, is an important discussion on the issues educators and administrators should consider when developing accessible online assessments. The authors begin by describing the various challenges in developing universally designed assessments and technology-based accommodations for learners with special needs. The chapter explains how universal design principles can be applied to the features of online assessment and alerts educators to the complexities that special learners experience when completing an online assessment. The authors conclude by offering suggestions to improve existing online assessment tools and the accommodations for such tests.

The last chapter is this section, Chapter VI, focuses specifically on the best strategies for using online discussions as part of a course assessment. Katrina A. Meyer provides a thorough literature review of all the current research related to best practices, assesses online discussion, and develops a model for educators to consider in the application of their online courses. Meyer distinguishes the differences between face-to-face interactions and online discussions and then demonstrates how educators can use the data from online interactions to improve the overall quality of teaching and learning in the course.

Section III details the challenges in online assessment and measurement that administrators and educators may face when developing and using online assessments in the classroom. In Chapter VII, Clark J. Hickman, Cheryl Bielema, and Margaret Gunderson discuss the importance of developing a comprehensive system for designing, developing, and delivering online assessment or evaluation. The authors first clarify the parameters of student and program evaluation, noting the process that educators should follow. The authors offer a systematic model for selecting the right assessment tool for the right environment. The chapter identifies the specific challenges that occur during the process of designing an effective assessment strategy.

Richard Schuttler and Jake Burdick reinforce the idea of creating a systematic assessment strategy in Chapter VIII. They begin by discussing the modalities of online education and explain why a different approach for assessing learning in an online environment is needed. Using the University of Phoenix as their model, Schuttler and Burdick demonstrate how a unified system of assessment can be applied across disciplines by outlining a step-by-step process. The authors use the business program to illustrate their model and to demonstrate how unification begins by designing a comprehensive structure and consistent curriculum. The authors explain how this model can be applied and integrated.

Chapters IX and X both focus on the legal implications of online assessment. In Chapter IX, Bryan D. Bradley offers a straightforward discussion of the legal issues that educators may face when using online assessments. He describes specific legal issues that arise in online assessment and outlines the core criteria for creating legally defensible exams. Bradley also explains how legal issues can influence the reliability and validity of testing. This chapter also provides important information for establishing secure testing and for preventing unfair bias in testing.

In Chapter X, Robert R. Hunt's analysis focuses on the legal issues that affect test developers. His chapter begins by examining how test developers can ensure the copyright of tests and assessment. This chapter also examines whether online assessment can satisfy the legal thresholds for security necessary to secure test copyright and trade secrets. The chapter provides examples in the professional certification context and notes why legal practices in assessment are important. Hunt also explains in detail the legal rights test developers have in the distribution of their assessments. Hunt concludes with an analysis of how various trade-secret laws have been applied to online assessment.

Chapter XI is perhaps one of the most thorough discussions of the issues associated with providing accommodations and ensuring validity for test takers with special needs. Eric G. Hansen and Robert J. Mislevy begin by giving a detailed overview of developing accessible designs for assessments but then raise the question of how accessibility features may undermine the validity of specific tests. This chapter explores in detail an evidence-centered assessment design (ECD) model as a conceptual framework that can assist in defining the requirements of accessible computer-based tests. This chapter explains how specific features, built in to provide accessibility, can indeed challenge the validity and reliability of the test outcome. The authors conclude by providing research and suggestions to improve accessible designs to meet the challenges of ensuring validity.

Section IV is designed to provide administrators and educators with strategies to secure online assessments. These chapters offer examples of issues that arise in an online context. In Chapter XII, Eric Shepherd, John Kleeman, and Joan Phaup provide a clear method assessing the security for delivering computer-based assessments. The chapter identifies the most common forms of assessment and the security issues that can arise for each specific format. The authors then identify the consequences of not addressing the specific security issues associated with each format and offer suggestions that educators or administrators can use to reduce these security issues. This chapter offers a very illustrative view of how to respond to potential security issues in the online context and proves to be a very comprehensive resource of information.

From the corporate perspective, Jamie Mulkey and John Fremer of Caveon Test Security discuss securing and proctoring online assessments from a corporate view in Chapter XIII. The authors describe how the rise of cheating has affected online assessments. They review the guidelines specific to computer-based testing and note how test developers need to examine the test itself, the authentication of the test taker, and finally, the process that ensures the security of the testing. The authors identify a taxonomy of cheater-types and offer solutions to address security problems associated with each cheating style. The authors conclude by discussing current methodologies being studied to improve test security for high stakes testing.

More specifically, Bernadette Howlett and Beverly Hewett provide an analysis of the literature on securing and proctoring online tests. From their review in Chapter XIV, Howlett and Hewett identify the key components educators must implement to ensure the security of their online assessments. The authors offer a historical overview of cheating and how traditional-based methods can be applied to the online context. The authors offer advice related to improving the instructional design of online assessments as well as technological solutions that educators can employ to ensure greater security. This chapter offers practical advice for educators who want to improve the ways in which their online tests are secured and proctored.

To address issues of academic misconduct, Brian F. Fox explains the role of a university's student services in improving the security of online assessment. In Chapter XV, Fox explains ways in which a student services department could be employed to assist faculty and administrators in ensuring a code of conduct for online assessment and plagiarism. Fox lists various strategies that can be used in conjunction with the course to ensure that students understand the importance of academic integrity in the online learning environment. This chapter is useful to educators seeking ways to reinforce academic codes of conduct in the classroom.

Finally in the last chapter, Boris Vilic and Marie A. Cini discuss the role of user authentication in online assessment and describe effective practices such as systematic course design, the use of portfolios, and the development of academic integrity policies. Vilic and Cini examine the broader context of how technological solutions can be used to improve user authentication. The chapter offers an exploration of the resources readily available and those that are in the process of development. An overview of these tools and how they can be used is given. The authors also explain the legality of using such tools within the context of an academic environment. This chapter concludes with a discussion of both the ethical and legal considerations in ensuring authentication.

Conclusion

There is little doubt that as online programming for education and workforce development training continues to expand, the challenges of online assessment and measurement will continue to be a key topic of discussion. For this reason, it is our hope that this book will give readers a comprehensive overview of the topic and serve as a foundation of material for future exploration. This book defines the topic from several perspectives, outlines the best practices based upon the most current research literature, and identifies the most important challenges that educators, administrators, and test developers face in designing, developing, and delivering online assessment. As the first text in our series of books on this topic, *Online Assessment and Measurement: Foundations and Challenges* offers suggestions and advice from many of the experts in the field and serves as a catalyst for future research of the topic.

References

Anderson, R., Bauer, J., & Speck, B. (2002). *Assessment strategies for the online class: From theory to practice.* San Francisco: Jossey-Bass.

Baker, E., & O'Neil, H. (1994). *Technology assessment in education and training.* Hillsdale, NJ: Erlbaum.

Brown, S., Race, P., & Bull, J. (1999). *Computer-assisted assessment in higher education.* London: Kogan Page.

Serban, A. M., & Friedlander, J. (2004). *Developing and implementing assessment of student learning outcomes.* San Francisco: Jossey-Bass.

Acknowledgments

It may be hard to believe, but the genesis of this project grew out of a series of questions. In April 2003, Dr. Mehdi Khosrow-Pour, Editor-in-Chief at Idea Group Publishing sent me an email asking if I would be interested in working on another project with the group. Since I was already involved in a research project on online assessment, I told him that it would have to wait until next year. Following this exchange, I resumed my work and soon realized that the type of information I needed on online assessment and measurement was not readily available, so I posted a query on a few listservs, asking for reading recommendations on the subject.

Most of the respondents had few titles to offer, and many writers wanted *me* to give them copies of the information I had compiled. One of the respondents who did offer several suggestions was Scott Howell. Scott provided me with several titles, but admitted that he was also looking for a good publication on the subject. As Scott and I discussed the limitations in research literature, somewhere along the way we talked about putting together our own book on online assessment and measurement. Recalling the earlier request from Idea Group Publishing, I mentioned to Scott that we might have a publisher. And so the project began.

When Scott posted the CFP for chapters, we received numerous responses, and in turn asked Idea Group Publishing if they would be willing to pursue three books on the topic of online assessment and measurement as a series. We had received so many outstanding submissions that it made sense to us to pursue this format. When Dr. Khosrow-Pour agreed, we asked David Williams to become a member of the team. Through Idea Group Publishing's work and support, *Online Assessment and Measurement: Foundations and Challenges* became the first volume of the series. The next two books in the series have already been submitted for publication.

We would like to thank Dr. Mehdi Khosrow-Pour for his belief that our prospectus was worthy of publication and for his willingness to pursue the expansion of our initial project. His enthusiasm for the series motivated us throughout the process. I would also like to thank Jan Travers and Michele Rossi, our development editors, who tolerated our incessant e-mails of queries and comments regarding the status of the project. They kept the project on schedule, and we appreciate their patient reminders of all our deadlines. They were also very supportive and helpful throughout the process.

Special thanks goes out to all the staff and publishing team at Idea Group Publishing, whose assistance and efforts throughout the process of this project made us realize the value of collaboration and teamwork. The ongoing professional support from the editorial staff at Idea Group Publishing made this publication possible.

In addition to thanking the staff at Idea Group Publishing for all of their support and assistance, I personally want to thank Scott for all his work in the organization and recruitment of authors to participate in this project. As far as I am concerned, Scott kept the project moving. He is a great project leader and I enjoyed working with him. I also would like to thank David for his willingness to take on the third book in the series.

We both are so impressed with the quality of work our authors have presented and wish to express our gratitude for their excellent contributions to this book. We have learned so much from their insights and research and appreciate their commitment to participate in this project. It has been a privilege for us to work with each of them. Many of the authors that have written chapters are experts in the field of online assessment and measurement. This book also brings together many of the primary researchers of this topic. We are very pleased with the development of this first volume.

We also wish to acknowledge the work and assistance of the reviewers of the chapters. We appreciate their time and participation in the process of selection and revision. We appreciate their suggestions and advice in terms of making final decisions regarding the chapters and other materials needed to complete this project. Having additional readers for these chapters was very important to ensure the accuracy of information and verification of citations.

Scott and I would also like to thank our colleagues at Brigham Young University and Kent State University – Geauga. The support, assistance, and encouragement of our co-workers helped us complete this project. My library staff and Scott's graduate assistants helped us with a variety of tasks, all necessary to complete this project. We appreciate all the time and efforts of these individuals who make our work so much easier.

Finally, we wish to acknowledge the love and support of our families and friends. We greatly appreciate their patient understanding and tolerance during the process of completing this project.

Mary Hricko, Editor

Section I

Understanding Online Assessment and Measurement

Chapter I

Assessment is as Assessment Does:
A Conceptual Framework for Understanding Online Assessment and Measurement

Jeanette M. Bartley, University of Technology, Jamaica

Abstract

With the increasing adoption of advanced education technologies, such as Internet-based communications, there are greater demands for more effective, flexible, interactive, customized and just-in-time online instructional and assessment systems. As institutions rush to digitize, virtualize and globalize their campuses, there remains the significant issue of the ability to measure outcomes as a means of determining the credibility of technology-mediated learning experiences. This chapter addresses selected concerns including the distinction between online assessment and traditional assessment; the relevance of authentic and alternative assessment; the practical considerations relating to security, integrity, validity, and reliability of assessment; emerging principles of good practice relating to quality indicators, benchmarks and criteria for effective online assessment; and the relative significance of online assessment trens to various drivers and stakeholders.

Introduction

With the increasing adoption of advanced education technologies beyond print, audio, and television media to more sophisticated forms of interactive electronic communication networks over the last two decades, the traditional orientation to teaching and learning is intensely challenged (Berge, 2000; Harisim, 1990; Rogers, 2000). This phenomenon has created an increasing demand for more effective, flexible, interactive, customized, and just-in-time instructional systems to keep pace with knowledge explosion and socioeconomic trends of a service-oriented global marketplace (Palloff & Pratt, 2001; Phipps & Merisotis, 2000; Whitis, 2001). Several authors suggest that the most extensive developments during the 1990s may have taken place in the area of teaching on the Internet, with electronic access to course materials (Berge, 2000; Drummond, 2003; Mason, 1998; Moore, Winograd, & Lange, 2001).

Mason (1998) observed that communication, whether synchronous or asynchronous, and whether one-to-one or many-to-many, has become easier, with the availability of Web-based conferencing systems providing Windows interface to messaging. The ability to provide asynchronous, interactive learning activities has also become the signature characteristic that sets Internet-based distance education apart from most of the other technologies (Phipps & Merisotis, 2000; Liang & Kim, 2004; Palloff & Pratt, 2001; Whitis, 2001).

There are increasing references in the literature to the digital generation and information age, reflecting how today's students perceive their learning environments as boundless, despite traditional expectations (Barone, 2003; Palloff & Pratt, 1999; Tapscott, 1998). "They tend to use physical space differently than prior generations, and they blur the boundaries between physical and cyber space" (Barone, 2003, p. 42). These behaviors have immediate implications for campus architectural design and for all aspects of student services, including policies relating to the definition of a course and how ownership of learning is determined and assessed. In this digital age, students' access to information and their expectations of when, where, how, and how fast they learn, are motivating faculty members to change their methods of instruction, interaction responsiveness, and approaches to assessment (Barone, 2003; Bober, 1998; Drummond, 2003; Palloff & Pratt, 1999).

Instructors in the online learning environment have been utilizing Internet-based technology to deliver courses in several ways, such as by materials on interactive or static Web sites, electronic textbooks, electronic whiteboards for graphical illustrations, bulletin board or discussion forums for posting lecture materials and stimulating asynchronous or synchronous conversations, and electronic mail for individual and group messages (Bober, 1998; Conrad & Donaldson, 2004; Palloff & Pratt, 2001; Rogers, 2000). A continuum of courses is emerging, from complete to partial delivery online, with varying degrees of interactivity and increasing use of course-authoring software or courseware packages, including gradebooks to assess learner progress through the creation of tests, quizzes, and polls (Palloff & Pratt, 2001).

Boettcher and Conrad (1999) identified three types of online courses: (a) Web courses, with material placed on a Web site, but little or no interaction between students; (b) Web-enhanced courses, with both face-to-face meetings and Web-delivery; and (c) Web-centric courses, which are interactive and delivered exclusively on the Web. One implication is that all online courses are not offered exclusively at a distance, in terms of the physical separation of learner and instructor in space and time for a significant portion of the course (Palloff & Pratt, 2001).

As these online instructional systems evolve in the digital age, through various integrated (hybrid) and nonintegrated (exclusive) technology-mediated formats, there is a growing interest among educators, researchers, and business persons in the development of related educational assessment processes involving the use of technology (Bauer & Anderson, 2000; Boettcher & Conrad, 1999; Hartley & Collins-Brown, 1999; Morley, 2000). The issue of assessment is central to the wider debate of the acceptance of distance education programs by accrediting bodies and higher education institutions (Morley, 2000; Serwatka, 2002; Sun, 2002). The ability to measure outcomes is therefore a major factor in facilitating the credibility of online courses, even as institutions rush to digitize, virtualize, and globalize their campuses (Drummond, 2003; Mason, 1998). The practical and ethical challenges relating to the validity of results and legitimacy of educations programs are forcing academicians to create unique testing alternatives (Morley, 2000; Sun, 2002). One could conclude that online learning and online assessment are impacting the pedagogical evolution in higher education (Mason, 1998; Palloff & Pratt, 2001).

In discussing research on technology and higher learning, Ehrmann (n.d.) found that changing a course involves shifts to unfamiliar materials, creation of new types of assignments, and inventing new ways to assess student learning. These course changes also foster shifts in roles and relationships, and communication styles. As the mechanism of learning paradigms is changed, so should the assessment delivery method, because the established techniques do not work as well for the different formats (Drummond, 2003; Liang & Kim, 2004; Palloff & Pratt, 2001; Serwatka, 2002; Sun, 2002).

There is also an increasing drive among schools and corporations towards measuring performance, improving results, and controlling costs (Lazerson, Wagener, & Shumanis, 2000; Weisburgh, 2003). Measurement has become increasingly important in K-12 and higher education, as schools have to comply with No Child Left Behind requirements, while administrators, parents and real estate brokers are keen for the latest test results (Weisburgh, 2003). The availability of assessment software to address these tasks is leading to assessment services becoming one of the fastest growing software niches, both in the corporate and in the educational markets (Mason, 1998; Weisburgh, 2003).

The educational research literature is replete with accounts of the virtues of technology-mediated learning such as Internet-based and computer-assisted applications in expanding access, convenience, and cost effectiveness to diverse populations, anytime and any place (Baker 1999; Bourne, McMaster, Rieger, & Campbell, 1997; Bull, 1999; Phipps & Merisotis, 2000). Concepts including the notion of the classroom, time on task, role and responsibility of the teacher and learners are also being redefined, to reflect distinctions between virtual and real-time, asynchronous and synchronous interactions (Mason, 1998; Morley, 2000; Palloff & Pratt, 1999). The question that arises is no longer whether but how these concepts are specifically applicable in monitoring, informing and assessing academic performance and progress in the online learning environment.

The topic of online assessment and measurement is therefore timely and important in terms of discussing continuous improvement in quality of learning, as various institutions grapple with how best to implement the assessment processes for online learning (Drummond, 2003; Mason, 1998; Sun, 2002). There is even deeper significance in the decisions regarding the development of comprehensive online assessment processes that articulate student learning objectives and program outcomes; facilitate student reflection, preparation, and achievement; and guide improvement and accountability in teaching–learning processes and overall program delivery (Bober, 1998; Drummond, 2003; Graham, Cagiltay, Lim, Craner, & Duffy, 2001; Juwah, 2003; Mason, 1998).

One underlying assumption of this topic is that "assessment is as assessment does," implying that the term is defined by its function. As aptly expressed by Boud (cited by Juwah, 2003): "Assessment must perform double duty; not only does it assess content, it must also prepare learners for future learning" (introduction). Juwah also commented that, for assessment to be effective, it should perform several functions and cites the following:

Act as a motivator for learning (Boud et al., 1999; Cowan, 1998); promote deep learning—in which the learner engages with (a) the learning materials and resources, (b) other learners and (c) tutor/facilitator (Marton & Saljo, 1984); contribute to the development of skills (Boud et al.,1999; 2002; Gibbs,1992; Ramsden,1992); be cost effective and sustainable (Boud, 2002). (Juwah, 2003, ¶2)

Approach of This Chapter

This chapter proposes that the evolution of a conceptual framework for online assessment and measurement is largely determined by prevailing notions about online teaching–learning processes, and the extent to which the quality of learning can be measured within context of this technological medium. The emerging concepts integral to the topic were traced through a brief, comprehensive review of current educational research literature, to identify the dominant issues, philosophies and perspectives, emerging best practices, trends, and the current state of the debate.

Selected concerns and issues include the distinction between online assessment and traditional assessment, the relevance of authentic and alternative assessment, the practical considerations relating to security, integrity, validity, and reliability of assessment; emerging principles of good practice relating to quality indicators, benchmarks, and criteria for effective online assessment; and the relative significance of online assessment trends to various drivers and stakeholders.

The literature review was conducted by purposeful sampling of research articles drawn from leading educational journals, professional educational organizations, agencies, and other credible online sources, including major distance learning bibliographic databases. The articles were selected primarily through an Internet search by the use of keywords or phrases to identify best practices and diverse perspectives related to various aspects of online assessment, and to shortlist research articles that focused on trends or conceptual issues of assessment and online learning. Several terms in the literature, including online learning, e-learning, virtual learning, networked learning, Web-based and computer-mediated learning, and technology-assisted distance learning were used interchangeably to refer to learning that involves interaction among learners and between learners and instructors, using Internet communication technology (Baker, 1999; Goodyear, 2002; Graham, Scarborough, & Goodwin, 1999; MacDonald, 2002; Twigg, 2001).

Online Assessment of Learning and Assessment of Online Learning

The terms *online assessment of learning* and *assessment of online learning* are used interchangeably in the literature, given that online learning includes both Web-centric and Web-enhanced courses. There is a technical difference between the terms, depending on the relative emphasis on the assessment tools or the learning interactions. The phrase "online assessment of learning" denotes the use of scientifically based electronic tools (technology mediated) to test or measure learning outcomes, in both face-to-face and distance-learning environments. The emphasis here is on the online characteristic of the assessment tools, such as computer software, conferencing systems, or Internet-based applications, though the learning process can be either conventional or online (Flint, 2003; Graham et al., 1999; Mason, 1998; Weisburgh, 2003). On the other hand, the phrase "assessment of online learning" denotes the use of scientifically based tools to test or measure learning, which occurs exclusively within the online environment. The emphasis here is on the online context of learning through interactions across and between learners and instructors, which are mediated by Internet-based technology, including e-mail, computer software, or conferencing systems and Web-based applications (Bober, 1998; Flint, 2003; Goodyear, 2002; Mason, 1998). The assessment tools can be either conventional (manual) or online (electronic), though the latter usually prevails for purposes of effective design.

This distinction of terms has impacted the nature of the literature search, given the relative emphases. For the most part, this chapter focuses on online assessment in the online learning environment, with respect to the use of electronic tools to assess Internet-based learning interactions, where instructors and learners are separated by distance. However, there are cursory references to manual assessments and face-to-face interactions, where applicable.

Background

Several authors have commented that although a great deal has been written on the advantages of online teaching, there have been few quantitative evaluations, with little known on how assessment is implemented in online classrooms and how to use computer-mediated tools to monitor and inform performance and progress (Bourne et al., 1997; Bull, 1999; Liang & Kim, 2004). It is possible that the dimension of assessing online learning has been overlooked, given the wider debate about the legitimacy of teaching–learning mediated by technology and the degree of quality learning in Internet-based education (Bourne et al., 1997; Morley, 2000). Moreover, the focus of many studies of assessment has been on the appropriateness, effectiveness, or acceptability of the program itself, rather than on assessing learner achievement (Drummond, 2003).

Most of the research on distance learning in general, and online learning in particular, is anecdotal; flawed or lacking in adequate theoretical base, empirical basis, or conceptual framework; and refer to few instructional, assessment, or evaluative models (Bober, 1998; Drummond, 2003; Merisotis & Phipps, 1999). Research on technology-mediated learning and predictions about its future may be even more difficult because of the rapid change in technologies and academia's ingrained aversion to risk (Whitis, 2001). Thus, the challenge is in trying to find out what online assessment is and how it works, at the same time that one is exploring the new context, criteria and psycho-cultural expectations that will fundamentally alter the very concepts being studied.

In exploring the emerging conceptual framework, it was also important to review the use of key terms and phrases as are used in the educational literature on online assessment. Given the context of controversy and debate, there are no single acceptable definitions or right ways of doing assessment.

Concept of Assessment

To most students and teachers, the term *assessment* is traditionally associated with the concepts of tests, grades, reports, and standards. Over the last 2 decades, the term continues to be the subject of intense debate among educational institutions and corporate and government agencies, with various schools of thought emerging in the effort to define the nature and function of assessment in education (Angelo, 1999; Angelo & Cross, 1993; Bull, 1999a; National Research Council, 2001). The body of literature reveals that there is an assessment movement in education, which has been evolving through cycles of reform and expansion (Herman, Aschbacher, & Winters, 1992; Kulieke et al., 1990; Lazerson et al., 2000; National Research Council, 2001). There have been marked changes from the culture of testing towards a culture of multiple assessments; from the focus on the single behavioral or cognitive attributes to include multiple dimensions of intelligence—abilities and skills; from simple measures to incorporate more complex measures on a continuous basis; from exclusive individual assessment to group-process assessment; from paper–pencil to authentic assessments; from the themes of standardization, accountability, and certification to the integration of assessment into instructional practice (Herman et al., 1992; Kulieke et al., 1990; Twigg, 2001).

The following definition was offered at the American Association for Higher Education (AAHE) Assessment Forum, in 1995:

> *Assessment is an ongoing process aimed at understanding and improving student learning. It involves making our expectations explicit and public; setting appropriate criteria and high standards for learning quality; systematically gathering, analyzing and interpreting evidence to determine how well performance matches those expectations and standards; and using the resulting information to document, explain, and improve performance. When it is embedded effectively within larger institutional systems, assessment can help us focus our collective attention, examine our assumptions, and create a shared academic culture dedicated to assuring and improving the quality of higher education.* (T. A. Angelo, as cited in AAHE, 2003a, p.7)

Assessment has also been defined broadly, to include all activities that teachers and students undertake to get information that can be used diagnostically to alter teaching and learning (Black & William, as cited by Liang & Kim, 2004). Central to these definitions is the notion of systematic process of gathering and interpreting information, to provide feedback. Angelo and Cross (1993) defined assessment as the multidimensional process of appraising the learning that occurs in the classroom before and after assignments are graded, with the feedback used to improved teaching and, hence, student learning.

Alexander Astin suggested that the term *assessment* can refer to two very different activities: (a) the mere gathering of information (measurement), and (b) the use of that information for institutional and individual improvement (evaluation; Astin, as cited by St. Cloud State University Assessment Office, 1999). Nevertheless, the terms *assessment* and *evaluation* are often used interchangeably in the literature, although assessment may be considered a more comprehensive process, because it also involves the dimensions, capacity, quantity, or amount of something ascertained by measuring.

Assessment has been described as a form of communication involving a number and variety of sources (MacAlpine, 2002; Weisburgh, 2003). MacAlpine explained that as a communication form, assessment may be directed to the students (feedback on their learning), to the lecturers (feedback on their teaching), to the curriculum designer (feedback on the curriculum), to the administrators (feedback on the use of resources), and to the employers (feedback on the quality of applicants). Weisburgh observed that in its simplest state, assessment involves a four-step process: "(a) the person being assessed performs some task; (b) the system evaluates the result; (c) the evaluation is compared with some standard of performance; and (d) the evaluation results in some communication, which can be a report, advice, or a chart" (Weisburgh, 2003, p. 27).

Although assessments used in various contexts and for differing purposes often look quite different, they share certain common principles (National Research Council, 2001). One such principle is that assessment is always a process of reasoning from evidence. By its very nature, however, assessment is imprecise to some degree. Assessment results are only estimates of what a person knows and can do.

Every assessment, regardless of its purpose, rests on three pillars: a model of how students represent knowledge and develop competence in the subject domain, tasks or situations that allow one to observe students' performance, and an interpretation method for drawing inferences from the performance evidence thus obtained. (National Research Council, 2001, p. 2)

Assessment is as Assessment Does

One interpretation of the phrase *assessment is as assessment does* is summarized in the first of AAHE's Principles of Good Practices for Assessing Student Learning: "Assessment is not an end in itself but a vehicle for educational improvement. Its effective practice, then, begins with and enacts a vision of the kinds of learning we most value for students and strive to help them achieve" (AAHE, 2003b, principle 1). Many educators concur that assessment is a central or integral part of instructional practice (Hartley & Collins-Brown, 1999; Herman et al., 1992; Kulieke et al., 1990; Liang & Kim, 2004).

Liang and Kim (2004) suggested that an effective assessment is an instructional event that describes, and promotes students' best performance across time and uses a range of methods. Beyond the knowledge of pedagogical content, achieving higher academic standards is influenced by the teachers' ability to determine what students really know and can do and where the learning gaps are, so that they can target instruction to fill the gaps. John Biggs observed that "what and how students learn depend to a major extent on how they think they will be assessed" (as cited in Maki, 2002, slide 3). In this sense, assessment strategies convey, implicitly or explicitly to students, the range of competences associated with a subject and function as preparatory tools for future learning.

The National Research Council (2001) reported that:

Every assessment is also based on a set of beliefs about the kinds of tasks or situations that will prompt students to say, do, or create something that demonstrates important knowledge and skills. The tasks to which students are asked to respond on an assessment are not arbitrary. They must be carefully designed to provide evidence that is linked to the cognitive model of learning and to support the kinds of inferences and decisions that will be based on the assessment results. (p. 47)

This broader function of assessment is reflected in the fourth of AAHE's Principles of Good Practice:

Assessment requires attention to outcomes but also and equally to the experiences that lead to those outcomes. Information about outcomes is of high importance; where students "end up" matters greatly. But to improve outcomes, we need to know about student experience along the way—about the curricula, teaching, and kind of student effort that lead to particular outcomes. Assessment can help us understand which students learn best under what conditions; with such knowledge comes the capacity to improve the whole of their learning. (AAHE, 2003b, principle 4)

In the literature, there is an important functional distinction between assessment *of* learning and assessment *for* learning. Elwood and Klendowski (as cited in Liang & Kim, 2004) used the term *assessment of learning* to denote assessment for the purposes of grading and reporting with its own established procedures, whereas the term *assessment for learning* denotes assessment whose purpose is to enable students, through effective feedback, to fully understand their own learning and the overarching goals. The technical implication is that assessment for learning requires teachers to examine processes and experiences and not to rely on the student's grade as the only goal of the assessment. "Rather, assessment performance is taken as a proxy for the student's status with respect to target instructional domain" (Popham, as cited in Liang & Kim, 2004 §Assessment as Part of Instructional Practice). Both types of assessment are important.

There is a vital link between assessment, learning, and teaching across environments (Herman et al., 1992; Liang & Kim, 2004; Rowe, 2004). Instructors can build in many opportunities to include students in the assessment of learning and then use the information to make beneficial changes for both learning and instruction. Liang and Kim (2004) observed however, that there is a challenge for instructors in creating these opportunities, especially for online instructors when the instructional environment and communication devices are different.

Moreover, the unique features of Web-based instructional environments open up a new frontier for online instructors to practice a more student-centered pedagogy. The use of technology itself for accessing and communicating course information is one main feature to be reflected in the design of learning activities and assessment practices (Sun, 2002).

Online Assessment in the Online Learning Environment

Mason (1998) discussed the phenomenon of the online course environment for learning in relation to three main elements of asynchronous group and individual messaging, access to course materials, and real-time (synchronous) interactive events. One of the important considerations for effective online assessments is to ensure that the tool incorporates these elements, fits the mode of delivery, and legitimately measures the desired outcome. One cannot simply transfer assessment activities from conventional teaching into the online environment, without consideration of the role of the Internet-based technology in the learning processes (Drummond, 2003; Morley, 2000; Sun, 2002). Regardless of the format, the tool must legitimately and honestly measure the desired outcome (Ehrmann, n.d.; Morley, 2000). For the online environment, this means carefully selecting among available synchronous and asynchronous assessment tools, fostering the development of new tools, and ensuring that the results are acceptable to meet the standards for certification, degrees, and the workplace (Bourne et al., 1997; WGU Virtual University Design Team, as cited in Morley, 2000; Sun 2002).

It has been proposed that one of the main advantages of using assessment software over manually assessing performance is primarily the savings in cost and time (Dowsing, Long, & Craven, 2000; Weisburgh, 2003). Scoring and evaluating tests used to take a lot of manual effort, whereas software can dramatically reduce, or even eliminate, the manual effort, and results can be instantaneous (Weisburgh, 2003). Through the Internet, assessments can be delivered wherever or whenever they are needed. Data warehouses can track responses for longitudinal studies by individuals or groups (Weisburgh, 2003). As the software tools increase in sophistication, with advances in computer adaptive testing (e.g., Michigan State's CAPA) and intelligent tutors (e.g., Carnegie Mellon), then online assessment is becoming more feasible and easier to manage (Twigg, 2001). Dowsing et al. (2000), however, cautioned that the reduction in costs may be illusory in the short term for computer-based assessments, because the cost of software development and maintenance and the costs of retraining and reorganizing administration may be considerable.

Mason (1998) also observed that the ease with which students can submit assignments electronically and take self-tests and even examinations online has led many institutions to exploit the technology to globalize their courses and in some cases to relieve tutors of the more tedious aspects of marking. "Web-in-a-box software customized for education offers forms for easy creation of multiple choice tests as well as assignment submission systems and record keeping facilities" (Mason, 1998, § C, Online Assessment). He indicated that there are also a number of firms marketing software for developing more complex assessment processes, involving surveys and collecting other forms of data from users.

The majority of assessments used in the online learning environment are in the asynchronous environment, where the assessment is completed in delayed time, outside the presence of an instructor (Bourne et al., 1997; Mason, 1998; Morley, 2000; Palloff & Pratt, 1999). Online asynchronous tools may involve alternating interactions between instructors and individual students or entire groups through computer conferencing software and modem or network connections (Brem, 2002; Morley, 2000). These assessments can take many different forms, from traditional examinations of written assignments, case studies, research projects, and multiple choice examinations, to alternative measures such as portfolios, student diaries, or journals to assess higher order abilities (Bourne et al., 1997; Herman et al., 1992; Morley, 2000; Muirhead, 2002).

Synchronous assessment models also play an important role in legitimizing the distance education process because dishonesty is minimized and the instructor has continual management of the testing environment (Morley, 2000). These models include any form of testing where the instructor and students are interacting in real-time during the assessment. Online synchronous assessments may be mediated by two-way interactive conferencing systems with telephone or video, chat rooms, or text messaging tools, through modem or network connections (Morley, 2000; Palloff & Pratt, 1999). One typical exam format involves asking students one question at a time, similar to oral exams, and requiring them to type in answers within a limited time frame (Kouki & Wright, as cited in Morley, 2000). Accreditation agencies prefer this method of synchronous testing because the teacher has significant interaction with the remote students during the examination (Morley 2000).

In the quest for new, unique online testing alternatives, while preserving the value of assessment results, some institutions have opted for combined formats involving on-site student-service center approach. In this design, all of the courses are completed remotely, with the testing done on site at a regional exam or student-service center (Morley, 2000; Palloff & Pratt, 2001). Nonetheless, it is the availability of credible, real-time conferencing tools that will facilitate online assessment for more remote students in the future (Mason, 1998; Morley, 2000).

One implication of incorporating these technology-mediated models is that online learning highlights the need for a new set of teaching–learning assessment decisions and concerns (Brem, 2002). Some of the more frequent questions heard on many campuses adopting online learning approaches include how do you conduct online exams, how do you prevent plagiarism and cheating in online assessments, and how do you verify student identity online (McNett, 2002; Rowe, 2004). These logistical considerations regarding security, integrity, and validity are often based on the notion that online assessment must follow the conventional assessment methods (Holzen, Runyon, & Heeler, 2001). However, one cannot overlook the significance of the interactive, collaborative nature of online medium, which redefines the context for teaching and learning and, therefore, impacts the facilitation of related assessment processes. One underlying aspect of the conceptual debate therefore is whether or not online learning and its related assessment processes also represent a paradigm shift in education (Bourne et al., 1997; Holzen et al., 2001; Palloff & Pratt, 1999; Sims, Dobbs, & Hand, 2001).

Implications of a Paradigm Shift for Online Learning and Assessment

In examining the question of whether education has been revolutionized by the Internet, Drummond (2003) found affirmative results, as one cannot ignore the current impact of the Web in representing information and the requirements for Web access through computers and Internet connection. Other authors refer to the pedagogical revolution or evolution of online learning and the need for a new paradigm (Mason, 1998; Palloff & Pratt, 2001). Sims et al. (2001) commented that the paradigm shifts embrace both pedagogy (from instructivist to constructivist) and technology (classroom to online). Bauer and Anderson (2000) also distinguished between the traditional teacher-centered behaviorist model, with objective testing as the basis for both formative and summative evaluation, and the more contemporary constructivist model, with individual participation and group projects, which are initiated and controlled by learners.

More recently, there has been the use of the term *pedagogical reengineering* to describe the change in online pedagogy from one that is teacher centered to one that is focused on learner activity (Collis & Moonen, as cited in Liang & Kim, 2004). However, the caution is that no research has focused on the interpretation of pedagogical reengineering to online classroom assessment.

Twigg (2001) observed that much has not changed in the online environment, as the vast majority of online courses are organized in much the same manner, as their campus counterparts. She concluded that the higher education paradigm, honed and perfected for hundreds of years, has served us well.

> *The courses are developed by individual faculty members, with some support from the IT staff, and offered within a semester or quarter framework. Most online courses follow traditional academic practices ("Here's the syllabus, go off and read or do research, come back and discuss."), and most are evaluated using traditional student-satisfaction methods. This is hardly surprising, since most online courses are offered by traditional institutions of higher education.* (Twigg, 2001, preface)

Yet, over the last decade there has been growing concern among higher education institutions about the limitations of traditional approaches to assessment (Barone, 2003; Juwah, 2003; Mason, 1998). The National Research Council (2001) suggested that the time is right to rethink the fundamental scientific principles and philosophical assumptions that underlie current approaches to educational assessment. These approaches have been in place for decades and have served a number of purposes quite well, but the world has changed substantially since the approaches were first developed, and the original foundations may not support the newer purposes to which assessments may be put. The National Research Council (2001) recognized that advances in the understanding and measurement of learning bring new assumptions into play and offer the potential for a much richer and more coherent set of assessment practices.

Many traditional assessment practices consist of essay problem-type examinations and growing reliance on multiple-choice questions, for easy designs (Herman et al., 1992; Juwah, 2003; Lynch, 2001). However, it is argued that these assessment practices cannot adequately test for critical thinking, creativity, reflection, and authentic learning (Conrad & Donaldson, 2004; Drummond, 2003; Lewis & Johnson, as cited in Juwah, 2003; Lynch, 2001). Recent developments in assessment are advocating alternative and diverse assessment methods, including peer assessment, portfolio, reflective journaling, self-assessment, and performance-based assessment, which are deemed to be constructive, authentic, contextualized assessment, and to promote deep learning and skills development (Conrad & Donaldson, 2004; Herman et al., 1992; Juwah, 2003).

Mason (1998) concurred that there needs to be a rethinking of current assessment procedures in higher education. These procedures are particularly ill-suited to the digital age in which using information is more important than remembering it, and where reusing material should be viewed as a skill to be encouraged, not as academic plagiarism to be despised. Mason maintained that many online courses are leading the way in devising assignments and assessment procedures, which reflect the call for higher education to teach information technology literacy, team working ability, and knowledge management skills. Current approaches to teaching and learning in higher education are becoming dominated by the importance of interactivity in the learning process, the changing role of the teacher from sage to guide, the need for knowledge management skills, and for teamworking abilities and the move towards resource-based rather than packaged learning (Mason, 1998).

In the general literature on educational assessment, there are also references to the growing trend of classroom assessment (Angelo & Cross, 1993; Cross & Steadman, 1996; Flint, 2003). Classroom assessment is a formative approach, with the aim of improving the quality of student learning rather than providing mere evidence for evaluating or grading students. Angelo and Cross (1993) stated that "as the college classroom changes, we have an opportunity to closely monitor and modify the teaching/learning process within the classroom. Classroom assessment provides a compelling model for realizing this opportunity" (p. 304).

Nature and Context of Online Assessment

The medium is a significant consideration in the delivery of quality-based online learning and assessment. It influences how content is accessed and used, defines the teaching–learning processes and outcomes (Liang & Kim, 2004; Palloff & Pratt, 2001). Therefore, it is important that the approach to online assessment reflects the nature and type of the online teaching–learning environment (Conrad & Donaldson, 2004; Liang & Kim, 2004; Morley, 2000; Palloff & Pratt, 2001).

The online medium is used for retrieving class content, subject-related information, and student–teacher interactions in Web-based instruction (Sherry, Bilig, & Jesse, as cited in Liang & Kim, 2004). Students are usually required to participate in some way to complete an online course (Palloff & Pratt, 2001). As more courses go online, there is increasing quantification of class participation, instead of the conventional class attendance, and more professors require electronic submissions on floppy disks, by e-mail, or on discussion boards (Bauer & Anderson, 2000).

In the online environment, it becomes necessary to change one's concept of the classroom to that of the online course room (Flint, 2003). As suggested by Barone (2003), this reconceptualization of the learning environment is important in making the transition from learning in a physical space, such as the classroom or the library, to learning in a student-centered learning environment in cyberspace. "The course is not the container; teaching 'space' is not a physical place; personal does not mean 'in person'" (Barone, 2003, p. 42). Barone also suggests that the technology enables the design of learning situations that actively engage and guide learners while allowing them to choose the style of the learning experience and to organize the knowledge outcomes.

Moving courses from the traditional classroom to an online setting fundamentally shifts human interaction, communication, and learning paradigms (Robles & Braathen, as cited in Liang & Kim, 2004). In distinguishing between online instruction and traditional face-to-face classroom instruction, three key components of content expandability, content adaptability, and visual layouts can be identified as being more extensive for online instruction (Jung, as cited in Liang & Kim, 2004). Greater emphasis for successful online teaching is also placed on three types of interaction: academic interaction, collaborative interaction, and interpersonal interaction (Juwah, 2003; Jung, as cited in Liang & Kim, 2004). Other references in the literature include four levels of online interaction in terms of the nature of communication among learners, with instructors, with the content, and with the technology (Gunawardena, 1992).

The online discussion tool is one of the most significant components of the online learning environment and is increasingly used for assessment purposes. An online discussion is identified by the use of a computer-mediated conversational environment, and in some aspects it seeks to mirror face-to-face classroom discussions by developing norms about conversation shifts, number and quality of posts, and peer collaboration (Brem, 2002; Palloff & Pratt, 2001). Online conversations may be synchronous, through real-time chat or instant messaging, or asynchronous, by means of a listserv or bulletin board. The ease and convenience of participation in asynchronous conversations—anytime from anywhere—are preferred considerations, especially for courses across global time zones (Palloff & Pratt, 1999). The online dialogue may be in text format only or enhanced with images, animations, hyperlinks, and other multimedia formats (Brem, 2002). Tools for online conversation are becoming increasingly sophisticated, popular, and available, and this increases the appeal of using online discourse as a source of archival data (Brem, 2002; Taylor, 2002).

The role, responsibilities and interrelationships of instructors and students in the online learning process are also affected by the incorporation of technology for teaching and learning. Without a teacher being physically present, Web-based teaching requires new and appropriate instructional practices built on a unique relationship between learners and instructors, as mediated by the technology (Liang & Kim, 2004). Instructors need to be willing to give up a degree of control and allow the learners to take the lead in learning activities, and students also have to be oriented to their new roles and the ways in which learning occurs online (Drummond, 2003; Palloff & Pratt, 2001). Consequently, the related assessment methods should embrace and reflect the nature of the learner-centered and technology-mediated environments. Lynch (2001) offered five examples of how the related assessment techniques are translated by instructors to the online environment in terms of (a) giving up control of the assessment to the student, (b) reassessing outcome evaluation beyond testing, (c) real-world application, (d) project-based learning assessment, and (e) incorporating student reflection as assessment.

It is also essential to recognize that one type of assessment method does not fit all situations; neither is there any single best assessment approach or design (Morley, 2000; The National Research Council, 2001). The online environment offers the opportunity for individualized and flexible testing design, instant feedback, self-paced learning, recursive learning, and ease in creating individual learning profiles (Sun, 2002). Online assessment technology also allows for faster rate of identifying errors, according to predefined assessment criteria, than do human examiners, although it is limited in the interpretation or classification of those errors (Dowsing et al., 2000).

As with any other pedagogical tools, Internet-based assessment tools can be effective in their enhancement of the teaching and learning process only when they are used within a context in which teaching and learning goals are clearly articulated (McAlpine, 2002; Sun, 2002). The ultimate challenge to educators is how to develop an appropriate, authentic, reliable, and ethical online assessment methodology that measures learning, engages the learner, is integrated into the learning process, and promotes further learning (Conrad & Donaldson, 2004; Drummond, 2003; Juwah, 2003).

Issues, Controversies, Problems

In the general literature on online assessment and measurement, there are areas of concern and challenge that can be readily identified as issues, controversies and problems. Most concerns are generic to the larger concept of assessment and arise at a time when education stakeholders are questioning whether current large-scale assessment practices are yielding the most useful kinds of information for informing and improving education (National Research Council, 2001). Over the past two decades, the larger movement in educational assessment has been focusing on setting challenging academic standards and measuring students' progress in meeting those standards, thereby increasing the role of educational assessment in decision-making (McAlpine, 2002; McMillan, 2000; National Research Council, 2001)

Online assessment is greatly affected by issues such as the validation of learning and the verification of student assessment, which have been areas of discontent in distance education since its inception (Conrad & Donaldson, 2004; Liang & Kim, 2004; Morley, 2000; Palloff & Pratt, 2001). The resolution of these issues can be directly related to the acceptance of distance education delivery by accrediting bodies and higher education institutions (Morley, 2000; Serwatka, 2002; Sun, 2002). As with accountability for all teaching practices in general, Internet-based teaching needs to answer questions of test reliability, validity, equity, consistency, transparency, security, accessibility, and appropriateness, among other criteria, as required by accrediting bodies in evaluating distance learning (Sun, 2002). The issues of credibility and validity are also compounded by controversies regarding the appropriateness of current standardization approaches for the online context, the practical technological challenges in an online medium, and the inherent debates among stakeholders in pursuing individual agendas for assessment.

The ensuing discussion of issues, controversies, and problems related to online assessment and measurement is presented within a theoretical and practical framework. The theoretical issues relate to the conceptual definition and interpretation of assessment as part of the instructional process and to the implications of assessment for and of learning. The practical issues relate to the logistical and technical considerations in the process of designing strategies, collating, and communicating data during various stages of assessment involving online technology.

Rethinking online assessment. Some educators advocate the power of online assessment to facilitate changes in teaching and learning, with technology as a catalyst for rethinking approaches to assessment (Bull, 1999a; Mason, 1998; Palloff & Pratt, 2001). Given the current demand for reforms and transformation in educational assessment, the goal is to promote online assessment as scholarship, by examining how Web-based technology develops or transforms thinking (Maki, 2002). Consistent with the proposals for assessment as a scholarship of teaching (Litterst & Tompkins, as cited in Maki, 2002), it would be important for online faculty to channel their intellectual curiosity into exploring relationships between pedagogy and student learning; to focus on the ways in which students integrate; to draw upon and use the knowledge, abilities, habits of mind, ways of knowing, and problem solving that characterize those who work in a discipline.

There are also several implications for managing the format and type of online assessments, as derived from the following suggestions by Bull (1999a) regarding the potential impact of computer-assisted assessment (CAA) within institutions. Immediate advantages of CAA for online learning are that the tools can be used to enhance the quality and speed of feedback to students while facilitating the increased awareness of academicians regarding their students' progress and deficiencies. In terms of quality assurance, CAA can become a catalyst for institutions to reconsider their existing online assessment methods, given the wealth of monitoring and evaluative data to be readily obtained during the process (Bull, 1999a).

On the other hand, Bull (1999a) indicated that the implementation of CAA may present challenges to existing organizational structures, despite its potential for greater collaboration between support and academic staff. Similarly, the initial implementation of online assessment processes may be challenging, as disparate groups are forced to find ways of working together; but such collaboration is essential to the process of fostering pedagogical and technological advances.

Validating online assessment: The message and the medium. Of particular relevance to the issue of validating online learning outcomes is the comment that "the awareness of the distinction between the evaluation of the effectiveness of the learning and the functionality of the medium is crucial as access statistics and the number of messages do not indicate student learning"(Graham et al., 1999, § F.) Though access to the technology is an integral component of success, Graham et al. (1999) cautioned against placing too much emphasis on quantifiable outcomes such as the marks gained in the subject or the number of times students used the system. They contended that it is the content of the messages that displays students' understanding of the theory, and the qualitative feedback from students that indicates the value of their learning experience. Though marks (quantifiable scores) indicate ability to conceptualize introductory theory, they may not reflect the educational process through which students have progressed. On the other hand, Graham et al. (1999) suggested that qualitative feedback may reveal particular benefits with students' learning outcomes, such as improved understanding of the material (tangible) and a sense of belonging to a group (intangible).

Graham et al., (1999) indicated also that there are problems associated with online quantitative assessments, which are compounded by the difficulty in measuring and controlling particular variables of student learning outcomes. They argued that it is difficult to establish a randomly selected control group online without inherent self-selection and creating inequitable opportunities among educational providers: "Direct comparison of student grades in online and conventional programs, is of limited use due to the discrepancy between the two pedagogical regimes" (Graham et al., 1999, § F.)

Another immediate consideration in validating online assessment arises from comparison to conventional assessment, in that the differences in modalities may also affect what and how different abilities are being measured during the assessment experience. In accessing online assessments, learners will require skills in submitting assignments as attachments or other software applications, in navigating the Web and participating in chat room or discussion forums, and downloading files or operating a media player. Students are expected to be adept at using technology as part of their learning mechanisms (Pain & Le Heron, 2003). As emphasized by Dowsing et al. (2000), there should be harmony between the technological nature of the subject matter and the learning and assessment practices for computer-based assessments. An online assessment experience therefore extends beyond achieving levels of proficiency in course content, to include technological competence in using the delivery system. The interpretation of the message for learning is therefore impacted by the medium of delivery and this should be reflected in all assessment strategies.

Appropriateness of assessment techniques: Alternative and authentic assessment. There is increasing deliberation about the appropriateness and design of multiple-choice questions and short-answer questions, which are two popular online assessment strategies (Conrad & Donaldson 2004; Sims et al., 2001). Multiple-choice questions have been heavily used with computer-based tools for easy programming to enable automatic marking, easy retrieval of test scores, and to provide instant feedback to learners, especially those in classes with large online enrollments (Conrad & Donaldson, 2004; Lynch, 2001). These testing strategies have facilitated individualized testing designs, self-pacing, and ease in creating individual learning profiles (Sun, 2002). Yet, there is the argument that these strategies are limited in scope and seem to contradict the contemporary approaches to learning, which advocate active participation, self-assessment, and reflection by learners (Lynch, 2001; Sims et al., 2001).

The prevailing view in the literature is that the more essential media to demonstrate competency in the online environment is written assessments (Conrad & Donaldson, 2004; Liang & Kim, 2004). "Good writing is synonymous with good scholarship and must be accorded a high place in assessment" (Bauer & Anderson, 2000, p. 67). Many authors contend that the online engaged learning environment requires authentic and alternative assessment methods that draw upon learners' prior experiences and real-world situations while measuring the depth of critical thinking and reflection through open-ended questions or culminating research papers and projects (Conrad & Donaldson, 2004; Palloff & Pratt, 2001; Whitis, 2001). Examples of authentic and alternative assessment strategies for online engaged learners include discussion analysis, activity rubrics, team assessment, and reflective self-assessment (Conrad & Donaldson, 2004).

Bauer and Anderson (2000) proposed that assessing the adequacy and timeliness of participation, through the quantity of student writing, is an important indicator of the level of trust between instructor and students in a learning-centered environment. Regular and significant amounts of writing encourage the student to continuously reflect on and discover particular topics, especially the student who is uncomfortable with speaking before a live audience. The regularity of writing also becomes a channel for the instructor to interact with students and to develop an awareness of their diverse and personal styles. Appropriate assessment strategies for continuous writing and reflection may include electronic portfolios, journals, and peer reviews (Bauer & Anderson, 2000; Juwah, 2003; Muirhead, 2002).

Reliability of online assessment results. Scoring a multiple-choice test does not require complicated judgment, though human judgment is a factor in phrasing questions and deciding the most appropriate response (Herman et al., 1992). Greater concerns arise however with alternative and authentic assessments, such as electronic portfolios, journals, multimedia presentations, and other individualized tools for creative and higher order online assessments, which invite a wider range of possible responses to be judged on the basis of the underlying quality or process involved (Herman et al., 1992; Muirhead, 2002).

The extended or open-ended responses are usually more complex; they are beyond right or wrong judgments and require well-defined, structured assignments with clear, meaningful objectives and instructions that are relevant to the course goals and specific scoring guidelines, rubrics, or criteria (Bauer & Anderson, 2000, Conrad & Donaldson, 2004; Whitis, 2001). When guidelines are vague or unstated, the assessments are less likely to be reliable in terms of being consistent, fair, and accurate in judging learner responses between one assessment activity and another or among a group of raters (Herman et al.,1992).

Although many instructors report that alternative assessment methods require much time and effort to develop and integrate into the curriculum, the benefits of having clear, articulated assessment criteria are worth the effort (Conrad & Donaldson, 2004; Muirhead, 2002). The criteria are important for clarifying the instructional goals, setting the performance standards, and defining priority outcomes. The criteria may include rating scales, checklists, numerical and qualitative scales (Herman et al., 1992).

Defining and measuring quality in online discussions. Conrad and Donaldson (2004) suggested that the greatest challenge in assessing an online engaged activity is determining the quality of thought expressed. There is the absence of social cues and body language and facial expressions, as in face-to-face assessment, to guide the online instructor in monitoring and responding to learners' questions, comments, and asides (Alessi & Trollip, as cited in Liang & Kim, 2004). The online discussion response is usually limited to written text with static images, and this may be compounded by learners' inexperience with this medium or their inability to respond (i.e., type) as quickly as they would in verbal conversations (Phipps & Merisotis, 2000). Other related practical challenges include the volume of postings and the limitations of the assessment software in interpreting the quality of the communication or related errors.

The literature refers to new software tools, such as discussion analysis tools, which are emerging to assist with this task, so that an instructor does not have to read every discussion posting in order to adequately evaluate each learner's depth of critical thinking. One such tool is the ForumManager, which evaluates patterns in online interactions (Jeong, as cited in Conrad & Donaldson, 2004).

Even with these tools, there is much more research required, as very little is still known about the intricacies involved in measuring online discussions (Spatariu, Hartley, & Bendixen, 2004).

Assessing online group learning. Another challenge facing online learning involves the incorporation of group assessment at a distance (Morley, 2000). Group projects are considered to be an essential part of assignments for adult learners, as representative of the working world and wider society (Graham et al., 1999; Becker & Dwyer, as cited in Morley, 2000). However, the main concerns tend to relate to determining whether group members participate in equal shares in terms of the quantity and quality of writing. One popular strategy is the use of analytic electronic portfolios to facilitate the grading of each section on an individual scale, along with peer and self-assessment (Bauer & Anderson, 2000). Students may be asked also to create journals of their reflections of the process, explanations of how they met course objectives, discussions of their personal learning, and critiques of the work of others and the quality of peer relationships (Bauer & Anderson, 2000; Juwah, 2003; Muirhead, 2002).

Recent advances in online groupware, such as Lotus' LearningSpace and IBM's ConferenceWare, have simplified issues such as scheduling problems, monopolization of the conversation, and impersonalization, which have plagued online group processes and slowed its acceptance (Morley, 2000). However there remains a challenge when the message threads are sometimes difficult to follow, as this can affect the quality of interpreting the student responses and create biases in assessment results. It helps therefore to assess postings over the length of a course, to establish the pattern of learning (Bauer & Anderson, 2000).

Writing skill as a confounder in online assessment. Liang and Kim (2004) found in their research that it was difficult to accurately measure how actively engaged students are with assignments and discussion, especially those with poor writing skills. For many educators, students are assessed for growth and work production based on the students' own writing. Even their online presence is verified by their written responses (Liang & Kim, 2004; Palloff & Pratt, 2001). Sometimes it takes a while to distinguish between the student's lack of preparation or engagement and the student's poor writing skill. In this context the variables of writing and the assessment of one's writing skill may be confounded with one's performance-based assessment. Other times, there may be miscommunication between and among learners and instructors caused by the writing process or the writing style (Liang & Kim, 2004).

One observation is that instructors have to make the effort to keep discussions on track and to frame their comments in a way that does not stifle the discussion. Very few instructors recognize the extent to which their own writing and writing style may influence interaction and learning in the online environment (Liang &

Kim, 2004). Morley (2000) also referred to the subjectiveness that may be associated with grading reports and research papers. In assessing written interactions between learners and instructors, it is important, therefore, to create a rubric for assessing content, expression, and participation (Bauer & Anderson, 2000, Conrad & Donaldson, 2004).

Learner autonomy and role-shift between learners and instructors. Many proponents of online learning advocate that online courses should be extremely interactive, engaging, and require learners to be independent and reflective, while relying on collaborative learning and peer feedback, to encourage team-work and participation for course assessment (Conrad & Donaldson, 2004; Liang & Kim, 2004; Palloff & Pratt, 2001; Whitis 2001). In this context, the online instructor may be perceived as a facilitator, consultant, and promoter of self-directed learning (Bauer & Anderson, 2000; Liang & Kim, 2004). Online learners are therefore expected to be autonomous in initiating the learning process and being responsible for reading the material, exploring the links, partaking in the discussion, asking questions, choosing to learn the objectives, setting aside the time to learn, and selecting a layout for presenting learning outcomes (Liang & Kim, 2004).

Many online learners may have to learn how to be autonomous, and for some learners this will involve a steep learning curve. As online learners take time to adjust to their roles of being in charge of their own learning, their instructors also require time to identify and facilitate the processes (Liang & Kim, 2004). The challenge is in designing online assessment strategies that are appropriately interactive and engaging, while incorporating the element of increasing learner autonomy over time.

Classroom assessment techniques in the online environment. Given the emerging trend of classroom assessment in the wider field of education, it was instructive to review three main issues identified by Henderson (2001) relating to adapting classroom assessment techniques (CATs) to asynchronous learning networks (ALN). Henderson proposed that a CAT may have a different impact in the ALN setting than in face-to-face or in-class settings. One issue is that teachers do not always know the kind of physical learning environment experienced by ALN students, as there may be variations in environment due to students traveling; trying to connect via a hotel telephone; in a quiet office; or at home, trying to deal with a busy household.

A second issue highlighted by Henderson (2001) is that ALN students may be at different stages in a course, at the point of the CAT. Unlike face-to-face CATs given during specific class periods to students who have participated in same class activities, some ALN students may be at the beginning of a particular unit while others have finished. This variation has implications for the design and format of the CAT. Henderson (2001) also addressed the difficulty in generating anonymous responses in ALN, as instructors can usually identify the names of students who send feedback via e-mail or preaddressed surface mail. The alternative consideration is to have mail or e-mail sent to a third party who removes all identifiers, or to use online survey tools that keep responses anonymous (Pain & Le Heron, 2003).

Time and logistical challenges. Online learning and assessments, which rely on computer-based technology, require a lot of effort in preparation, review, and implementation (Bourne et al., 1997; Bull, 1999a; Pain & Le Heron, 2003). Innovative assessments, including computer simulation, microworld, or other interactive applications, require time and resources to develop, although they are worth the long-term gain in terms of challenging learners to construct mental models (Bull, 1999; McNett, 2002). There is also the challenge of keeping up with changes and upgrades in online assessment software and testing environment. Henderson (2001) referred to the practical challenges of having to adapt course assessments to fit online software templates rather than having the opportunity to create new question types or test formats to match the unique requirements of a course. Bull (1999a) noted that time is required to invest in designing appropriate tests with challenging and effective questions for meaningful feedback and to master new forms of computer-assisted assessments as the technology advances.

Pain and Le Heron (2003) observed that there will be problems, no matter how good the software, hardware platforms and institutional support. Much depends therefore on the quality of preparation, damage control, recovery from difficulties, and ultimately clarity of related roles and responsibilities. Technical problems may include the occasional unexplained crash, installation and network difficulties, browser differences, and password problems, resulting in stress on both learners and instructors. Bull (1999a) also commented on the heavy time investment and cultural shift required for preparing computer-based assessments as disadvantages or risks associated with using the technology. "The crucial issue, for both staff and students, is credibility—can the system be trusted, will it work as expected, and can the information collected (test answers) and the information reported (results) be relied upon?" (Pain & Le Heron, 2003, conclusion).

Controversies and Problems

Many references in literature highlight cheating and plagiarism as two of the most controversial issues in online assessment (Morley, 2000; Rowe, 2004; Pain & Le Heron, 2003). Several writers use these terms to cover all forms of academic dishonesty, including unauthorized collaboration and misrepresentation during the assessment process (Le Heron, 2001; McNett, 2002; Rowe, 2004). Some concerns have been attributed directly to the nature of online technology and course design, especially in reference to interactions at a distance that do not occur in real time or with face-to-face contact. These concerns are central to the debate relating to the acceptance of online learning and assessment. Other concerns pertain to the broader nature of assessment, including the conventional approaches to academic integrity, given that many of the problems regarding plagiarism and authentication also occur in the traditional classroom (Illinois Online Network, 1999). Dowsing et al., (2000) proposed that the major problems with automating any assessment process relate more to consistency, completeness, and ambiguity of the assessment criteria. The following outline of controversies will focus on the diversity of perspectives and solutions emerging in the literature or practices.

Cheating in online assessments. The term *cheating* has been used to describe all deceptive or unauthorized actions, such as taking materials into tests or quizzes, looking at others' answers, breaking the rules governing assessment, or taking another person's identity (Illinois Online Network, 1999; Le Heron, 2001). Rowe (2004) cited various statistical references to studies of cheating and the high percentages, exceeding 70%, of high school and college students who had cheated at least once in their school experience, many without being caught. Most university regulations condemn the practice of cheating and threaten disciplinary procedures (Le Heron, 2001).

It is often argued that the very nature of online technology increases the opportunities or temptation for cheating, given the sense of separation and distance, with the decreased likelihood of being able to monitor the student (Illinois Online Network, 1999; Rowe, 2004). Rowe also highlighted the relative ignorance, denial, or poor judgment of instructors and administrators of the possibilities for cheating online, especially in regard to multiple-choice and calculation questions. Other grounds for cheating online may arise from difficulties in ensuring that students take tests or quizzes simultaneously, retaking of assessment multiple times due to technology glitches or password theft, and receiving unauthorized help due to access to other electronic media and handheld devices, inadequate authentication measures, or the inability to verify that the student is acting alone during assessment (Le Heron, 2001; Rowe, 2004).

Yet, there is the proposition that cheating should not be a major problem if the online course is well designed, learner centered, or that online assessment is continuous so that it is less cost-effective to cheat (Palloff & Pratt, 1999; Rowe, 2004). There may also be ambiguity about what constitutes cheating, if there is no distinction between the inherent practice of collaborating in a learning community to extend knowledge and understanding and the requirement to submit independent work in courses that encourage peer group discussion and debate (Le Heron, 2001).

The proactive suggestions for improving online assessment include less reliance on objective testing or increasing the use of randomized objective testing; using online chats for oral tests; using authentic and alternative approaches such as electronic portfolios, small group projects, simulations, case studies, and self-reflections; giving only take-home tests or quizzes or proctored tests or quizzes; using student photo ID and Web-cams during tests; setting time and access limits for taking tests (Illinois Online Network, 1999; McNett, 2003; Pain & Le Heron, 2003; Rowe, 2004; Serwatka, 2002).

However, Rowe (2004) cautioned that in seeking to eliminate one problem, one may inadvertently create or compound others. For example, the use of a combination of online and traditional paper-and-pencil testing may dilute the problems of cheating online, but not eliminate it. The use of proctors and testing centers to simulate traditional assessment approaches involves expenditure on personnel, facilities, and identification cards, and the travel arrangements and other logistics could be perceived as an imposition on the student and being at cross purposes with the legitimate reasons for taking online courses (McNett, 2002; Rowe, 2004). The use of randomized testing requires a very large pool of questions and can also be problematic in terms of grading students fairly when they randomly select questions of differing levels of difficulty. Rowe commented also that group projects can reduce cheating if students monitor one another, but suggested that group projects are not appropriate for many subject topics and learning skills.

Some writers have maintained that assessment should be continuous, so it is less cost-effective for students to cheat (Kulieke et al., 1990; Palloff & Pratt, 2001; Bork, as cited in Rowe, 2004). However, Rowe countered that continuous assessment requires considerable work on the part of the instructor in setting up a course, whereas students have less opportunity to study and digest the material at their own paces, as reflective of self-education. He believed that it may also foster a climate of distrust, suggesting that students cannot be trusted to learn without constant testing.

Rowe (2004) argued therefore for final examinations online, instead of continuous assessment, as the latter almost inevitably overemphasizes a student's short-term memory, when the purpose of learning is to foster long-term memory impacts on students. However, McNett (2002) advised that multiple opportunities for assessment should be structured into the course, in addition to, or in place of, final examinations. Testing should never be the only means by which the abilities of students are assessed (Conrad & Donaldson, 2004; Illinois Online Network, 1999; Holzen et al., 2001; McNett, 2002; Palloff & Pratt, 2001).

Le Heron (2001) proposed that technologically-based objective assessment has allowed for the reduced occurrence of cheating. Teaching staff and the vast majority of honest students find it very demotivating to see students achieve success by cheating. Le Heron observed that in cases where students submit work as a group or have the opportunity to develop work with others before handing in their *own* work, there tends to be only a small percentage who put their energy into "beating the system" by cheating. Suggestions for alleviating cheating include collaboration between institutions, the swapping of question banks, the sharing of best practice, though Pain and Le Heron (2003) cautioned that these activities require not only good planning and continuity of staff, but also assume an environment in which institutions collaborate rather than compete.

Plagiarism in online assessments. The issue of plagiarism is another major concern, as examination results become more important for jobs and personal advancement (Dowsing et al., 2000; Le Heron, 2001). Plagiarism is described as the reproduction and presentation of others' work, without acknowledgement, or the attempt to receive credit for the ideas or words of others. This may involve verbatim copying or unacknowledged paraphrasing significant parts of a paper (Illinois Online Network, 1999).

There are the familiar arguments that online (computer-based) technology, for file copying and transfers, makes plagiarism simple and easy. On the other hand, the computer technology in the form of commercial services and software packages enable faster and more efficient automatic checking of sophisticated plagiarism patterns than do manual, paper-based systems (Dowsing et al., 2000; McNett, 2002). This checking facility has been a deterrent in some instances.

Suggestions to alleviate plagiarism include the instructor's familiarization with each student's academic writing ability through continuous assessments, individual interviews, or debriefings with students concerning their papers, assigning special format or style requirements for references and bibliography, use of tracking services, and familiarization with active Internet and term-paper sites (Illinois Online Network, 1999; McNett, 2002). More proactive suggestions for constructing plagiarism-resistant assignments include requiring the incorporation of unique resources or requiring a perspective involving analysis and synthesis, rather than the repetition of facts (McNett, 2002). Preventative strategies include helping students learn how to properly cite work, providing and enforcing clear instructions and expectations, requiring students to submit source materials, annotated bibliographies, and rough drafts or incremental stages of assignments (Foothill College, 2003; Illinois Online Network, 1999; McNett, 2002).

Security and confidentiality in online assessments. The decision to use automated or online assessment strategies, usually require serious consideration about the security implications of transmitting biographical details and students' scripts remotely or centrally, across distance and time (Dowsing et al., 2000;

Palloff & Pratt, 1999; Whitis, 2001). Rowe (2004) observed that the problems of educational security are as common as are the problems of information security.

Whitis (2001) cited a national survey in which over 64% of Americans were very concerned or somewhat concerned about confidentiality online (p. 58). Concerns related to authenticating the identity of students, protecting confidential transmittals including personal data and conversations, and electronic tracking of activity and interactions. There are always concerns regarding the risk of widespread security leaks within institutional settings, intensified by the number of persons who have easy access to data. One suggestion for restricting access is the use of a centralized service for online assessments to minimize the number of copies, processes, and persons required, especially in using particular software or handling test results (Dowsing et al., 2000).

The concern regarding protection of confidential transmissions is of particular relevance to the increasing use of online discussions for assessment of group processes, peer feedback, and interactive research projects (Brem, 2002; Palloff & Pratt, 1999; Rowe, 2004). Online discussions present new opportunities for higher order critical thinking skills and reflection, but also present new concerns and considerations, especially regarding the ethics of monitoring, storing, and retrieving information in discussion forums (Brem, 2002). For example, participants may not be aware or may forget that their conversations are being unobtrusively monitored or will be preserved for future retrieval and use as research data. Brem indicated that online exchanges of information require the same levels of protection as face-to-face exchanges, though this can be more complicated to achieve (Brem, 2002).

Proactive Solutions and Recommendations

With newer forms of academic dishonesty emerging, as newer software packages are developed, there are genuine and significant concerns about the pursuit of online learning as an effective delivery system by various educational institutions (McNett, 2002; Rowe, 2004). The prevailing countermeasures are inadequate and unsatisfactory in several ways, so online assessment in distance-learning programs should be done with caution until more progress is made on the technical development of countermeasures (Rowe, 2004). Fortunately, the same technologies, which offer new opportunities for cheating, also offer new ways to detect and prevent cheating (McNett, 2002).

To eliminate academic dishonesty, several institutions are opting to control the assessment situation by moving towards proctored examinations within specified time frames at testing facilities, including alternative arrangements for remote students at a library or corporate human-relations department (McNett, 2002; Rowe, 2004). Yet, the technology has also been advanced by incorporating course management systems to allow administration of a proctored examination online or the use of Web-logging software to track students' movement on course sites. Another option is the use of question banks for randomizing the sequence of questions, which can also be created with several course-delivery software packages. Despite the extensive effort required initially to create a large pool of questions, the computer tools facilitate storage and retrieval over many years. Instructors could also invite students to submit good test questions as electronic files for future examinations. In this way, assessment would be incorporated with varied formats, as a community learning experience (Barone, 2003; Bourne et al., 1997; McNett, 2002).

Holzen et al. (2001) argued that the current efforts in online assessment need to move beyond the rhetoric that says that assessment should be used as a teaching tool and not as an evaluation mechanism. They promoted online assessment as an interactive mentoring opportunity, which would enable students to evaluate their own progress through the course materials and provide feedback on course content areas that need further enhancement and development. Proactive suggestions include the use of open-book quizzes and tests as means of promoting learning and representing only a small component of the overall assessment strategy for the online course. The assessment strategy would include a cumulative process of diverse array of assessment methods and opportunities to determine the student's understanding of the learning outcomes and provide for relearning and reassessment (Holzen et al., 2001).

More recent emphases in online assessment literature are advocating a shift in priority from policing students to facilitating creative design of assessment to accommodate collaboration, continuous reflection and use of other's ideas without plagiarism (Drummond, 2003; Le Heron, 2001; Pain & Le Heron, 2003). Holzen et al. (2001) listed examples of minute papers, concept maps, and analytic memos. Other writers suggested the use of electronic portfolios or digital scrapbooks for customized assessment, the use of threaded discussions, and conferencing software for peer reviews and reactions, storytelling and with the view of personalizing the subjects and providing opportunities for researching related areas of interest (Juwah, 2003; Kulieke et al., 1990; Muirhead, 2002). The ultimate aim is to encourage students to learn by doing, within a supportive environment of reflection, exploration, interactive stimulation, collaboration, feedback, and reinforcement (AAHE, 2003b; Conrad & Donaldson, 2004; Drummond, 2003)

The overall recommendations garnered from the literature are that because online learning allows for flexibility in delivery time and place, then instructors should use the technology to facilitate a more student-centered and authentic approach to assessment and feedback (Conrad & Donaldson, 2004; Drummond, 2003; Liang & Kim, 2004; Muirhead, 2002). Despite the current practices of on-site proctored examinations, it is recommended that conventional tests and quizzes be only a small component of the overall assessment strategy for online engaged learning (Conrad & Donaldson, 2004; Palloff & Pratt, 2001). In modifying conventional tests to incorporate higher order thinking and language skills, instructors will have to establish a clear rubric indicating grade allocation for specific criteria such as originality, creativity, application, synthesis, and evaluation of subject content, and where appropriate the criteria for presentation style including spelling, grammar and punctuation (Bauer & Anderson, 2000; Conrad & Donaldson, 2004; Muirhead, 2002). Students could be invited to assess an example of what is considered to be a model or poor sample of performance items, while multiple-choice tests could include short answer items, to enable students to demonstrate critical thinking in justifying choices.

With emphasis on providing individual feedback, instructors also need to create effective and efficient strategies and comprehensive guidelines for using their time and energy in providing feedback about assessment to large numbers of students (Drummond, 2003; Foothill College, 2003). Proactive strategies may include establishing a Word document with standard, canned responses, then adding the personal touch; setting up a Web form with selectable common issues and providing room for a personal response; offering opportunities for students to post potential examination questions, and conducting peer reviews of drafts and projects (Foothill College, 2003).

Other practical considerations include greater emphasis in assessing the application of content instead of the presentation or verbiage of content, assessing in ways that reflect diversity of learning styles, and creating tests to accommodate for universal learning. Recommendations for performance assessment designs include frequent, small assessment tasks rather than infrequent large tests, and the use of an assessment portfolio, where each different type of assessment plays a different role, rather than one or two types of assessments (Conrad & Donaldson, 2004; Herman et al., 1992; McNett, 2002).

Although further research is required on the intricacies involved in using online discussions as an assessment tool (Spatariu et al., 2004), there are practical considerations for developing measurement strategies already emerging in existing research studies. In one study, Spatariu et al. (2004) used four general categories of analysis: (a) levels of disagreement, (b) argument structure analysis, (c) interaction based, and (d) content analysis. In an earlier study, Hara, Bonk, and Angeli (2002) used five dimensions of analysis: (a) student participation rates; (b) electronic interaction patterns; (c) social cues within student postings; (d) cognitive (e.g., clarification, inference, judgment) and metacognitive (e.g., personal awareness, task knowledge, and strategic knowledge) aspects of students' postings; and (e) depth of processing ranging from surface to depth. Conrad and Donaldson (2004) also pointed to analysis results including the average number of discussion entries per participant, the level of interactivity, richness of the discussion, and depth of the discussion. One important implication of these research investigations is that the quality of online discussions to be assessed in the future will be directly affected by the clarity and precision of instructional guidelines relating to the multiple dimensions and levels of analysis possible.

In the literature, there were several examples for more creative and flexible assessments that extend beyond traditional examinations and term papers. These examples include video review and reactions, Web design or art production (e.g., pictures, diagrams, photos, videos), WebQuests, storytelling, service learning, research interviews, case studies, and peer critiques (Conrad & Donaldson, 2004; Herman et al., 1992; Morley, 2000; Holzen et al., 2001). These forms of nontraditional assessment serve to increase opportunities for creativity and flexibility in allowing learners to choose when and how they complete assignments and provide some degree of freedom for learners to pace themselves relatively faster or slower than others taking the course. Learners would also be able to choose whether or not to explore topics in more depth, through links to related Web sites. Of significance too are the types of assessment, which provide optional remediation for students who are falling behind, or allow learners to take the assessment more than once to improve their score.

Serwatka (2002) observed that with the use of practice quizzes there has been improved performance on randomized tests, as the questions do not appear in the same order. Even where these tests allow for open book, the students do not have time to look up the answers to each question. The tests are timed and automatically submitted at the end of the time limit.

Principles of good practice. Any discussion on proactive solutions for online assessment needs to include the emerging principles of good practice and benchmarking. This is a recurring theme in the literature on online learning, as proponents (Baker, 1999; Barone, 2003; Chickering & Ehrmann, 2000; Graham et al., 2001) emphasize the need for consistent application of the principles to ensure quality of learning experiences and to establish benchmarks for success in Internet-based distance education. Phipps and Merisotis (2000) affirmed that this extraordinary growth of technology-mediated distance learning in higher education has prompted several different organizations to develop principles, guidelines, or benchmarks to ensure quality distance education. Baker (1999) offered a collation of quality indicators, standards, principles, guidelines, and benchmarks for distance education in general and technology-assisted (electronic-based) learning in particular. Baker's sources were drawn from several leading organizations and agencies, including the American Association of Higher Education (AAHE), the American Council on Education (ACE), Western Interstate Commission for Higher Education (WICHE), the North Central Regional Educational Laboratory (NCREL), the Open University–UK, the Commonwealth of Learning (COL), and the Global Alliance for Transnational Education (GATE).

In outlining the guidelines for ensuring the best quality of learning management processes and practices for technology-assisted distance learning, Baker (1999) emphasized the use of standards to reflect assessment of learning that is frequent and timely; appropriate and responsive to the needs of the learner; in various forms such as written and oral assignments, self-assessment, demonstrations, and exams; and is competency-based (p. 6). In a separate category, Baker referred to quality standards for authentic assessment of learning through faithful representation of the contexts encountered in the field of study or in real-life tests faced by adults; engaging and important problems and questions; nonroutine and multistage tasks and real problems; self-assessment; trained assessor judgment; and the assessment of habits of mind and patterns of performance (p. 6).

Chickering and Ehrmann (2000) suggested that if the power of the new technologies is to be fully realized, they should be employed in ways consistent with the seven principles for good practice. A subsequent qualitative study conducted by Graham et al. (2001) provides a practical lens for adapting these seven principles to the online environment. Although the authors of this study caution against making generalizations, the lessons learned are instructive for all practitioners in the online learning environment. In implementing good practices of assessment and evaluation in the online environment, one should therefore consider the following provisions:

1. Encouraging student–faculty contact by instructors providing clear guidelines for interaction with students

2. Encouraging meaningful cooperation among students by providing well-designed discussion assignments

3. Using active learning techniques by requiring students to present course projects

4. Giving prompt feedback by providing at least two types of feedback: information and acknowledgement

5. Emphasizing time on task by providing course deadlines

6. Communicating high expectations by providing challenging tasks, sample cases, and praise for quality work

7.   Respecting diverse talents and ways of learning by allowing students to choose project topics, and incorporating diverse views into online courses (Grahamet al., 2001)

McMillan (2000) asserted that an understanding of fundamental assessment principles can be used to enhance student learning and teacher effectiveness while incorporating technology. He observed that as technology advances and teachers become more proficient in using technology, there will be increased opportunities to use computer-based techniques (e.g., item banks, electronic grading, computer-adapted testing, computer-based simulations), and Internet resources. However, McMillan cautioned against the mindless use of these resources, without consideration for adequate evidence of reliability, validity, and fairness.

Future Trends

Two decades ago, the Educational Testing Service president Anrig (1985) observed that current forms of standardized testing serve important accountability and institutional needs. He indicated that these needs would continue to exist in the future, as would the current array of achievement, admissions, and licensing tests. Anrig (1985) also proposed that

> *advances in cognitive psychology and technology, however make possible new kinds of measurement instruments. This new generation of tests will have three functions: (a) it will serve individuals more than institutions, (b) it will aim primarily at helping individuals learn and succeed rather than simply yielding scores for institutional decision making, (c) it will guide instruction and self-development on a continuing basis rather than compare performance among test takers. ...This new generation of tests will be helping measures, enabling individuals to keep pace with rising standards in education and the workplace. They will capitalize upon electronic technology for their development, design and delivery. (pp. v-vi)*

Anrig's vision of the future is unfolding in this decade, though it would seem that the systemic and systematic changes required for the advance of the new generation of assessment techniques within an online environment are still at an early stage (Bull, 1999b). This early stage is identified as one of three stages, characterized by the limited exploitation of the potential of the technology, with the focus on replication of paper-based testing methods (Bennett, as cited in Bull, 1999b). The second stage is characterized by a relative increase in exploitation of technology in terms of greater automation in test generation and marking, with focus on summative assessment, facilitated by the integration of student performance tracking features; the third stage reflects the use of a full range of interactive simulation and modeling coursewares with the integration of medium-specific testing methods (Bennett, as cited as Bull, 1999b).

The educational paradigm is therefore shifting from a culture of testing in terms of standardization and certification to a culture of assessment reflected by multiple dimensions and the integration of assessment and learning (Herman et al., 1992; Kulieke et al., 1990). In this context, multidimensional assessments will enhance the power and diversity of active and authentic learning, create multiple sources of information to support instructional decision making, and help students become more reflective and capable learners (Holzen et al., 2001; McNett, 2002).

In proposing a vision of possibilities, the National Research Council (2001) commented that although it is always risky to predict the future, it appears clear that advances in technology will continue to affect the world of education in powerful and provocative ways. Many technology-driven advances in the design of learning environments, which include the integration of assessment with instruction, will continue to emerge and will reshape the terrain of what is both possible and desirable in education (p. 283).

The CEO Forum on Education and Technology (2001) suggested that as schools integrate technology into the curriculum, the method of assessment should reflect the tools employed in teaching and learning. The implication is that if the technology tools help to expand the range of possibilities by making it easier for students or faculty to create, share and master knowledge, then that technology tool should also be used for the related assessment of the learning processes and outcomes. The real challenge therefore is to harness the advantages of the Internet-based technology for stimulating synchronous or asynchronous conversations, electronic messaging, peer collaboration and global access to information across time and distances, with the goal of integrating assessment within the learning processes.

In a general search for related articles on the Internet, it was observed that the major trends in online assessment were primarily described in context of the paradigm changes or emerging standards in online learning or e-learning. Taylor (2002) referred to the current trends as the second wave of e-learning and explained it primarily in terms of advances in the development of online content, technology, and services.

Taylor (2002) maintained that regardless of the sophisticated technology features or flashy graphics and images, it is the quality and relevance of the content to the business issue and learner objectives that are most important. There is a growing movement towards the development of learning objects among content providers, which has been enabled by the growth and adoption of the guidelines provided by the standards initiative. The breaking of content into searchable learning nuggets is allowing new configurations in unique ways based on learners' needs and objectives. Barron (2002) suggested that the use of learning objects will enhance the speed in uploading learning content, the simplification of content authoring using template-based tools, and the ability to draw from a pool of learning objects to customize learning content for multiple audiences. The result is the ability to offer more personalized learning experiences, in response to the needs of individuals (Barron, 2002; Taylor, 2002).

There are significant implications for online assessment, if the nature of and approach to the content covered in the learning process are similarly reflected in the nature of and approach to assessment of topics and skills. As the content becomes more personally relevant and accessible, the related assessment outcomes will be more learner-centred, customized and managed by the learner (Bauer & Anderson, 2000; Liang & Kim, 2004). The emerging assessment measures will focus on more authentic, creative, reusable modular formats, with flexible and collaborative interaction responses to the learning objects (Barron, 2002; Drummond, 2003; Juwah, 2003). The use of electronic portfolios, online discussion forum and interactive data-sharing projects emerge as appropriate tools for online learning and assessment (Barone, 2003; Brem, 2002; Mason, 1998; Morley, 2000; Weisburgh, 2003).

Taylor (2002) observed that the e-learning technology continues to evolve as suppliers search for every advantage possible in a tight technology spending market and merge or combine forces, streamline and refine their product offerings, and introduce new features in an attempt to attract buyers. Wilson (2001) commented that the increasing range of representational and modeling tools including 3-D animation and digital TV are affecting the notions of reality and possibility in the learning process. These tools in combination with sophisticated wireless and mobile communication systems have led to virtual worlds that increasingly allow for rich experience and interaction anytime, anywhere. The advances in learning management systems (LMS) and content management systems (CMS) and all the variants also offer an unprecedented opportunity to track, monitor, quantify, and analyze performance variables (Taylor, 2002).

One positive implication of these trends is that online assessment measures will increasingly reflect the digital possibilities of archivability, searchability, replicability, and hypertext linkability (Wilson, 2001). The digital tools will also facilitate self-assessments and peer reviews and feedback in the integration of interactive learning and assessment.

In the literature, there are also increasing references to the emerging provision of blended learning support services, in terms of offering myriad combinations of asynchronous technology-mediated learning experiences with synchronous traditional-instructor led training in new and creative ways to satisfy diverse learner needs and preferences (Taylor, 2002). This hybrid format provides increasing access to technology while allowing for direct interactions to support the learner.

The implication for online assessment is primarily in terms of the need for multiple approaches, including both asynchronous and synchronous formats. Morley (2000) cautioned that the selection of any assessment method must include practical considerations for the nature of the subject matter, terminal consequences for dishonesty, cost associated with the method, and the layout of the course. The selection criteria should foster a match of the assessment method with the personality of the course and institution, and ensure a consistent measure of competence. With primary reference to the distance education context, Morley (2000) endorsed the provision of a repertoire of assessment alternatives to ensure that learners are challenged and remain involved with the course.

A multitiered assessment structure has been proposed by Riel and Harisim (as cited in Bober, 1998). They emphasized that for the design and structure of any networked learning or online learning assessment, there must be consideration for the size of the learning groups, the types of leadership provided, and the tasks assessment movement. Bober (1998) commented that Riel and Harasim's proposal is a fairly complex evaluation system that includes systematic observation, user feedback (through self-report surveys or usage logs), and examination of extant computer data. The proposal also included the recommendations for the collection of attitudinal data (e.g., students' confidence with both the content and the technology and their comfort with collaborative learning strategies).

For assessments relating to social interaction, the main considerations must include analyses of e-mail or forum messages and account for patterns of individual participants, response times, and varying levels of involvement. Finally, in assessing individual learning outcomes, the underlying considerations must address changes in participant interactions, increases in skill or knowledge, and the quality and type of skill and knowledge transfer (Bober, 1998).

Angelo (1999) described the growing acceptance by faculty and administration of the approach to assessment as if learning matters the most, although assessment for accountability also matters. Angelo observed that much of this acceptance is driven by the desire to improve student learning, even when there is the lack of solid evidence of any learning improvement. Angelo's explanation for the lack of evidence was that most assessment efforts have been implemented without a clear vision of the nature of higher or deeper learning, and without an understanding of how assessment can promote such learning. He proposed that the piecemeal attempt arising from a mechanistic, additive model of assessment, be replaced by a transformative, assessment-as-culture model (Angelo, 1999).

A similar model could be adapted by the online learning assessment movement. Angelo's three step approach is generally applicable: (a) the development of a clear vision facilitated by assessment for learning communities, (b) a new mental model of assessment as cultural transformation, and (c) research-based guidelines for effective assessment practice. The underlying four pillars of transformative assessment are important preconditions: (a) shared trust, (b) shared visions and goals, (c) shared language and concepts, and (d) shared internal guidelines (Angelo, 1999). The transformative assessment model will require institutional commitment and alignment throughout the institution at all levels, to provide the relevant insight into learning processes and outcomes (Barone, 2003).

The transformative model is worth further research investigation for more specific application and relevance to online learning and assessment practices undertaken by the participants. Barone (2003) cited a few examples of institutions such as the National Learning Infrastructure Initiative (NLII) of EDUCAUSE, and the University and Washington, which are implementing transformative assessment projects. She noted that there is the prevailing notion that transformative assessment could play the same role that technology has performed in transforming teaching and learning.

In observing the future trends in the content, technology, and services for online learning and assessment, one can detect the underlying emphasis on the evolution of the technology as a distinguishing feature in the assessment movement (Barone, 2003; Drummond, 2003; Taylor, 2002; Wilson, 2001). However, the focus on the technology tool, by itself, does not automatically denote the desired quality and appropriateness of the assessment process.

More critical for future decisions will be the learner's ability to use the online technology to access and process information, communicate with instructor and peers, and demonstrate their mastery of learning (Morley, 2000; Sun 2002). In confirming the achievement of learning outcomes, which are authentic, reliable, and creative, one has to ensure the integrity of the learner's performance, the relevance of measures, the security of personal information, and the appropriate use of technology tools (Drummond, 2003; Ehrmann, n.d.; Sun, 2002). The distinctive feature of online assessment using online tools will be the flexibility and individualization of the process in facilitating feedback and interaction, and the learner's control of the pace of assessment for learning and diverse learning styles (Barron, 2002; Palloff & Pratt, 2001; Sun, 2002; Taylor, 2002). Effective decisions about using online technology tools are essentially teaching and learning decisions (Barone, 2003; Brem, 2002).

Conclusion

Online course delivery is in its infancy, and the current paradigms for an assessment system tend to be based on fluid caveats rather than set criteria (Porter, as cited in Bober, 1998). Yet, one cannot simply transfer assessment activities from the conventional learning environment into technology-mediated environment without setting guidelines for the anticipated impact of the technology on content, interactions, roles, and responsibilities (Barone, 2003, Drummond, 2003; Liang & Kim, 2004; Morley, 2000; Sun, 2002). In this context, online course and program developers will have to lead the pedagogical evolution by becoming change agents and innovators working from the ground up, and expending more time and resources for ongoing experimentation with new technologies and educational methods (Bober, 1998; Mason, 1998; Palloff & Pratt, 2001). The process becomes even more complex as the technologies constantly change (Whitis, 2001).

Several authors have proposed that as the mode of communication and learning paradigm shifts, the assessment practices in online environment should use a model that reflects this shift, to direct teaching and promote learning (Bober, 1998; Brem, 2002; Liang & Kim, 2004; Macdonald, 2002; Morley, 2000; Palloff & Pratt, 1999). They observed that technology advances have provided opportunities for online assessment to be more learner centered, to promote self-directed learning, and to increase learner autonomy. They also recommended the practice of assessment for learning as a means of cultivating student ownership, affecting effort and achievement, eventually. The transformative assessment model (Angelo, 1999) is worth further consideration in developing the vision and sense of community regarding online assessment for online learning.

Within the context of online assessment becoming more learner centered, there is the underlying notion of assessment being defined by its function, or being a means to an end: assessment for learning. This reinforces the expression that assessment is as assessment does, because assessment is not an end in itself. Regardless of the mode for assessment (online or conventional), the general perspective emerging in the literature is that assessment is a central part of instructional practice (Hartley & Collins-Brown, 1999; Herman et al., 1992; Kulieke et al., 1990; Liang & Kim, 2004).

As an instructional practice, online assessment has to incorporate multiple approaches and a diverse array of assessment methods to accommodate various learning styles, and provide continuous feedback on learning performance, with opportunities for relearning and reassessment (Bauer & Anderson, 2000; Holzen et al., 2001; Liang & Kim, 2004; The National Research Council, 2001). Some current observations, however, indicate that the organization of many online courses is not much different from its face-to-face counterpart (Twigg, 2001). Without a total paradigm shift for online assessment, it is not surprising that many of the popular questions about the feasibility of online assessment reflect the concerns about authentication, security, cheating, and plagiarism that also occur in the traditional classroom. Some aspects of these concerns have been related to the limitations of the online technology tools, especially at a distance, with the loss of visual clues.

Proponents of online assessment for online learning contend, however, that if the full potential of the online technology is exploited fully for more interactive, creative, authentic, reflective, and constructive assessment, then many of the concerns will be mitigated (Illinois Online Network, 1999; McNett, 2003; Palloff & Pratt, 1999; Rowe, 2004; Serwatka, 2002). The culture of online learning assessment will also shift from a focus on objective testing and policing students towards more meaningful, collaborative, customized, and multitiered assessments (Conrad & Donaldson, 2004; Holzen et al., 2001; McNett, 2002; Morley, 2000).

There is a growing collection of publications from leading educational organizations and agencies on assessment guidelines for distance education in general (Baker, 1999; Phipps & Merisotis, 2000). Much more research is needed, however, on the specific and consistent application of principles of good practice for online assessment (Graham et al., 2001). As new trends emerge in online learning, reflecting the use of more sophisticated technology features, there is the greater need for appropriate standards to ensure the validity, reliability and high quality of the content, tools, methods of online assessment systems and procedures. As they are being assessed, online learners have to be challenged yet encouraged to remain involved in the learning experience (Morley, 2000).

References

American Association for Higher Education. (2003a). Assessment frequently asked questions (FAQs). Retrieved June 10, 2004, from *http://www.aahe.org/assessment/assess_faq.htm#define*

American Association for Higher Education. (2003b). Nine principles of good practice for assessing student learning. *AAHE Assessment Forum*. Retrieved June 10, 2004, from *http://www.aahe.org/principl.htm*

Angelo, T. A. (1999, May). Doing assessment as if learning matters most. *AAHE Bulletin.* Retrieved July 2, 2004, from *http://aahebulletin.com/public/archive/angelomay99.asp*

Angelo, T. A., & Cross, K. P. (1993). *Classroom assessment techniques: A handbook for college teachers* (2nd ed.). San Francisco: Jossey-Bass.

Anrig, G. R. (1985). The redesign of testing for the 21st century. *Proceedings of the 1985 ETS Invitational Conference, Educational Testing Service,* Princeton, NJ.

Baker, K. (1999, March). *Quality guidelines for technology-assisted distance education* (Proj. Rep.). Community Association for Community Education (CACE) and the Office of Learning Technologies (OLT) of Human Resources Development Canada (HRDC). Retrieved June 15, 2004, from *http://www.futured.com/pdf/distance.pdf*

Barone, C. (2003, September-October). The changing landscape and the new academy. *EDUCAUSE Review, 40-47.* Retrieved June 27, 2004, from *http://www.educause.edu/ir/library/pdf/erm0353.pdf*

Barron, T. (2002, May). Learning object approach is making inroads. *ASTD Learning Circuits* [Electronic version]. Retrieved August 4, 2004, from *http://www.learningcircuits.org/2002/may2002/barron.html*

Bauer, J. F., & Anderson, R. S. (2000, Winter). Evaluating students' written performance in the online classroom. In R. E. Weiss, D. S. Knowlton, & B.W. Speck (Eds.), *Principles of effective teaching in the online classroom. New Directions for Teaching and Learning* (84, 65-71). San Francisco: Jossey-Bass.

Berge, Z. (2000, Winter). Components of the online classroom. In R. E. Weiss, D. S. Knowlton, & B.W. Speck (Eds.), *Principles of effective teaching in the online classroom. New directions for teaching and learning,* 84, 23-28. San Francisco: Jossey-Bass.

Bober, M. (1998, November). Online course delivery: Is meaningful evaluation possible? *Distance Education Report, 2*(11), 1-3.

Boettcher, J. V., & Conrad, R. M. (1999). *Faculty guide for moving teaching and learning to the Web.* Los Angeles, CA: League for Innovation in the Community College.

Bourne, J. R., McMaster, E., Rieger, J., & Campbell, J. O. (1997, August). Paradigms for on-line learning: A case study in the design and implementation of an asynchronous learning networks (ALN) course. *Journal of Asynchronous Learning Networks, 1*(2). Retrieved June 4, 2004, from *http://www.aln.org/publications/jaln/v1n2/v1n2_bourne.asp*

Brem, S. (2002). Analyzing online discussions: Ethics, data and interpretation. *Practical Assessment, Research & Evaluation, 8*(3). Retrieved June 21, 2004, from *http://pareonline.net/getvn.asp?v=8&n=3*

Bull, J. (1999a). Computer-assisted assessment: Impact on higher education institutions. *Educational Technology & Society, 2*(3). Retrieved June 14, 2004, from *http://ifets.ieee.org/periodical/vol_3_99/joanna_bull.html*

Bull, J. (1999b). Conclusion: A glimpse of the future. In S. Brown, P. Race & J. Bull (Eds.), *Computer-assisted assessment in higher education.* London: Kogan Page.

CEO Forum on Education and Technology. (2001). School technology and readiness—Key building blocks for student achievement in the 21st century: Assessment, alignment, accountability, access, analysis (Forum Rep.). Retrieved June 4, 2004, from *http://www.ceoforum.org/downloads/report4.pdf*

Chickering, A. W., & Ehrmann, S. C. (1996, October). Implementing the seven principles: Technology as lever, new ideas and additional reading. Retrieved January 10, 2004, from *http://www.tltgroup.org/programs/seven.html*

Conrad, R. M., & Donaldson, J. A. (2004). *Engaging the online learner: Activities and resources for creative instruction.* San Francisco: Jossey-Bass.

Cross, P., & Steadman, M. (1996). *Classroom research: Implementing the scholarship of teaching.* San Francisco: Jossey-Bass.

Dowsing, R. D., Long, S., & Craven, P. (2000, March). The effectiveness of computer-aided assessment of IT skills. Paper presented at the *International Conference on Learning with Technology*, Temple University, Philadelphia, PA. Retrieved June 14, 2002, from *http://l2l.org/iclt/2000/papers/134a.pdf*

Drummond, C. M. (2003). Authentic learner assessment in an online environment: Using instructional design techniques to create an assessment model for an introductory computer science course. (Doctoral dissertation, Capella University, 2003.) Retrieved June 24, 2004, from *http://wwwllib.umi.com/dissertations/fullcit/3112977*

Ehrmann, S. C. (n.d.). Asking the right question: What does research tell us about technology and higher learning? Retrieved June 15, 2004, from *http://www.learner.org/edtech/rscheval/rightquestion.html*

Flint, W. (2003, May). Classroom assessment techniques for online instruction. RP Group eJournal of Research, Planning, and Practice, Publication 1. Retrieved June 10, 2004, from *http://www.rpgroup.org/publications/eJournal/Volume_1/volume_1.htm*

Foothill College. (2003, April). Best practices in online teaching and learning: Outcomes from the round-tables. Retrieved June 10, 2004, from *http://www.foothillglobalaccess.org/main/best_practices_printer_friendly.htm*

Goodyear, P. (2002). Teaching online. In N. Hativa & P. Goodyear (Eds.), *Teacher thinking, beliefs and knowledge in higher education* (pp. 79-101). The Netherlands: Kluwer.

Graham, C., Cagiltay, K., Lim, B. R., Craner, J., & Duffy, T. M. (2001, March–April). Seven principles of effective teaching: A practical lens for evaluating online courses. *The Technology Source* [Electronic version]. Retrieved June 3, 2004, from *http://ts.mivu.org/default.asp?show=article&id=839*

Graham, M., Scarborough, H., & Goodwin, C. (1999, May). Implementing computer mediated communication in an undergraduate course: A practical experience. *Journal of Asynchronous Learning Networks, 3*(1). Retrieved June 24, 2004, from *http://www.aln.org/publications/jaln/v3n1/v3n1_graham.asp*

Gunawardena, C. N. (1992). Changing faculty roles for audiographics and online teaching. *The American Journal of Distance Education, 6*(3), 58-71.

Hara, N., Bonk, C.J., & Angeli, C. (1998, March). Content analysis of on-line discussion in applied educational psychology courses. *Society for Information Technology and Teacher Education 98.* Washington, DC.

Harasim, L. (Ed.) (1990). *Online education: Perspectives on a new environment.* NY: Praeger.

Hartley, J. R., & Collins-Brown, E. (1999). Effective pedagogies for managing collaborative learning in on-line learning environments. *Education Technology and Society, 2*(2). Retrieved June 20, 2004, from *http://ifets.gmd.de/periodical/*

Henderson, T. (2001, September–October). Classroom assessment techniques in asynchronous learning networks. *The Technology Source.* Retrieved June 14, 2004, from *http://ts.mivu.org/default.asp?show=article&id=908*

Herman, J., Aschbacher, P., & Winters, L. (1992). *A practical guide to alternative assessment.* Alexandria, VA: Association for Supervision and Curriculum Development.

Holzen, R. von., Runyon, D., & Heeler, P. (2001). Assessment in online courses: Practical examples. *Proceedings of ALN Conference 2001.* Retrieved June 4, 2004, from *http://www.aln.org/conference/proceedings/2001/ppt/01_vonholsen.ppt*

Illinois Online Network. (1999, December). Beating cheating online. In *Pointers & clickers, technology tip of the month*. Retrieved June 24, 2004, from *http://www.ion.illinois.edu/Pointers 1999_12.html*

Juwah, C. (2003, January). Using peer assessment to develop skills and capabilities *USDLA Journal 17*(1). Retrieved June 27, 2004, from *http://www.usdla.org/html/journal/JAN03_Issue/article04.html*

Kulieke, M., Bakker, J., Collins, C., Fennimore, T., Fine, C., Herman, J., et al. (1990). Why should assessment be based on a vision of learning? Retrieved June 10, 2004, from *http://www.ncrel.org/sdrs/areas/rpl_esys/assess.htm*

Lazerson, M., Wagener, U., & Shumanis, N. (2000). What makes a revolution: Teaching and learning in higher education, 1980-2000. Retrieved August 1, 2004, from *http://www.stanford.edu/group/ncpi/documents/pdfs/5-11_revolution.pdf*

Le Heron, J. (2001). Plagiarism, learning dishonesty or just plain cheating: The context and countermeasures in information systems teaching. *Australian Journal of Educational Technology, 17*(33), 244-264. Retrieved July 31, 2004, from *http://www.ascilite.org.au/ajet/ajet17/leheron.html*

Liang, X., & Kim, C. (2004). Classroom assessment in Web-based instructional environment: Instructors' experience. *Practical Assessment, Research & Evaluation, 9*(7). Retrieved June 10, 2004, from *http://PAREonline.net/getvn.asp?v=9&n=7*

Lynch, M. M. (2001). Evaluating student mastery and program effectiveness. *DEOSNEWS Archives, 11*(12). Retrieved July 10, 2004, from *http://www.ed.psu.edu./acsde/deos/deosnews/deosnews11_12.asp*

MacAlpine, M. (2002, February). Principles of assessment (Computer-assisted Assessment Centre Bluepaper No.1). Retrieved June 15, 2004, from *http://www.caacentre.ac.uk/dldocs/Bluepaper1.pdf*

Macdonald, J. (2002, August). Developing competent e-learners: The role of assessment. Paper presented at the *Learning Communities and Assessment Cultures Conference* organized by the EARLI Special Interest Group on Assessment and Evaluation, University of Northumbria at Newcastle.

Maki, P. (2002). Aligning teaching, learning, and assessment methods. Presentation at AAHE 2002 National Conference on Higher Education, Community of Practice. Retrieved July 9, 2004, from *http://www.aahe.org/nche/2002/Cops/assessing/activity_4/1*

Mason, R. (1998, October). Models of online courses. *The ALN Magazine, 2*(2). Retrieved July 4, 2004, from *http://www.aln.org/publications/magazine/v2n2/mason.asp*

McMillan, J. H. (2000). Fundamental assessment principles for teachers and school administrators. *Practical Assessment, Research & Evaluation, 7*(8). Retrieved July 2, 2004, from *http://PAREonline.net/getvn.asp?v=7&n=8*

McNett, M. (2002, May-June). Curbing academic dishonesty in online courses. Retrieved June 24, 2004, from *http://www.ion.illinois.edu/Pointers 2000_05/default.asp*

Merisotis, J., & Phipps, R. (1999). What's the difference? Outcomes of distance vs. traditional classroom-based learning. Retrieved June 24, 2004, from *http://www.siue.edu/TLTR/GrpB2.htm*

Moore, G. S., Winograd, K., & Lange, D. (2001). *You can teach online: Building a creative learning environment.* Boston: McGraw Hill.

Morley, J. (2000, January). Methods of assessing learning in distance education courses. *Education at A Distance, 13*(1). Retrieved June 4, 2004, from *http://www.usdla.org/html/journal/JAN00_Issue/Methods.htm*

Muirhead, B. (2002). Effective online assessment strategies for today's colleges and universities. *Educational Technology & Society, 5*(4). Retrieved July 31, 2004, from *http://ifets.ieee.org/periodical/vol_4_2002/ discuss_summary_october2002.html*

National Research Council. (2001). Knowing what students know: The science and design of educational assessment.Retrieved June 15, 2004, from *http://books.nap.edu/books/0309072727/html/R1.html#pagetop*

Pain, D., & Le Heron, J. (2003). WebCT and online assessment: The best thing since SOAP? *Educational Technology & Society, 6*(2), 62-71. Retrieved June 16, 2004, from *http://ifets.ieee.org/periodical/6-2/7.html*

Palloff, R. M., & Pratt, K. (2001). *Lessons from cyberspace: The Realities of online teaching.* San Francisco: Jossey-Bass.

Palloff, R. M., & Pratt, K. (1999). *Building learning communities in cyberspace.* San Francisco: Jossey-Bass.

Phipps, R., & Merisotis, J. (2000, April). Quality on the line: Benchmarks for success in Internet-based distance education (Study ED444407). Retrieved January 14, 2004, from *http://www.ihep.com/Pubs/PDF/ Quality.pdf*

Rogers, D. L. (2000). A paradigm shift: Technology integration for higher education in the new millennium. *Educational Technology Review, 1*(13), 19-33. Retrieved June 24, 2004, from *http://www.aace.org/dl/files/ETR/ ETR11319.pdf*

Rowe, N.C. (2004). Cheating in online student assessment: Beyond plagiarism. *Online Journal of Distance Learning Administration, 7*(2) [Electronic version]. Retrieved July 31, 2004, from *http://www.westga.edu/~distance/ojdla/summer72/rowe72.html*

Serwatka, J. (2002, April). Improving student performance in distance learning courses. Retrieved June 10, 2004, from *http://www.thejournal.com/magazine/vault/A4002B.cfm*

Sims, R., Dobbs, G., & Hand, T. (2001). Proactive evaluation: New perspectives for ensuring quality in online learning applications. Retrieved March 5, 2003, from *http://www.ericit.org/olweb-cgi/fastweb?search*

Spatariu, A., Hartley, K., & Bendixen, L. (2004, Spring). Defining and measuring quality in online discussions. *Journal of Interactive Online Learning, 2*(4). Retrieved June 30, 2004, from *http://www.ncolr.org/jiol/archives/2004/spring/02/index.html*

St. Cloud State University Assessment Office. (1999). Definitions of assessment. Retrieved June 10, 2004, from *http://condor.stcloudstate.edu/%7Eassess/*

Sun, J.R. (2002, March). E-assessment: The design of assessment activities in technology enhanced and distance learning. Paper presented at the *Ohio Learning Network Conference 2000*, Columbus, OH. Retrieved March 11, 2004, from *http://www.oln.org/conferences/OLN2002/pdf/Sunpaper.pdf*

Tapscott, D. (1998). *Growing up digital: The rise of the net generation.* New York: McGraw Hill.

Taylor, C. R. (2002, October). E-Learning: The second wave. *ASTD Learning Circuits*[Electronic version]. Retrieved August 4, 2004, from *http://www.learningcircuits.org/2002/oct2002/taylor.html*

Twigg, C. A. (2001). Innovations in online learning: Moving beyond no significant difference. Retrieved June 20, 2004, from *http://www.center.rpi.edu/PewSym/mono4.html*

Weisburgh, M. (2003, October–November). Assessing the market. *Upgrade Magazine* [Electronic version]. Retrieved June 7, 2004, from *http://www.siia.net/upgrade/archive/1011_03/1011_03.pdf*

Whitis, G. R. (2001). *A survey of technology-based distance education: Emerging issues and lessons learned.* Washington, DC: Association of Health Centers.

Wilson, B. G. (2001). Trends and futures of education: Implications for distance education. Submitted to *Quarterly Review of Distance Education.* Retrieved June 5, 2004, from *http://carbon.cudenver.edu/~bwilso/TrendsAndFutures.html*

Chapter II

Defining Online Assessment for the Adult Learning Market

Betty Bergstrom, Promissor, USA

Jim Fryer, Promissor, USA

Joel Norris, Promissor, USA

Abstract

Learning for many adult professionals culminates in some form of assessment. Doctors, cosmetologists, private detectives, insurance agents, real estate brokers, plumbers, contractors, electricians, nurses, roofers, dietitians, and many others are trained in a variety of programs from community colleges to graduate programs, and increasingly their education and testing is conducted wholly or in part via e-learning platforms. Online assessment provides educators with the ability to measure learning needs, assess the results of learning activity, and speed learners toward professional credentialing by leveraging the efficient delivery framework of the Internet. In this chapter, we describe several low- to high-stakes online test delivery methods, which, when coupled with e-learning systems, can offer education providers with valuable opportunities for enriching the validity and credibility of their programs. We discuss the relationship of online assessment models in the context of e-learning, as well as distinguish between the types of testing technologies used to deliver them. We conclude by discussing the advantages and disadvantages of offering assessments online and the future of online assessment systems in general.

Introduction

E-learning is a form of technology-enabled instruction designed to increase learners' knowledge and skills so they can be more productive, find and keep high-quality jobs, advance in their careers and have a positive impact on their employers', families' and communities' success. E-learning takes place online—via the Internet or through some form of courseware—and typically offers users a variety of content presentation modes as well as resources for retrieving class materials, researching subject-related information, and facilitating peer-to-peer interactions. Without requiring an instructor to be physically present, e-learning has been viewed as revolutionary because it explodes the concept of a personalized instructor-to-student relationship and allows the education process to be individualized rather than institution based.

The rising popularity of e-learning has paralleled extraordinary increases in domestic and international Internet access. Over the past 10 years, the number of Internet sites and users has grown from thousands to millions. As of 2004, three out of four Americans have Internet access in the United States. Of Americans living in households with a phone line, 74.9% have home access to the Internet. This amounts to 204 million Americans out of the projected 272 million who are at least two years old (Nielsen/NetRatings, 2004). Placed in historical context against other media forms, the Internet has become as pervasive as radio or television, in a fraction of the time. The impact of readily available home and work Internet access has been, and will continue to be, a substantial contributor to the availability of e-learning technologies for many people without access to traditional forms of education. According to the U.S. Department of Commerce (2002), although ethnic minority groups in the United States—particularly African Americans—were among the last to use the Internet in the early 1990s, their usage has grown significantly over the last 10 years, and individuals with disabling conditions have also been connecting in growing numbers.

By eliminating mandatory clock-hour attendance in favor of performance and outcome-based measures, and by emphasizing customized learning solutions over generic one-size-fits-all instruction, the self-service potential of e-learning is particularly appealing to these nontraditional learners. Laurillard (1993) stated that one of the major roles for distance learning is to promote self-directed learning and increased learner autonomy. More recently, Collis and Moonen (2001) coined the term *pedagogical re-engineering* to describe the change in online education from an environment that is teacher centered to one that is focused on learner activity. But as theorists and educators at many levels have come to embrace online technology as a valuable and important teaching tool, the issue of integrating assessment into online learning has remained widely unaddressed. One of the most important characteristics of an e-learning environment should be its ability to adapt instructional design to meet individual needs and to evaluate the knowledge acquisition and retention rate of its learners. E-learning platforms that purport to reliably measure learner knowledge or competence must not only follow well-established theories of measurement and testing, but incorporate some form of assessment into their design. Unfortunately, mechanisms for easy, active tracking and evaluation of user learning are missing from many e-learning platforms today.

Online assessment provides distance educators with the ability to measure learning needs, assess the results of learning activity, and speed learners toward professional credentialing by leveraging the efficient delivery framework of the Internet. Given the array of products and services available to fulfill these needs, understanding the role of assessment in e-learning is critical to making informed decisions about what types of online testing systems should be considered appropriate for specific online learning programs. Our goal in this chapter is to describe several low- to high-stakes online test delivery methods that, when coupled with e-learning systems, can offer distance education providers valuable opportunities for enriching the validity and credibility of their programs. We discuss the relationship of online assessment models in the context of e-learning as well as distinguish between the types of testing technologies used to deliver them. We conclude by discussing the advantages and disadvantages of offering assessments online and the future of online assessment systems in general.

Types of Online Assessments

Online assessment is a method of using the Internet to deliver, analyze, and report exam content and, when appropriately used, it can greatly enhance the efficacy of online learning (Bergstrom & Lopes, 2003). It is important to keep in mind that assessment can and does take place throughout the entire learning process, not just at the end. Black and William (1998) defined assessment broadly to include all activities that teachers and learners undertake to get information that can be used diagnostically to alter teaching and learning. When viewed in this way, assessment is not limited to just assigning scores, grades, or some other form of credentialing to learners in the form of an exam. Rather, assessment persists throughout the learning process and can encompass teacher observation, learning discussions, group collaboration, and analysis of educational or training work.

The idea of administering assessments online is attractive because it assists e-learning providers in overcoming the barriers of time, location, and cost of test delivery. But to administer online assessments effectively, it is essential that they interface with the delivery components of an e-learning system and integrate with the overall pedagogical objectives of a well-realized learning program. In addition, a not-inconsequential aspect to any online testing program is the development of appropriate business rules for test delivery, which can include well-developed registration processes, sophisticated item delivery options, robust score reporting features, and powerful data retention capabilities. By examining how the test delivery process can be used for different purposes and at different levels, we can begin to understand how online assessment can aid adult learners in focusing on specific learning outcomes and enhance online education practices.

Table 1 contrasts several different types of online assessments against the following evaluation characteristics: the function, outcomes, and stakes of the assessment; the method used to deliver the assessment; the assessment's monitoring requirements; and the user access restrictions to the assessment.

The content of an e-learning environment dictates the functions or uses of a corresponding assessment. Understanding the goals of an e-learning system, then, cannot only help us determine the purpose of its related assessments but, subsequently, how those assessments are delivered. Assessments can be classified into three broad categories, according to their general use. They can be used prior to, during, and following learning (Swearingen, 2004) as

- diagnostic assessments
- formative assessments
- summative assessments

Diagnostic assessments identify learner strengths and weaknesses. These assessments can be used to identify specific personality characteristics or traits (e.g., motivation for success, personality type), determine suitability for a particular type of job or trade, or allow individuals to self-assess their ability to complete a task or demonstrate knowledge of a particular subject area. Delivered as knowledge practice tests, diagnostic assessments administered

Table 1. Assessment and online delivery use matrix

	Practice	Diagnosis	Course Evaluation	Continuous Course Assessment	Screening /Placement	End of Course or Benchmark Assessment	Credentialing
Function	Diagnostic	Diagnostic	Formative	Formative	Formative	Summative	Summative
Outcomes	To help an examinee become familiar with test or item formats	To discover the strengths and weaknesses a person may have in a content area	To evaluate the effectiveness of course, curriculum, instructors, materials or facilities	To assess learner progress and provide feedback to learner	To decide if a person qualifies for a particular program or to place the learner properly within a program	To provide a course grade or final assessment	To aid in the decision to license or certify an individual
Stakes	Low	Low to Medium	Low to Medium	Low to Medium	Medium	Medium	High
Delivery Method	Web-based	Web-based or secure site	Web-based or secure site	Web-based or secure site	Web-based or secure site	Web-based or secure site	Secure site
Monitoring Requirements	Non-proctored	Proctored or Non-Proctored	Non-Proctored	Proctored or Non-Proctored	Proctored	Proctored	Proctored
Interface and Access Restrictions	Unrestricted	Restricted or unrestricted	Restricted or unrestricted	Restricted or unrestricted	Restricted	Restricted	Restricted

Table heading: **Assessment and Online Delivery Use Matrix**

prior to a training or education program can be used to identify learner strengths and weaknesses or, in the form of screening tests, be used to determine proper course placement. Diagnostic tests can also be used by employers to identify individual training needs and improve the skills of employees in the workplace. Because diagnostic tests require the collection and retention of learner information for a prescriptive purpose, diagnostic assessments frequently include mechanisms for user data collection as well as detailed reporting tools.

Formative assessments take place during the learning process. Formative assessments involve the delivery of multiple-choice or short-answer quizzes administered at the end of a textbook chapter, learning module, or other learning benchmark in a course or training program. Feedback is almost always provided during or following the delivery of these assessments, and opportunities for self-remediation may also be made available. Formative assessments can provide teachers with data that can be used to guide an individual student's progress, improve the curriculum, or serve as the starting point for a remediation loop that prescribes specific learning modules based on their assessment outcomes. Formative assessments are usually considered low-stakes exams and good question writing, including the creation of detailed prescriptive feedback, is critical to these assessments' successful use.

Summative assessments frequently take place in the middle or end of a learning or evaluation program and can be used for grading, certification, and high-stakes evaluation. Summative certification, licensure and some cognitive ability tests are administered with the purpose of identifying the best candidates to be awarded some form of credential. Summative assessments are almost always delivered in a proctored environment. They are often considered "high stakes" because their outcomes have an impact on a learner's ability to advance in a course, receive some form of accreditation, or receive permission to professionally practice a learned skill. As a result, summative assessments typically call for candidates to authenticate their identity in advance of the assessment by requiring them to demonstrate proof of identity or class enrollment status.

Just as in any other learning environment, the goals of an assessment in the context of e-learning have important implications for how it should be administered. Out of both practical and pedagogical necessity, assessment should always be at the center of e-learning curriculum design to ensure quality online instruction. In turn, the goals and purposes of an assessment frequently have a direct impact on what and how learners learn. For example, a diagnostic practice test or a survey designed for personal development or to capture a candidate's opinion is typically easier to deliver via a Web browser then a licensing exam for a real estate or insurance agent. More care in test administration with regard to security, validity, reliability, and access is required for higher stakes assessments, perhaps even requiring the test to be delivered in a secure, dedicated test site rather than through a Web browser at home or in the classroom.

Online Assessment Delivery Methods

The delivery methods for online assessments can be divided into two broad categories: *Web-based assessments* and *secure site assessments*. Web-based assessments use Web interfaces such as Internet Explorer or Netscape to deliver test content to end users via the Internet. These tests can be delivered quickly and easily, any time, any place. This approach to test delivery is appropriate for low-stakes tests in which the sponsor has not made a large investment in test content and wherein cheating would have little impact or no benefit to the test taker. Secure site tests take place at high-stakes test centers or proctored third party locations such as community colleges or conference centers and go much further than browser-based tests with regard to managing the security of test content. Each of these delivery methods is described in detail.

Web-Based Assessments

With the wide availability of course management software (courseware), the World Wide Web is becoming increasingly attractive as a mechanism for the delivery of assessments, especially when directly associated with e-learning.

Web-based assessments are designed from the ground up to make the greatest use of Internet technology. Web-based assessments are written on a standards-based application server, such as Java 2 Enterprise Edition (J2EE) or .NET, and are designed from the outset to be accessed over the Internet by a Web browser. Many students now make use of self-service online learning products in their day-to-day studies, leveraging this technology into the realm of assessment brings assessment practice and use closer to the learning environment of students.

Within the last 5 years, a variety of e-learning providers, and even high-stakes certification and licensure sponsors, have begun to look to Web-based testing for the delivery of mid- to low-stakes examinations such as practice tests and quizzes. Distinct from high-stakes computer-based test (CBT) delivery methods, which frequently require some form of localized clientserver network at a secure testing site for test administration, Web-based testing requires only a common Web browser and active Internet connection for test delivery. Many sophisticated methods of test delivery can now be supported via the Web, and increasingly, Web-based testing systems are offering many of the same test-delivery and security benefits as a dedicated test center.

The advantages of offering testing services over the Web are many and varied. For e-learning providers who cannot afford or otherwise deliver tests through a high-stakes channel of testing centers, the World Wide Web is a powerful and often less expensive way to extend quality testing services to the home, workplace, or remote testing venues. Offering tests over the Web enables test developers to use a wide array of multimedia technologies that can enhance a test's authenticity or improve its accessibility. As the preferred method of digital data transfer, the Web offers near-instant access to test data, which allows test results and test-taker feedback to be made available on demand. Web-based test delivery can also permit organizations to take a much more active hand in the creation and administration of their testing services, and offers a relatively accessible way to integrate testing with e-learning objectives. As with any CBT, a Web-based test eliminates the need for test booklets and answer sheets.

Currently, the most popular use of Web-based assessments is for diagnostic and formative testing. Diagnostic tests such as quizzes and surveys can be administered easily via the Web to measure knowledge or competency prior to course or training delivery. For assessments such as these, Web-based assessment is ideal. Offering the capability for real-time synchronous delivery of test content and instructional feedback, Internet connectivity must be maintained throughout the entire test, and questions and responses must be continuously relayed back and forth from the test taker to the assessment provider's Web server. Another popular use of online technologies is for the electronic submission of written assignments that can streamline administrative processes; the results of which can aid the calibration of instructional methods to meet individual learning needs.

Nonproctored, unsecured online assessments do not require that an observer be present to monitor the test-taker's progress, nor does it require controls to be used that prevent test takers from accessing any information other than the test itself. Nonproctored, unsecured Web-based assessments are commonly used for self-tests such as the Houghton Mifflin College Division's online practice tests for accounting (*http://college.hmco.com/accounting/needles/principles/ 9e/students/ace/index.html*). Both professors and students are able to access these online tests by pointing to a Web address provided in their textbooks. Practice tests such as these are widely available and can be offered as standalone practice tests or be offered as a means of online chapter review and self-assessment.

Another example of a nonproctored, unsecured practice test for adult professionals is the Association of Registered Diagnostic Medical Sonographers (ARDMS) test. ARDMS offers short, 30-question practice versions of these practice tests via the Web. All of the practice exam questions are representative of the corresponding content outlines of the real ARDMS examinations. To mimic the pacing and experience of a real ARDMS exam, the time allotted to answer the test questions in the practice exam is the same as it is for the real exam, which is about 1 minute per question (30 questions in 30 minutes). The questions included in these practice exams are randomly selected from a database of test items, so no two practice exams will be exactly the same.

Figure 1. Houghton Mifflin College Division ACE practice test in accounting

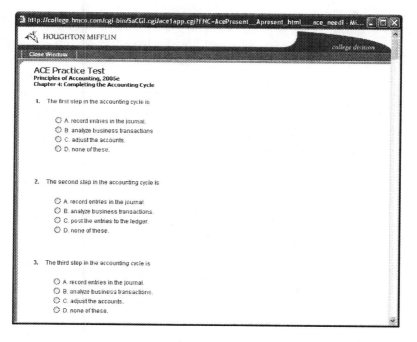

Figure 2. Splash screen for the ARDMS Cardiovascular Principles and Instrumentation practice exam hosted by Promissor

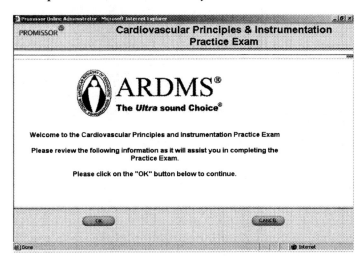

Candidates can take any test as many times as they like to learn more about what they need to know to pass the actual ARDMS credentialing examinations. After they finish a practice exam, candidates will receive a test report that lists the content areas that they may wish to further investigate (e.g., see *http:// www.ardms.org/practiceexams/index.htm*).

Web-based tests can easily be offered under proctored conditions for mid-stakes tests that may require some form of monitoring. Under this Web-based assessment model, test takers are not allowed to begin a test until an assigned proctor enters a login ID and password, which must be authenticated by the testing system. By design, proctored online testing fully supports test session monitoring. Because a proctor must authorize the start of all tests by entering a personal ID and password, it provides an opportunity to verify test-takers' identities prior to administration as well as effectively control access to the test. By monitoring the examination during the testing period, the proctor can also prevent cheating and help guarantee that test takers take the exam without influence from outside sources.

Online proctored testing can offer a more efficient, cost-effective approach for test delivery than dedicated test center administration. Such a system, for instance, is ideal for classroom or education use. Proctors may be third party contractors, designated staff, or even teachers. Delivered under the right conditions, proctored online tests can be used for both mid-stakes and high-stakes assessment. For testing in remote locations where test volume does not warrant a dedicated test center, proctored online tests may, in fact, be the only viable testing method available.

Web-based testing can be made even more secure if the test taker is required to download and install an application to prevent them from accessing other applications or viewing other Web pages during the test. This type of secure Web-based testing offers the convenience of testing online while preventing test takers from compromising the credibility and reliability of a Web-based assessment. Many vendors today offer applications like these that provide one or more of the following security benefits:

- Hide all navigation controls, menu options, and toolbars from the browser display area
- Disable all browser shortcut keys and right-click mouse menus
- Prevent test takers from accessing other applications
- Prevent test takers from printing questions on the screen
- Present a display area that encompasses the entire screen and prevents test takers from resizing or moving the display frame to access other applications on the desktop
- Offer a single means of closing the application that, when selected, effectively terminates the testing session
- Guarantee that all test items are cached on the test taker's machine only for as long as they appear on the screen

Assessment delivery applications such as these automate important security safeguards that protect the value of an organization's assessment. The purpose of these safeguards is to prevent test takers from using the assessment computer for any other purpose other than taking the assessment from the time the test session begins until it terminates. If required, these features can protect test content for low-stakes or practice tests in the office or classroom, or be used to facilitate assessments in the workplace for training evaluation.

In all of the previous cases, depending on the level of security and reliability required, test items may be downloaded from the Internet prior to test delivery and administered offline via a proprietary delivery application, or the entire test may be delivered over the Web in real time. Like nonproctored, unsecured assessments, these tests can be immediately scored upon completion and the results sent to a central data repository for consolidation and reporting purposes.

Secure Site Assessments

Without question, Web-based testing has profoundly changed the way tests are developed and delivered, but offering high-stakes tests via the Web concerns test developers and educators alike, and not without good reason. As online testing technology continues to evolve, two key factors that continually hound the feasibility of offering high-stakes tests via the Web are reliability and security. For high-stakes tests that result in some form of credentialing or licensure, Web-based delivery via e-learning is often not viewed as feasible for many programs today.

A test is considered "high stakes" when there are social, legal, or professional consequences linked to the test result—for the test candidate, the testing program, or the groups involved in the testing. Most licensing and certification tests are considered high stakes, and because of the stakes involved, administering them to the Web is not a simple matter. As the social and professional ramifications of these assessments can be extensive, licensure and certification tests must not only be developed to meet rigorous psychometric standards for measurement quality (e.g., reliability, equivalence of alternate forms) but delivered in a highly secure manner.

Many types of high-stakes summative assessments must be delivered at secure test sites. The nature of these sites vary, based on the type of the assessment delivered. Academic summative assessments may be delivered in a proctored classroom or computer center. Professional certification or licensure examinations may be delivered in highly secure, professionally owned and operated testing facilities. Because online summative assessments are usually linked to some form of credentialing, the data collected during the assessment is usually transmitted to and from the test taker's workstation in a much more sophisticated and secure manner than in diagnostic and formative online assessments. Furthermore, the integration and interpretation of data in the summative assessment process requires a high degree of skill, psychometric sophistication, and education. Following test delivery, summative assessment results may need to integrate with a number of different systems; they may be securely transmitted to credentialing organizations for licensure or certification evaluation, testing organizations for item analysis, and e-learning providers for end-of-course notification.

For many decades, credentialing examinations were given via paper and pencil in proctored testing environments to ensure the highest security possible, and for the most part, high-stakes CBTs have adhered to the same policies and rules established for paper-and-pencil test delivery. Today, high stakes CBTs are typically delivered by a variety of domestic and international vendors in dedicated test centers. Usually these tests are administered "off-line" or asynchronously, meaning that an electronic test file is sent to a test center prior to administration, but the test does not rely on a real-time or synchronous connection to the Internet for delivery. Many of these sites—even those that are established for temporary testing—rely on a local area network (LAN) to administer tests as opposed to relying on a direct connection to the Internet for assessment delivery. This ensures that the test is administered without interruption, on a secure network, under the watchful eye of a dedicated proctor (and possibly video camera), in a very secure environment.

The facilities where a secure site test take place typically have well-defined practices and procedures for test administration and always require proctors to authenticate test takers' identities as well as monitor their activities during an assessment. The sponsors or vendors who provide high-stakes testing services under these conditions may operate dedicated centers for this type of assessment delivery, or subcontract with community colleges, training centers, or other third parties to administer and proctor assessments on their behalf. Before the examination begins, the files necessary for test administration are sent via the Internet to the test center or test station. These files include all of the test-taker information—possibly including the test taker's photo or even fingerprints—and the entire test (including the pool of questions if the test is randomly or adaptively administered). A live Internet connection is required only at test download and upon completion of the test for return of results.

For high-stakes testing purposes, sponsors also want their test-delivery technology to have features for effective test proctoring. For example, it is important to control the timing of high-stakes tests. Just as in the Web-based assessment systems described previously, a test administrator or proctor frequently must authorize test initiation by entering a personal ID and password into the test-delivery system. Another important concern with high-stakes delivery is how test content and results are sent back and forth between the test site and the testing system's servers. Because sending test data over the Internet could risk theft through wiretapping, router snooping, or other means, this data should be encrypted to maintain security. In addition, test data should also be stored in an encrypted and unreadable format.

For the highest level of security, the sponsors or vendors providing high-stakes testing services should have full control over the testing facility, its test administrators, and its proctors. Although this control helps prevent cheating and the theft of test content, by far the most important factor is the security provided by the test proctors. Highly effective test supervision, when combined with secure testing technology, will produce the highest levels of test security. Ideally, such proctors should be directly employed by the sponsor or testing vendor, thoroughly trained and certified in proper security procedures, and closely supervised to ensure that their performance meets defined security standards.

Secure site testing is especially well suited for the use of assessments that make use of item response theory (IRT). For example, computerized adaptive tests (CATs) that tailor difficulty to the ability level of each examinee can be efficiently delivered through this medium. In this process, IRT technology would be used to select which items are given so that they are of appropriate difficulty for each examinee. In this way, online assessment can also offer the potential for assessing abilities and skills not easily assessed by paper-and-pencil methods.

An example of a high-stakes assessment CAT delivered via the Internet to secure test sites is the Council on Certification of Nurse Anesthetists (CCNA) Exam. The CCNA is charged with protecting and serving the public by assuring that individuals who are credentialed have met predetermined qualifications or standards for providing nurse anesthesia services (Markway, Vitek, & Wolkober, 2004). The certification program began in 1945 and the National Certification Examination, which tests the knowledge of entry-level nurse anesthetists, was moved to a CAT format in April 1996.

When using a CAT, each candidate takes an individualized test administered on a computer. Candidate competency is continually assessed online, and the difficulty of each item administered is targeted to the current ability estimate of

Figure 3. Council on Certification of Nurse Anesthetists certification information Web page

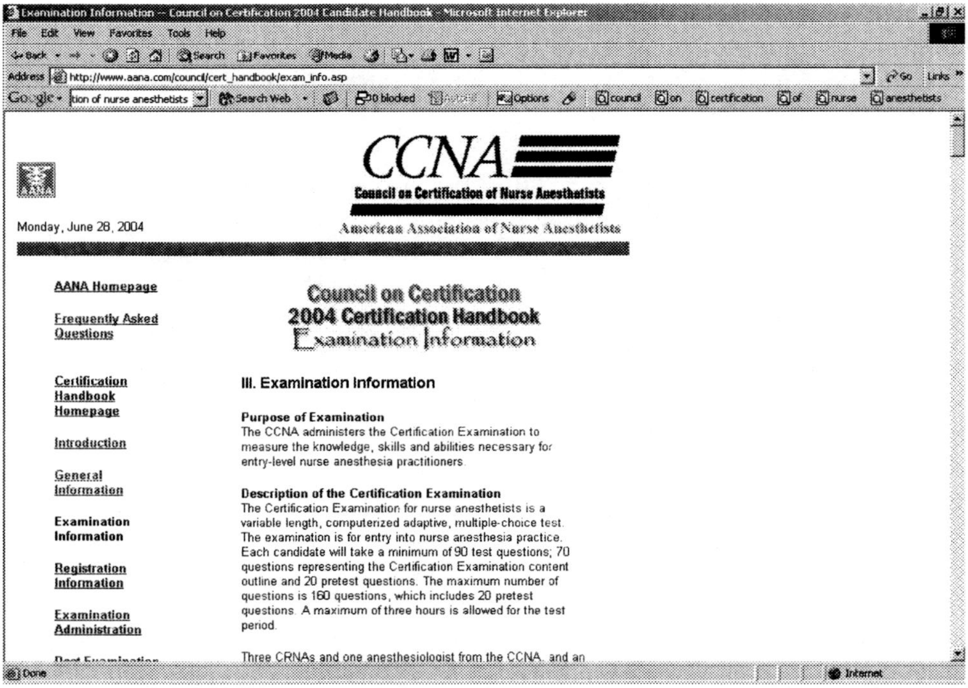

the candidate or to the pass point (Bergstrom & Lunz, 1999). Adaptive testing requires that the test file pushed via the Internet to the secure test centers contain a large number of items and all of the test specification that enable this sophisticated test to be properly administered. Each candidate thus receives an individualized test that is content balanced and of appropriate difficulty. The CCNA provides online information for candidates to help them prepare for the examination (e.g., see *http://www.aana.com/council/cert_handbook/ exam_info.asp*).

Combining E-Learning with E-Assessment: City & Guilds

With over 500 qualifications, City & Guilds is the leading provider of vocational qualifications in the United Kingdom (*http://www.city-and-guilds.com*). The Smartscreen e-learning component (*http://www.smartscreen.co.uk/*) provides online assessment information including projects, syllabus details, careers service, recommended Web-sites and unit-specific support materials. For example, e-Quals Software Development Level 2: Smartscreen provides

- Scheme of work
- Lesson plans and identification of key skills within lesson plans
- Sample question paper
- Sample assignment and software for sample assignment
- Answer for assignment
- Worksheets for selected Level 2 units (overview)
- Glossary of terms
- Levels chart
- Career progression chart
- Useful web sites
- Recommended reading list

In addition to e-learning components, City & Guilds offers e-assessment. In order to complete the qualification, learners go to one of over 3,000 City & Guilds centers to take their qualifying test.

Utilizing Different Delivery Modes: Training and Assessment in the Insurance Industry

In the United States, to work in the Insurance industry, it is necessary to pass an examination for the line of business that you wish to sell. Licensing in the insurance industry is regulated at state level as opposed to the federal level. States vary on eligibility requirements for the individual. The universal key, however, is passing the examination.

Some states mandate that the individual take a prelicensing education course. In those states, a preset curriculum (and often a required number of study hours) is mandated and the individual must receive the education through an approved prelicensing provider or school. This self-study course may be completed in the traditional textbook format with a proctored or nonproctored exam, or in many states, it may be completed online. Texas, for example, allows online prelicensing education (*http://www.insurancece.com/global/map/texas.html*).

Once individuals complete their online education, they may choose to practice for the licensure examination with an online practice test. These tests give candi-

Figure 4. Promissor insurance and real estate practice test frequently asked questions Web page

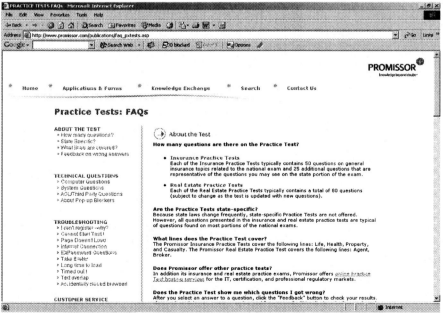

dates a chance to take questions that are similar to the actual licensure examination that they will be taking and to experience a computer based examination (*http://www.promissor.com/publications/faq_pxtests.asp*).

After preparation, the individual must sit for the licensing examination. Most jurisdictions have outsourced their examinations to national testing vendors, and examinees take their licensing examination in a proctored environment where the exam has been sent via the Internet. Individuals sign up for their licensure exam via the Web and then go to a testing center to take the examination in a secure environment.

After licensure, all jurisdictions require the licensee to take continuing education (CE) courses to maintain and renew the license. All states allow online education, though most require that the individual pass an examination (either nonproctored or proctored by a disinterested third party) to receive credit. Ohio allows individuals to earn credit through interactive online education without an examination. Those courses allow the individual to move through the program, answering questions, following links, and so forth as evidence that the program is completed. States such as Texas provide information about the examinations and give examples on their Web sites: (e.g., see *http://www.tdi.state.tx.us/agent/agcehome.html*).

Utilizing Different Delivery Methods: Training and Assessment in the Real Estate Industry

The real estate industry also utilizes an online learning, online assessment, and high-stakes credentialing model. The Association of Real Estate License Law Officials (ARELLO) is an organization composed of real estate regulatory officials from across the United States, Canada, Australia, Bermuda, Hong Kong, Philippines, South Africa, and Zimbabwe. ARELLO (*http://www.arello.net*) promotes better administration and enforcement of licensing and regulatory laws in member jurisdictions. This organization certifies online courses and its distance education certification is recognized by over 40 jurisdictions. ARELLO defines Internet courses as follows:

> *Internet courses are courses that require an Internet connection to complete. Internet or web-based courses are typically facilitated by web technologies that use a web browser as the primary means of*

content delivery and providing interactivity. Typically, Internet based courses store student progress information in a database housed on the provider's server. Internet courses can typically be run from any computer having an Internet connection, a web browser and any software needed to run that particular course. In an Internet based course, the content is delivered to the student in logical increments and assessments are performed at appropriate intervals using Internet or web technologies. (ARELLO.net, http://www.arello.net)

ARELLO has the following standards with regard to assessment. Diagnostic pretests are not required, but they are encouraged. Incremental (formative) assessments are typically considered to be quizzes given after each logical unit of instruction and are required at appropriate intervals throughout the course to obtain certification. These assessments should be designed to measure adequately whether mastery of the material has been achieved. Summative assessments are not required by ARELLO, although some jurisdictions require them. ARELLO also highly recommends that course providers develop banks of questions from which assessments can be drawn (ARELLO.net).

Disadvantages and Advantages of Online Assessment for Adult Learners

There are, of course, drawbacks to offering assessments online. Perhaps chief among them, particularly for e-learning organizations, is the issue of authenticity. As Naglieri et al. (2004) stated in the APAs Task Force Report for Psychological Testing on the Internet,

Tests can be placed on the Internet in a manner that suggests authority and conveys confidence, although many of these tests may have little to no documentation of reliability and validity, test-takers often ignore disclaimers that might appear, and self-administered tests can yield inaccurate interpretations. What is needed is considerably more accountability of the Internet site authors.... This would include, for example, documentation summarizing standardization samples, reliability, and validity as well as additional evidence such as equivalence of tests delivered on Internet and paper, uniformity of stimulus quality on different displays, and so forth, to ensure high quality test administration. (Naglieri et al., 2004, p. 65)

Many educators already feel challenged by the task and cost of producing high-end coursework delivered with reliable technology to students and may simply be unequipped to meet this level of standard. In the face of market pressure to offer distance education and e-learning courses, it can be very difficult to guarantee that users receive the same kind of protections obtained in traditional classrooms or training facilities with regard to certain types of assessments, particularly high stakes assessments. Similarly, test developers and publishing companies that enter into Internet testing programs may have difficulty ensuring that Internet tests are held to the same psychometric standards as traditional tests.

Another issue is support. Providing round-the-clock support to students who are now learning and testing at all hours can be expensive and time consuming. Many traditional educators feel that the face-to-face, iterative interaction with students is an important part of the learning process. While students can get immediate feedback from an online exam, educators do not always get the feedback of what topics students are answering incorrectly or are confused about. This raises a subsequent issue: The appropriateness and meaningfulness of the feedback following an assessment is often as important as the assessment itself. Assessments that fail to link meaningfully to outcomes can quickly be perceived as barriers instead of gateways to credentialing.

Loss of connectivity can be an issue with online assessment. Learners utilizing online courses and assessment need reliable Internet connectivity. Assessments must be designed so that loss of connectivity does not result in loss of data so that the test can be restarted at the exact point of interruption. Online assessment systems also have a drawback in that students who perceive themselves as possessing poor information technology (IT) skills may be disadvantaged. For example, a fair number of people enter professions in later life, and some of them may feel uncomfortable with computers. Assessing student readiness for online learning and online assessment is another important consideration facing e-learning and online assessment providers. It should not be assumed that students are ready and able to cope with new forms of learning and assessment. Extensive preparation and support may be required when introducing learners to these forms of education and testing (Naglieri et al., 2004).

There are also obvious advantages to online testing for busy adult learners. Online assessments can be made available 24 hours a day, seven days a week, from any computer with access to the Internet. Online assessments provide easy access to registration, online payment, information about the test, initial test administration requirements, retake test administration, certification, recertification, and results. Test takers may visit a Web site to register to take a test on a specific day, at a specific time, and at a specific test center. Test taker access may include the ability to review online transcripts, scores, and pass status (Bergstrom, Meehan, & Bugby, 2001)

Test-taker responses are recorded at the item level. A complete record of the test-taker's experience, including every item administered, the test-taker's response choice, and the amount of time spent on the item can be saved during each test. If the test is interrupted, upon restart the test can automatically return to the item completed just before the system interruption. Item-level timing and test-level timing accurate to the end of the previous item should be restored as well.

A new test with accompanying translations could be made available around the world almost instantly. Test publishers can download new tests to secure testing sites in a matter of moments, whereas other test developers can put tests on their Web sites and make them available to anyone with an Internet connection. Updating a test is also much easier. For example, revising a paper-and-pencil test requires printing and distributing new test forms and answer keys and printing new or revised test manuals, an expensive process that may take several months or years. Revisions of a test that appears on the Internet can be downloaded to testing sites around the world in a few minutes at virtually no cost.

In many paper-and-pencil testing and assessment programs, examinees typically receive their scores and interpretive reports a month or two after they take a test. Their answer sheets must first be mailed to the test publisher, where they are scanned and scored, and perhaps interpreted. Then reports are created, printed, and mailed back to the examinees. In an Internet setting, responses are recorded in computer files as examinees answer each item. Software that computes test scores and generates interpretive reports can be run as soon as the last item is answered, with examinees receiving feedback within a few seconds of completing the test. Internet testing is more scalable than paper-and-pencil testing.

Online assessment lends itself to a pedagogical approach to learning in which assessment is integrated with learning processes. In today's CBT world, multiple-choice, text-only items are increasingly being replaced with items that offer a rich array of presentation stimuli and response options, including pictures, line drawings, tables, maps, and equations, as part of the item stem or response option. Embedding audio files within an examination creates an entirely new dimension for CBT and expands its application to a wider range of assessments. Improvements in video technology have enabled test developers to use high-quality, full-motion video in an assessment environment. Interactive responses, such as hot spots, matching, constructed response, fill in the blank and essay items are also possible in a CBT environment. Online exhibits can be included to allow a test-taker access to Internet content (Bergstrom & Cline, 2003).

Conclusion

Formal vocational learning is no longer confined by institutional boundaries but can occur in multiple contexts including the workplace, the home, or in virtual learning communities. To meet the exigent needs of the nontraditional learners, testing and assessment remain an integral part of the learning process. A well-designed online assessment not only helps students assess their level of knowledge of course material but also gives the instructor a better idea of what students understand. These forms of education afford easier access and greater flexibility in adult learners' education options, but they also change what we expect out of education providers and assessment metrics.

For now, online assessment enables instructors to obtain feedback about how a learner is performing, as well as to evaluate the effectiveness of the e-learning environment. But in many ways, online assessment, like online teaching, is still very much in an embryonic state. Today, the most common online assessment strategies involve the use of computer communications as simply a transfer medium to submit and comment upon assigned work such as essays, submit and compile portfolios, and deliver traditional paper-and-pencil tests in a computer testing environment (i.e., multiple choice tests). As the mechanisms for learning paradigms change, so will our assessment delivery methods.

References

ARDMS Currently Available Practice Exams. Retrieved June 28, 2004, from the ARDMS Web site: *http://www.ardms.org/practiceexams/index.htm*

ARELLO.net Home Page. Retrieved June 28, 2004, from *http://www.arello.net*

Bergstrom, B., & Cline, A. (2003). Beyond multiple choice: Innovations in professional testing. *CLEAR Exam Review.*

Bergstrom, B., & Lopes, S. (2003, December). *The role of computer based assessment in on-line learning.* Presented at the Online Educa Berlin Annual Conference, Berlin, Germany.

Bergstrom, B., Meehan, P., & Bugby, A. (2001). *Certification testing on the Internet.* National Organization for Competency Assurance.

Bergstrom, B. A., & Lunz, M. E. (1999). CAT for certification and licensure. In F. Drasgow, & J. B. Olson-Buchanan (Eds.), *Innovations in computerized assessment* (pp. 67-92). Mahwah, NJ: Erlbaum.

Black, P., & William, D. (1998). Assessment and classroom learning. *Assessment in Education, 5*(1), 7-71.

City & Guilds Home Page. Retrieved June 28, 2004, from *http://www.city-and-guilds.com*

Collis, B., & Moonen, J. (2001). *Flexible learning in a digital age.* London: Kogan Page.

Continuing Education: Agents, All Adjusters, Providers. Retrieved June 30, 2004, from *http://www.tdi.state.tx.us/agent/agcehome.html*

Council on Certification: 2004 Certification Handbook Examination Information. Retrieved June 28, 2004, from *http://www.aana.com/council/cert_handbook/exam_info.asp*

Foster, D. (2001, August). Examinations: Using Tests Properly. *Certification Magazine* [Online]. Retrieved June 28, 2004, from *http://www.certmag.com/issues/aug01/contrib_foster.cfm*

Houghton Mifflin Ace Practice Tests. Retrieved June 28, 2004, from *http://college.hmco.com/accounting/needles/principles/9e/students/ace/index.htm*

Laurillard, D. (1993). *Rethinking university teaching: A framework for the effective use of educational technology.* London: Routledge.

Markway, P., Vitek, L., & Wolkober, L. (2004). Council on Certification of Nurse Anesthetists Candidate Handbook 2004. Retrieved June 28, 2004, from *http://www.aana.com/council/pdfs/2004%20CCNA%20CandidateHandbk.pdf*

Naglieri, J., et al. (2004). Report on APA's task force for psychological testing on the Internet. *American Psychologist, 59*(4).

Nielsen/NetRatings. Retrieved June 28, 2004, from *http://www.netratings.com/pr/pr_040318.pdf*

Promissor Practice Tests: FAQs. Retrieved June 28, 2004, from *http://www.promissor.com/publications/faq_pxtests.asp*

Smartscreen Home Page. Retrieved June 28, 2004, from *http://www.smartscreen.co.uk/*

Swearington, R. (2004). A Primer: Diagnostic, formative, & summative assessment. Retrieved June 28, 2004, from *http://www.mmrwsjr.com/assessment.htm*

US Department of Commerce (2002, February). *A nation online: How Americans are expanding their use of the Internet.* Washington, DC: Author.

Chapter III

Definitions, Uses, and Benefits of Standards

Eric Shepherd, Questionmark Corporation, USA

Abstract

This chapter explains the background and significance of specifications and standards that make it possible for content and management systems to work together. It examines why standards are needed and traces the challenges of developing them. It also explains the technical and operational standards that are currently available and the standards for exchanging data and communicating data. It describes the organizations that establish standards and specifications. I hope to give readers a working understanding of the principles behind today's use of technical interoperability standards as they relate to assessments and learning.

Introduction

Many people recognize the importance of standards and specifications to allow developers to create tools, content, and management systems that will work together. This interoperability allows solutions to work together and allows organizations not only to use generic systems but also to add best-of-breed modules when required.

As useful as standards are, the language relating to them can be confusing. This chapter provides readers with a working understanding of the principles behind today's use of technical interoperability standards as they relate to assessments and learning.

The chapter discusses the following topics:

- Background and significance of specifications and standards
- Differences between standards and specifications
- Challenges of developing standards and specifications
- Technical and operational standards currently available
- Standards for exchanging data
- Standards for communicating data
- Organizations that establish standards and specifications
- Key standards and specifications that are crucial to interoperability

Background

Many devices are based on standards that evolved and were developed decades ago; everything from traffic lights to electrical sockets to a vehicle's control system relies on standards to ensure people's safety and convenience. But these standards did not always exist. They had to evolve along with the technologies they regulate.

The first drivers of horseless carriages had no use for traffic lights, but as more and more people began to favor the automobile, a system needed to be established to dictate rights of way. When these lights first appeared at busy intersections, some drivers might have been annoyed, wondering why they should have to stop at a red light, but as traffic lights proved their worth, they came to be taken for granted. What would life today be like without them? A lot more dangerous!

No doubt these new contraptions created some confusion at first. The earliest traffic lights were based on railroads' stop and go signals; these signals were based on the flags used by signalers. Engineers used red and white lights to guide their trains, but the white light was changed to green after a red lens once fell out of a traffic light and caused an accident. Some cities initially chose the two-light signal based on the railroad system; others started to use a three-light system that could warn drivers that the light was about to turn from green to red.

And some cities didn't see the need to adopt traffic lights for decades after they were invented. What confusion this must have caused!

The history of the traffic light illustrates standards building upon existing practice and comparable standards, standards evolving in response to new issues and practices, and standards being adopted as needed rather than being mandated. The standards that are evolving today follow a similar process.

Specifications and Standards

The world of online assessments is just emerging from an early stage of setting standards. Innumerable discussions have taken place about how to enable one system to make contact with another and to ensure that the information exchanged between them is compatible and understandable.

There are seven criteria that may be used to judge the success of a specification or standard:

1. Durability: Will it stand the test of time?
2. Scalability: Can it grow from small to large?
3. Affordability: Is it affordable?
4. Interoperability: Will one system work with another?
5. Reusability: Can it be reused within multiple contexts?
6. Manageability: Is it manageable?
7. Accessibility: Can we all use it?

Using these criteria to judge the traffic light, we can see why it has become a successful standard. With standards and specifications relating to assessments and learning, we strive for these same qualities. These new standards and specifications follow three different styles:

1. Discussion of philosophies
2. Data models
3. Data interchange or application program interfaces (APIs)

Although the discussion of philosophies might not seem to provide much value, it does assist in framing the discussion and implementing systems that maintain interoperability. Although these philosophies could be regulated into a standard, this is unwise while systems are still evolving.

A data model describes the data elements and the relationships among them, so a student (data) might have a home address (data) and a college address (data) and might have attended previous colleges (data) and obtained various qualifications, skills, and degrees (data). Data, and the relationships between data elements, must be defined to provide the relevant meaning within a certain context. A prospective employer, for example, might want to see all the qualification data for a job candidate, and an association might care only about the address, a particular degree, and the money being paid.

Common data models allow disparate systems to derive a single meaning from the data that they share. For instance, a person might be referred to as a *student* in one system, a *learner* in another, and an *employee* in a third. Without a common vocabulary and data model, subtle differences can obscure the derived meaning. In the traffic light analogy, the white light and green light can be regarded as having the same meaning, although they are different colors. (No one said that this was going to be easy!)

Interoperability

If data resided within a single system, it would not interoperate. In order to interoperate, the data have to be shared between at least two systems. Data can be shared as batch (with data being bundled or packaged) or in real time (with data made available as they come into being). Real-time data sharing is preferable, as the latest data are always available. However, batch processing is easier to implement as the data can be checked, filtered, and rearranged offline without concern for the effect on the end users. If a real-time interface failed as a result of network congestion or system availability, the end-user experience would be frustrating.

Some systems use a hybrid approach by providing a batch of data and notifying the server to pick it up and process it. For example, a threaded discussion system could request information from a student management system to determine which threaded discussions should be available to a student, based on the courses the student is taking. The threaded discussion system could make a batch request each evening, or it could make a request on-the-fly as the student logs in.

Most modern interface specifications and standards focus on real-time data interchange. For example, some real-time interfaces that launch content and assessments by pushing data to the content and then receiving data back when completed include a protocol established by the Aviation Industry Computer-Based Training (CBT) Committee (AICC HACP) and the sharable content object reference model (SCORM API), developed by the Advanced Distributed Learning (ADL) Initiative.

Web services provide another form of real-time interface in which data can be pushed and pulled to and from other systems via the Internet or intranets. Web services offer a standard provided by the World Wide Web Consortium (W3C) to allow systems to exchange extensible markup language (XML) packaged data via the Internet using the simple object access protocol (SOAP). Web services are becoming the transport method of choice for the latest specifications and standards.

The challenge with Web services is that they are very configurable and, therefore, flexible. Using Web services does not make a system standards compliant, but it does indicate something about the way it expects to communicate with other systems. However, with a common transportation system in place, it is easier to define the packages that will traverse it. Web services are an important milestone in the pursuit of interoperability.

As Web services take front and center stage, more de facto standards will emerge. Consider, for instance, the usefulness of having a Web service to collect weather information at a training location. This standard is more likely to be provided by the weather-monitoring Web sites than by an industry consortium. The use of this Web service might become commonplace, and therefore become the de facto standard.

As real-time Internet-based transportation systems such as Web services become commonplace, work flow and practices will start to be defined to ease not only the data transfer between systems but also the processes that are maintained by the interoperability of the systems.

I have established that standards and specifications help one to understand the philosophies behind the interfaces—the data that need to be shared and the methodology to move the data from one place to another. But what are the benefits that can be derived from specifications and standards?

Benefits of Standards for E-Assessment and E-Learning

Specifications and standards that promote common data models and APIs are fostering the assessment and learning marketplaces in the following ways:

- The process of feature analysis, product design, and content design can be simplified.

- Specifications help vendors bring products to market more quickly, confident that they will interoperate with other systems and content to provide the solution desired by the customer.

- The overall market grows as systems become more robust; products specialize and diversify, and the technology finds new audiences.

- By using interoperable standards, two or more different pieces of software can be used together to create synergy. For example, a diagnostic assessment can route to the correct piece of learning. Standards help avoid requiring the users to log in multiple times, providing a more enjoyable user experience.

- Specifications and standards safeguard investment in digital content for both producers and consumers by assisting with portability, hardware, and operating system independence and reusability, all of which contribute to longevity.

- Content that can be driven from standards-compliant systems creates its own marketplace, which in turn stimulates production and a selection process that improves the quality of the available content.

With all these advantages, it would make sense to think that we should have had specifications and standards years ago. Indeed, some are as old as the Internet, but it is only now that these specifications and standards have matured sufficiently to be widespread. Only now do we have the infrastructure and understanding that can yield sufficient benefits to justify their use.

Challenges of Developing Standards

Although we recognize the value of specifications and standards, there are competing schools of thought about how these should be defined and refined. On one side are the idealists in search of the perfect model; on the other are the pragmatists who want results sooner rather than later. Academics and longer-term thinkers are normally in the former group, and business people and salespeople tend to be in the latter. It makes for an interesting debate.

The benefits and risks of each approach are shown in Table 1.

What we need are perfect standards quickly, but such a goal is elusive. Specifications and standards need time to mature.

Table 1. Benefits and risks of competing schools of thought

	Approach	
	Get it done soon	Get it done right
Benefits	<ul><li>Timely benefits are realized</li><li>People are engaged by the rapid process</li></ul>	<ul><li>It works as expected</li><li>Possible problems have been thought through and planned for</li></ul>
Risks	<ul><li>Can be difficult to maintain because there was not time to think through all of the issues</li><li>Might ignore similar standards and specifications</li><li>Might be overly simplistic and therefore handle only a limited set of uses</li></ul>	<ul><li>Happens slowly</li><li>Might miss the market requirement</li><li>Participants in the process can become disillusioned by the lack of results</li><li>Might be overtaken by a more pragmatic approach that becomes the de facto standard</li><li>Can become overly complicated and difficult to deploy</li></ul>

Another challenge with standards is the matter of patents and the ownership of the intellectual property (IP). There are two issues at stake:

- Who owns the specification so that it does not get reproduced and enhanced inappropriately (which could cause noninteroperability)
- Royalties for patents that have been included within standards

If a standards or specifications body requires the use of someone's Internet protocol, royalties might have to be paid to the owner. If royalties are to be mandated, it is essential that the standard's body or specification group has an open and transparent process to ensure that the standard or specification have not been manipulated by the owner of the Internet protocol to generate royalties inappropriately. This is a challenge that we have seen before. Between 1910 and 1923, the U.S. Patent Office granted approximately 75 traffic signal patents.

Another less obvious advantage of having publicly available standards and specifications is that they establish prior art. Prior art is a legal concept representing the body of public knowledge; it can be used to argue against a patent's validity. Prior art established by open organizations can be used to defend against patents invalidated by knowledge already in the public domain. As an example, there was a risk in the late 1990s that someone could have patented the methods to package assessment items into an XML schema, which would have restricted the ability of assessment content to flow between publisher and delivery system. The IMS Question and Test Interoperability (QTI) specification initiative ensured that packaging questions and tests remained open for all.

Business Drivers for Specifications and Standards

Over the past decade, many new applications for providing and supporting learning and training have emerged. Many of the top quality applications are specialized applications that address only a narrow range of the functions required by a comprehensive system. At the same time, many different models of how to create, deliver, and foster learning and measure knowledge and skills have evolved, and these can require very different functionalities and patterns of integration. There is no one right or best way to do things in the learning or measurement space, nor should there be. However, the emergence of a wide variety of applications that support e-learning and e-assessment has meant, at times, that one very useful system simply cannot talk to another.

One important response to this difficulty in communicating has been the emergence of a set of specifications and standards-building activities led by organizations such as the Institute of Electrical and Electronics Engineers (IEEE), the AICC, the IMS Global Learning Consortium, and the Advanced Distributed Learning (ADL) Initiative. These standards-building efforts have made great progress, and there is general acceptance of standards, such as the IEEE's Learning Object Metadata for describing learning resources, the IMS's QTI specification for allowing organizations to share assessment content, and the ADL's SCORM, which defines how learning resources and learning management systems should interoperate.

There has also been some criticism of these standards-building efforts, on the following grounds:

- They are moving too slowly for customers and vendors, and so proprietary Web services interfaces are being introduced.
- They are moving too quickly, which will lock systems into using older technologies.
- They are too complex and therefore difficult and expensive to implement.
- They do not deliver actual, functioning integration that has been tested.
- While certification helps, it is more important for vendors to demonstrate their commitment to integrations with other products.

Developing interoperability could be a pragmatic and evolutionary approach that takes advantage of market selection pressures. This evolution has provided de facto standards such as the Macromedia® Flash® player, Windows Internet Explorer, and standardized user interface standards, such as a search engine being queried by a text box and button.

Specifications vs Standards

Conformance or compliance testing distinguishes specifications from standards.

At the beginning of the standardization processes, users who have a problem consult with technologists. The technologists make proposals and produce specifications. They then prepare to build the technology just as a contractor might use a blueprint to build a house. However, the friendly technologists might interpret the finer points of the specifications differently. If only one person needs to interpret the document, all is well. But when 2, 3, 10, 100, or even 1,000 people need to write a system based on a specification, we need to ensure that they interpret the specification in the same way.

The first solution is to write a conformance statement and insist that all systems adhere to it, and then interpret the specification and the conformance statements. All the technologists conclude that their system conforms, so if other systems are incompatible, the problem exists within those other systems. Deadlock would result without the goodwill of the parties involved in the process. What if competitors are forced to interoperate based on a demanding customer? There is less goodwill in the process, and deadlock results.

Certification is a method to set criteria by which systems are tested to ensure compliance with the intent of the specification. Systems are tested against the criteria by a third party; specifications, conformance statements, and test criteria can be improved over time to ensure compatibility.

The process of moving from specifications to standards follows:

1. Users state requirements for systems to interoperate.
2. Technologists write a specification.
3. Programmers produce systems.
4. Technologists and programmers discuss the issues they encountered and update the specification, recommending practice and conformance documents.
5. Programmers test their systems with other systems.

6. Documentation is enhanced and best practices documents produced.

7. Specifications and best practice and conformance statements become robust.

8. Certification is put in place to ensure interoperability.

Specifications are not standards, but standards need specifications. Without specifications we cannot get past Step 2. It is not until certification, in Step 8, that standards acquire real meaning.

The Players in Technology Specifications and Standards

One notable fact about standards and specifications is that there are so many of them! Each serves its purpose, and each has an organization behind it to maintain the order and quality control that are essential for systems to interoperate effectively using standards and specifications. Here is the list of the key organizations in this arena:

Key Standards Related to E-Assessments and Learning

While listing the specification and standards players is enlightening, it does not help us make decisions or take action. Just as cities needed to decide when it was time to implement traffic signaling systems, organizations need to determine when specifications or standards are needed.

If an organization produces or buys systems or content and has no need for them to interoperate, then conformance with specifications and standards is probably not high on its list of priorities. However, if an organization produces systems that need to manage content, produces content that needs to be managed by other systems, or purchases content and systems that have to interoperate, then specifications and standards are extremely important.

It is essential, in this case, to determine what your interoperability requirements are and then seek out the specifications that will meet the organization's

Table 2. Key organizations in technology standards and specifications

Acronym	Name	Style	Comments
ADL	Advanced Distributed Learning *www.adlnet.org*	U.S. government initiative open to all those who wish to participate	Promotes, documents, and validates specifications; organizes events to test interoperability
AICC	Aviation Industry CBT Committee *www.aicc.org*	Aviation industry consortium open to all those who wish to participate	Produces specifications, known as AICC Guidelines and Recommendations (AGRs) and has a program to certify conformance for learning-related technologies
ALIC	Advanced Learning Infrastructure Consortium *www.alic.gr.jp*	Japanese consortium of private and public organizations	Promotes the adoption of e-learning in Japan by validating and documenting specifications
ANSI	American National Standards Institute *www.ansi.org*	Not-for-profit membership organization	ISO-recognized body that administers the U.S. voluntary standardization and assessment system; accredits the IEEE
ARIADNE	Alliance of Remote Instructional and Distribution Networks for Europe *www.ariadne-eu.org*	European foundation with an open membership policy	Produces and provides specifications and technology for online learning
BSI IST/43	British Standards Institution Committee that works in e-learning standards *www.bsi-global.com*	Not-for-profit UK organization open to all in the UK	Develops e-learning standards within Britain
CEN/ ISSS WS-LT	Comité Européen de Normalisation/Information Society Standardization System–Learning Technologies Workshop	CEN is an accredited standards body; ISSS WS is an open group that does not produce standards	Validates, modifies, and disseminates specifications for Europe
CLEO	Customized Learning Experiences Online *www.cleolab.org*	Closed consortium	Conducts research on technical and pedagogical issues related to SCORM
DCMI	Dublin Core Meta-Data Initiative *www.dublincore.org*	Open consortium	Produces and disseminates meta-data standards
EdNA	Education Network Australia *www.edna.edu.org*	Australian government-funded education initiative	Validates, documents, and disseminates specifications and standards to Australian educators

Table 2. (cont.)

Acronym	Name	Style	Comments
HR-XML	HR-XML Consortium *www.hr-xml.org*	Open consortium	Produces and disseminates data standards for competencies, recruiting and staffing, assessments, and other human resources applications
IEEE LTSC	Institute for Electronic and Electrical Engineers – Learning Technology Standards Committee *ltsc.ieee.org*	Accredited by ISO as a standards body	Deals with learning standards for the IEEE
IETF	Internet Engineering Task Force *www.ietf.org*	Large and open international community	Community concerned with the evolution of the Internet architecture and the smooth operation of the Internet Produces specifications and standards related to the evolution of the Internet architecture
IMS	IMS Global Learning Consortium *www.imsglobal.org*	Open consortium managed by Educause	An industry/academic consortium that develops specifications
ISO	International Standards Organization *www.iso.ch*	Membership open to every country's standards organizations	Creates standards using an open process based on consensus. ISO standards often become legally mandated around the world
ISO/IEC TC1 SC36	ISO International Electrotechnical Committee 1 (IT Standards) Subcommittee 36 (Standards for Learning, Education and Training)	Membership open to country's standards organizations	Produces accredited standards for learning, education, and training. CEN/ISSS and IEEE LTCS work closely with S36
JA-SIG	Java Administration–Special Interest Group *www.ja-sig.org*	Open consortium	Supports the free, open source, open standard portal for higher education
JCP	Java Community Process *www.jcp.org*	Open consortium	Ensures Java technology's standard of stability and cross-platform compatibility
OASIS	Organization for the Advancement of Structured Information Standards *http://www.oasis-open.org/*	Open consortium	Produces more Web services specification than any other organization

Table 2. (cont.)

Acronym	Name	Style	Comments
OKI	Open Knowledge Initiative *web.mit.edu/oki/*	Closed consortium of academic institutions	Creates and disseminates API specifications and reference implementations
Prometeus	Promoting Multimedia access to Education and Training in European Society *www.prometeus.org*	EU-sponsored open initiative	Comments on specifications and standards in order to build a common approach for producing and provisioning e-learning in Europe
SIF	Schools Interoperability Framework *www.sifinfo.org*	Open consortium	Develops and maintains specifications for managing data for schools and school districts
W3C	World Wide Web Consortium *www.w3.org*	Open consortium	Produces, maintains, and disseminates specifications, guidelines, software and tools for the World Wide Web. Focus is on HTTP, HTML, XML, and SOAP
WebDAV	Web-based Distributed Authoring and Versioning *www.webdav.org* Also *www.ietf.org*	Working group of IETF	Developed the specification for DAV which allows collaboration of authors over the Web
WS-I	Web Services Interoperability Organization *http://www.ws-i.org/*	Open industry effort	Provides guidance and recommended practices for developing interoperable Web services

objectives. Standards are more stable than specifications, so if there is a choice, standards offer the safest option in the long run.

Group standards and specifications can be grouped into the categories listed in Table 3.

If content needs to interoperate with a management system, Content Packaging and Launch and Track specifications or standards are most likely to hold the answer. If different management systems need to interoperate with a portal, then Data Definitions and Authentication should be considered.

With these broad categories established, we can relate the most popular and well-known standards and specifications to the categories in Table 4.

Table 3. Group standards and specifications categories

Type of Specification or Standard	Description
Authentication	Systems that allow authentication of individuals and others systems to potentially single-sign-on across various systems within a larger system
Content Packaging	Allows content and assessments to be packaged for simple transmission between systems containing sufficient information for the recipient system to run the content
Data Definitions	A schema, which is a collection of logical data structures, that defines such things as learner information, competencies, eligibility, qualifications, assessment items, and so forth
Data Transport	Describes how data can be moved from one system to another
Launch and Track	Allows content and assessments to be launched and tracked by a management system
Metadata	Allows content to be tagged to help management systems search for and discover properties about the content
Philosophical	Provides a framework for understanding and identifying critical system interfaces

Table 4. Most popular standards and speications categories

Supporting organization	Specification/Standard	Category
ADL	SCORM Content Aggregation Model	Content Packaging
	SCORM Run-Time Environment (RTE)	Launch and Track
	Sequencing and Navigation (SN)	Data definitions
AICC	AGR7	Content Packaging and file-based Launch and Track
	AGR10	Content Packaging and Internet-based Launch and Track. Probably the most widely used de facto standard for delivering assessments
HR-XML	Staffing Exchange Protocol	Data Transport
	Assessments Background Checking Benefits Enrollment Competencies Contact Method Date Time Education History Effective Dating Employment History Entity Identifiers Identifier Types ISO Utilities Job and Position Header Military History Organization Payroll Benefit Contributions Payroll Instructions Person Name Postal Address Provisional Envelope Specification Resume Stock Plan Interface Taxonomy Types Time Expense Reporting WorkSite	Data definitions

Table 4. (cont.)

Supporting organization	Specification/Standard	Category
IEEE	1484.1-2003 Learning Technology Systems Architecture (LTSA)	Philosophical
	1484.12.1: IEEE Standard for Learning Object Metadata	Meta-data
	1484.12.2: Standard for ISO/IEC 11404 binding for Learning Object Metadata data model	Meta-data
	1484.12.3: Standard for XML binding for Learning Object Metadata data model	Meta-data
	1484.12.4: Standard for Resource Description Framework (RDF) binding for Learning Object Metadata data model	Meta-data
	1484.11.1: Data Model for Content to Learning Management System Communication	Content Packaging
	1484.11.2: ECMAScript API for Content to Runtime Services Communication	Launch and Track
	1484.20 Learning Technology - Competency Definitions	Data definitions
	New - Digital Rights Expression Languages (DREL)	Data definitions
IETF	Lightweight Directory Access Protocol (LDAP)	Authentication
	WebDAV Distributed Authoring Protocol	Data transport
IMS	Abstract Framework	Philosophical
	Accessibility	Meta-data
	Competency Definitions	Data definitions
	Content Packaging	Content Packaging
	Digital Repositories	Data definitions
	Enterprise	Data definitions
	Learner Information	Data definitions
	Learning Design	Data definitions
	Meta-data	Meta-data
	Question and Test Interoperability (QTI)	Data definitions
	Resource List Interoperability	Data definitions
	Shareable State Persistence	Launch and Track
	Simple Sequencing	Data definitions
JCP	JSR-000168 Portlet Specification – *(enables interoperability between portlets and portals)*	Authentication
OASIS	Web Services for Remote Portlets (WSRP) *("plug-n-play" portlets with portals)*	Authentication
SIF	Schools Interoperability Framework	Data definitions Data transport

Specifications and Standards Groups Working Together

As with the pieces in a jigsaw puzzle, the participants in the creation of national and international standards all have something unique to contribute. Moving user requirements, bright ideas, and specification development through to becoming national and international standards is a somewhat random process. Technologists, intellectuals, pragmatists, users, and organizations each feel an affinity with and gravitate to a particular part of the specification or standards process. Each of these individuals and organizations is essential to completing the puzzle.

In some ways the specifications and standards bodies compete, but for the most part they collaborate. The competition between them tends to yield better specifications eventually, but there can be confusion at first because people do not know which specification will yield the greatest value. Competing interests, agendas, and egos make the process of producing specifications and standards quite lively and intellectually colorful. The people involved are doing a fantastic job by contributing their time freely for the advancement of specifications, standards, and e-learning and e-assessment.

With the ebb and flow of relationships, it is difficult to define the current status or future potential and synergies of specifications and standards organizations working together. It is valuable for organizations producing specifications to align themselves with standards organizations, so that the authors have the satisfaction of seeing their work become a standard. It is also valuable for standards organizations to have a feeder channel to think through the challenges during a specification writing process; this in turn eases the standardization process.

One challenge in the collaboration and handoff from one organization to another is who retains credit and copyright and who is allowed to take the specification forward. This will inhibit the smooth transition of specifications to standards, but it also provides healthy competition amongst specifications.

Conclusion

Purchasing power will ultimately drive people to standards. As governments and organizations determine specifications and standards, more vendors will produce compliant systems. As more compliant systems appear, potential customers will feel more comfortable in requiring support of specifications and standards—just as municipalities rely on traffic signals to help keep motorists safe.

Although they are not regulated by standards, the Institute of Transportation Engineers maintains the latest research on the optimum timing for red, yellow, and green lights. These calculations take into account the reaction time of the motorist, the speed of the vehicle, the comfortable deceleration rate, and the width and grade of the intersection. Calculations were first established in 1941 and were then updated in 1950, 1965, 1976, 1982, 1985, 1992, and 1999. The latest calculation looks like this:

$$y = t + \frac{V}{2a + 64.4g} + \frac{W + L}{V}$$

If there had never been a standard for traffic lights, this sophisticated mathematical equation could not have been created to enhance the safety and efficiency of driving a motor vehicle. I marvel that this simple yet essential user interface—something we experience nearly every day—relies on a complex calculation established through the use of a standard. It is the same with specifications and standards in e-learning and e-assessment. Specifications and standards provide a base of shared understanding that paves the way for further progress.

As the e-learning and e-assessment industries continue to evolve, the specifications and standards that are adopted now will become common practice and will eventually be recognized as best practice. As with traffic lights, electric power, and many other familiar systems, such standards will become an integral part of our lives; we won't even think about their origins or the painstaking work of the trailblazers.

Technology standards related to learning and the measurement of knowledge and skills will continue to affect our lives. An excellent way to track their progress is to visit the Learning Technology Standards Observatory (*http://www.cen-ltso.net*).

Recommended Reading

Allen, C., Nurthen, J., & Ensroth, K. (2002). Choosing SIDES: An introduction to staffing industry data exchange standards. Retrieved from *http://www.hr-xml.org/sides/HR-XML-2002-ChoosingSIDES1.ppt*

Brignano, M., & McCullough, H. (1981). *The search for safety: A history of railroad signals and the people who made them.* Pittsburgh, PA: Union Switch & Signal Division, American Standard.

Eccles, K., & McGee, H. (2001) A history of the yellow and all-red intervals for traffic signals (ITE report). Retrieved from *http://www.ite.org/library/yellowintervals.pdf*

Eduworks Corporation. (2002). White paper: What is the open knowledge initiative? Retrieved from *http://web.mit.edu/oki/library/OKI_white_paper_120902.pdf*

Institute of Transportation Engineers. (2004) ITS Standards. Retrieved from *http://www.ite.org/standards/index.asp*

The Masie Center. (2003). Industry report: Making sense of learning specifications & standards: A decision maker's guide to their adoption. Retrieved from *http://www.masie.com/standards/s3_2nd_edition.pdf*

Mueller, E. A. (1970). Aspects of the history of traffic signals. *IEEE Transactions on Vehicular Technology, VT-19*(1).

Sessions, G. (1971). *Traffic devices: Historical aspects thereof.* Washington DC: Institute of Traffic Engineers.

Sun Microsystems. (2002). White paper: E-Learning application infrastructure. Retrieved from *http://www.sun.com/products-n-solutions/edu/whitepapers/pdf/eLearning_Interoperability_Standards_wp.pdf*

U.S. Patent Database. Retrieved from *http://www.uspto.gov/patft/index.html*

Section II

Best Practices in Designing Online Assessment

Chapter IV

Ten Key Qualities of Assessment Online

Chris Morgan, Southern Cross University, Australia

Meg O'Reilly, Southern Cross University, Australia

Abstract

Student assessment belongs in the centre of our teaching and learning considerations—it is the engine that drives and shapes student learning. In online contexts, it is argued that although teaching and learning has been dramatically reconceptualised, assessment practices are lagging, and more likely to imitate conventional practices such as end of term exams that encourage rote learning and the dissemination of fixed content. The authors argue that it is essential for online educators to bring the same innovation to their assessment practices that they have to their other online teaching practices. Ten key qualities of good online assessment are offered for consideration and discussion, namely:

1. *A clear rationale and consistent pedagogical approach*
2. *Explicit values, aims, criteria, and standards*
3. *Relevant authentic and holistic tasks*
4. *Awareness of students' learning contexts and perceptions*

5. *Sufficient and timely formative feedback*

6. *A facilitative degree of structure*

7. *Appropriate volume of assessment*

8. *Valid and reliable*

9. *Certifiable as students' own work*

10. *Subject to continuous improvement via evaluation and quality enhancement*

Introduction

A traditional view of assessment is that it is a terminal event—something that *follows* teaching and uncovers how much has been learned. It is about educational testing and quantitative measurement. It takes the view that assessment is a science that is expressed in terms of efficiency, reliability, and technical defensibility—from the design of tests to the bell curve in which grades are apportioned. At another level, it also implies a view of knowledge itself— that knowledge is relatively fixed, finite and resides with the teacher.

Although there are many vestiges of this assessment tradition flourishing in universities today, there are also many new views of assessment that have emerged in the past 10 or more years. Far from being a terminal event, assessment is moving into the centre of our teaching and learning considerations. Research into student learning has consistently located assessment at the centre of students' thinking: how they spend their time, what they regard as important, and the kinds of learning approaches they adopt (Gibbs, 1992; Ramsden, 1997, Rowntree, 1977). We have come to reconceptualise assessment as the engine that drives and shapes learning, rather than simply an end-of-term event that grades and reports performance.

Online learning has created new opportunities for learning that require us to redesign our assessment practices (Alexander & McKenzie, 1998). We are challenged by a new medium, a broader and more diverse student population, new forms of interaction and dialogue, and potentially, new ways of knowing. Constructivist pedagogy has moved into the mainstream, and online learning, in its most potent form, is about the drama of the multiple meaning, the contrary viewpoint, the search for credible sources, and the elusive nature of "truth" in a postmodern world. Many teachers are also grappling for the first time with separation from learners, not to mention the complexities of cross-cultural dialogue, collaborative learning, and negotiated course content and assessment.

Ironically, however, online learning has promoted old thinking and old practices in assessment. While it is argued that assessment strategies such as objective testing and end-of-term exams, with accompanying pedagogies that emphasise dissemination of fixed content, have less relevance in these settings, many of these strategies now flourish with greater technological wizardry than ever.

This chapter articulates some of the most important values and practices that we believe should underpin assessment in online contexts across a broad range of disciplines, so that when we are making the shift to support assessment through online means, we can keep in mind the fundamental principles of quality teaching, learning, and assessment. The 10 key qualities that we put forward here for consideration have been drawn from our exploration of the processes of assessment in its many modes over a period of time (Carroll, 2002; Morgan, Dunn, Parry, & O'Reilly, 2004; Morgan & O'Reilly, 1999).

A Clear Rationale and Consistent Pedagogical Approach

From the beginning, it is important that you be clear about what you are hoping to achieve through your assessment tasks. What understandings are you hoping to encourage? What abilities are you hoping to develop in students? How do you want to shape and structure their learning? By focussing upon these questions, you are acknowledging the power of assessment to drive and shape student learning. If your assessment tasks have a clear rationale that is consistent with the values you espouse and the abilities and qualities you are seeking to develop in your students, then you are more likely to engender positive, meaningful learning experiences.

A second consideration is how consistently you support the assessment through your teaching. Is there a clear alignment between your rationale, the learning objectives, the content, the teaching and learning strategies and the assessment? Or does assessment appear almost as an afterthought at the end of the learning encounter? Do you weave your formative and into a course in a strategic way that maximizes deep learning? A consistent pedagogical approach in support of your assessment is vital in providing a consistent message to students regarding what you value.

If, for example, you value self-directed inquiry and critical thinking, but are assessing students by the use of online multiple-choice quizzes, then perhaps you are sending mixed messages to your students. Should they engage in deep inquiry or simply rote learn information to be fed back on cue? The assessment requirements, as the most powerful message of all, will no doubt supplant all others. Conversely, there is very little sense in opting for highly self-directed or constructivist assignments where regular discursive submissions are required online, if your teaching strategies are prescribed and directing students to adopt a fixed view. Unless we recognise assessment as the powerful learning tool it is, we may be missing valuable opportunities to shape student learning appropriately.

Explicit Values, Aims, Criteria, and Standards

Given that you have provided a clear explanation of purpose and rationale for assessment, an explanation of the aims and values of each assessment task will further help students understand the nature and quality of their expected performance. Why are your students being required to complete a certain task? How do these tasks relate to the goals of the subject? How do these assessment tasks relate to the students' own learning goals? What specifically, is going to be valued in their achievement of the tasks? An explanation of these relative values can assist learners to take charge of the extent of effort they might apply and to take responsibility for seeking a standard of completion that reflects their own personal values whilst responding to the stated requirements of assessment.

For example, when students are undertaking an online assessment, what is the relative importance of carrying out tasks requiring some technical competencies, as opposed to the value being placed on the higher order capabilities of critical thinking and the development of argument? Without your clear explanation of values and aims, how is the student to know the relative significance of the objectives of a subject in which they may be required to obtain information from the Internet, provide a critique and post the combined resource to a discussion area, as compared to the value you may wish to place upon the quality of resources discovered and the content of the students' critical commentary?

In addition, your description of the criteria against which students will be marked and the standards which they might expect to attain needs to be made explicit right at the start of the learning period. A detailed breakdown for standards of achievement can be readily negotiated with an online class, thus improving their understanding of qualitative differences between the standards. Students can then make an informed decision on their own aims to achieve a particular standard and this will better ensure against misunderstandings and disputes. Other markers with whom you may be sharing the workload can also confirm in advance a clear understanding of the requirements of each assessment task. In this way you are taking account of the requirements of the students, the subject, the faculty, and the institution.

Relevant, Authentic, Holistic

Because much of online learning is accessed by students from their homes and workplaces, it requires educators to actively reach out more into students' own worlds, work contexts, and life experiences. We find a wider range of student motivations and orientations to education, including those that tend towards intellectual and academic interests, personal challenge, self-improvement, social interaction, as well as vocational interest. Arguably, cohorts of online learners are much less homogenous than their classroom counterparts. The challenge therefore is to ensure that assessment tasks are diverse and relevant, and offer the broadest and most inclusive learning opportunities for students. The online environment also opens up student's access to disciplinary experts, thereby providing an authentic opportunity for discourse with noted specialists in the field having the potential to provide their input to the assessment process. Students may also benefit from cross-cultural dialogue that can bring global perspectives to their learning from across a diversity of geographic locations (Rimmington, O'Reilly, Gibson, & Gordon, 2003).

Authentic assessment tasks are particularly appropriate as a means of encouraging learners to engage with real-life issues and problems in their own worlds and workplaces. Authentic tasks strive to avoid fragmentary testing of atomised facts or competencies, preferring more complex holistic challenges such as problem scenarios, case studies, and projects in which the learner meaningfully participates. These also create opportunities for deep learning in which students are encouraged to develop higher order abilities such as problem solving and analytical thinking in applied settings.

Awareness of Students' Learning Contexts and Perceptions

Thoughtful planning of online assessments will include an awareness of students' context beyond the subject at hand, including competing assessment commitments in parallel subjects and prior learning experiences and knowledge that are brought to the learning encounter. What do you know of the existing knowledge that your students bring on commencement of study to your subject? What do you know of the competing assessment requirements being juggled by students taking other subjects at the same time as yours? Are you and your colleagues ensuring a spread of skills development across the program or are students expected to demonstrate the same skills (e.g., essay writing) over and over? Students who are overloaded with assessments or who are not sufficiently stimulated by a variety of assessment tasks or who are inadvertently out of their depth, may find themselves adopting surface approaches to learning.

Further, Ramsden (1997) alerted us to student's perceptions of assessment tasks, which may vary considerably from our own, and may prompt surface approaches to learning despite our most careful designs. He argued that if students feel there is insufficient time, or they have experienced insufficient support or engagement, or if they have been previously rewarded for reproducing strategies, or if their prior knowledge is insufficient, then they will feel constrained to use surface approaches.

Several strategies might be useful to address these concerns through utilising the benefits of the online environment. With the capability for rapid turnaround of students' work, you might decide to break down the assessment scheme into smaller components, which are due for submission on several occasions throughout the study period, rather than keeping two larger assignments. Continuous assessment as it applies to problem based learning is also possible to achieve when the class group is not too large. In both these situations, care should be taken to ensure that the smaller assessment tasks do not represent more assessment overall.

In fairness to students encountering online approaches for the first time, assessment weightings need to be spread across a number of methods to ensure there is no disadvantage to the technical novice. Media-rich learning activities and resources may be too slow for students to download and can be frustrating, stressful, and potentially inequitable when built into an assessment task. Screen dependent assignments can also be physically demanding, causing fatigue and eye strain (Kerka, Wonacott, Grossman, & Wagner, 2000). Consideration of fairness and equity needs to be ongoing in the design of online assessment.

An appreciation of student's prior knowledge can also be gained through early assessment. Where it is likely that a number of students have extensive prior knowledge within the cohort, a process of peer review might be useful to include. This peer review process would allow each student to learn the process of applying assessment criteria and providing constructive feedback in regards to the central elements of performance or understanding. Learning through assessment thus transforms a potentially perfunctory experience for some into a rich learning experience for all.

Sufficient and
Timely Formative Feedback

Formative assessment comprises all those activities designed to motivate, to enhance understanding, and to provide learners with an indication of their progress, in contrast to summative assessment, which reports on and grades student achievements. In online learning, formative assessment activities may take many guises, including dialogue with teachers and other students, feedback from assignments, and self-assessment activities which allow students to monitor their own progress.

How do online learners get developmental feedback on their progress? Is it sufficient and is it timely? In online learning, formative and summative assessment are ideally interwoven into a form of continuous assessment—each assignment building upon the former, with dialogue and formative feedback from one providing a scaffolding for students to develop their understanding and contributing to the quality of the next assessment task (Hogan & Pressley, 1998). This maximises the formative function of assessment by allowing students the greatest opportunity to develop their abilities and understanding, with ongoing and timely feedback. It also helps us to move away from the notion of assessment as the "terminal event" that follows teaching. Rather, formative and summative assessments work together as powerful drivers of learning that provide a useful structure, opportunities for rich interaction, break down summative assessment into manageable chunks, encourage and motivate students, and provide invaluable insight for learners into their progress.

Teachers often express a conflict in their roles as both formative guides and summative assessors. On the one hand we are attempting to guide and promote learning, and we may liken our role in online learning to a facilitator, mentor, or even a colleague of learners. Yet, in most instances we are then required to make unilateral decisions about their achievements. Our role suddenly shifts from advocate to judge. How do we find a balance between the competing demands of our role? Online learning readily lends itself to a range of strategies, including contract learning, self-assessment through auto-marked quizzes, and peer assessment which provide for great student input and negotiation in assessment products and processes. It also provides opportunities for teachers to maximise their role as facilitators of learning rather than just examiners of achievement. Both asynchronous exchanges with the class and synchronous events such as online office hours or small group feedback sessions enable these approaches (Rowe, 2003).

A Facilitative Degree of Structure

This is a phrase that Gibbs (1995) employed to describe the fluid balance we seek in online assessment between structured tasks and learner self-direction. Arguably, self-direction is not an innate quality in learners, particularly given the many years of teacher-centred experiences that have tended to socialise learners in the reverse direction. Rather, it is seen as a quality that can be fostered in learners by a progressive shift from teacher to learner control throughout a program of study, as learners incrementally acquire competence in information retrieval, goal setting, critical thinking, self-management, and self-evaluation (Brockett & Hiemstra, 1990; Candy, 1991). Thus, a "facilitative" degree of structure in online assessment is one that develops these abilities purposefully, with the explicit goal of self-direction, and seeks a balance between structure and self-direction at any given time, which is optimally helpful (Gibbs, 1995).

Finding a balance between structure and self-direction in online settings is most important. In early stages of a program, learners are often only beginning to come to grips with the online environment and its possibilities in terms of information networks, interactions and collaborations. Early encounters may be quite structured in relation to learning pathways and assessment outcomes. As learners progress, learning pathways and assessments become increasingly negotiable and self-directed. Pre-prepared inflexible course materials with a tendency towards content dissemination will work against this end goal, as they leave little room for personal exploration or construction of individual meaning.

In the online learning environment, we are now able to provide a number of supporting structures for students in the form of

- auto-marked quizzes which test core knowledge and understanding and provide immediate feedback,
- engagement with student peers in a non-competitive situation for mutual support and clarification of ideas,
- anonymous queries by students to seek reinforcement of fundamental concepts, and
- students' collaboration with teaching staff on the development of assessment criteria.

Where these structures are clearly associated with the goals of the subject and the program as a whole, students are better able to engage with a broad conception of learning. Students' increasing independence in the process of completing the requirements of each subject also improves their chances of a deep approach to learning (Prosser & Trigwell, 1999).

Understanding how to optimise the flexibility inherently available through online facilitation of learning and assessment is a complex question. On the one hand, through the capabilities of modern communication technologies, a great deal of mutual exchange can occur between students and staff to develop and support a student-centred assessment regime, thus addressing the diversity of student needs and prior knowledge. On the other hand, the greater the flexibility for students in terms of what, when, and how they agree to be assessed, the greater the pressure on staff to be available, negotiable, equitable, and accountable. The question of flexibility in assessment requires that we know to what degree can we sustain an approach to assessment that can take account of various needs of students, staff, faculty and any relevant accrediting bodies.

Appropriate Volume of Assessment

How much assessment is sufficient? If you accept that assessment is a powerful driver of learning, you may be tempted to increase your assessment load. Before taking this drastic action, and increasing your own workload, consider first the dangers of overassessment. Too much assessment in a subject creates anxiety and a surface or survival approach to learning (Ramsden, 1997). Although some believe that extensive assessment provides the necessary rigor to a subject, it may actually be having the reverse effect upon learners. It may also be impacting upon learners' capacity to meet assessment requirements for parallel subjects.

There is no universal formula for judging the appropriate volume of assessment, although there are often agreed standards within an institution or faculty. To begin, you should have a reasonably clear idea about the total time to be spent by students doing each of the various activities, such as reading, interacting online, researching, and preparing for assessments, and this should be communicated to students by way of a benchmark. Naturally students will differ according to abilities, prior learning, competing commitments, and so forth, but an approximation is a valuable guide for you and learners alike (Chambers, 1992, 1994).

Perhaps students who had previously only had experience of distance education with its frequent sense of isolation and an absence of opportunities for benchmarking progress, there may be a tendency to be carried away by the chance to interact with fellow learners, 24–7. Though this can be a wonderful doorway to shared understanding and a community of like-minded learners, providing an explicit indication of time to be spent on online assessment tasks can remind students of the decision they need to make to prioritise their time. This can be done informally online just as one would mention the need to move on in class.

Some teachers are prone to overassessing their students because they believe they have to assess everything in a subject. The challenge for teachers is to determine what really needs to be formally assessed, so that you are getting a balanced sample of the subject as a whole, but not creating a treadmill for students. As discussed earlier, authentic, holistic assessment is another strategy to consider when moving into the online context. Portfolios, projects, reflective journals, take-home examinations, problem-solving scenarios are all examples of ways to broadly sample student learning across a spectrum of topic areas, without overloading students. Consider individual assignment length as well. Assignments with large word limits will often encourage long-winded and rambling responses from students. Clearly, big is not necessarily better, if you are interested in rigorous and tight responses.

Valid and Reliable

Validity and *reliability* are terms commonly found in the lexicon of educational measurement. Validity poses the question of whether your assignments provide the truest picture possible of the particular knowledge and abilities being measured by the assessment task. Reliability poses questions about whether your assessment items can be marked with a high degree of consistency and relative objectivity, particularly if other markers are involved.

For validity to be high, the assessments should sample students' performance on each objective, an appropriate mix of assessment methods should be used, and assessment methods should be selected on the basis of providing the truest picture possible. For example, if you are testing learner's critical thinking skills, a series of short comments to the online forum may not successfully provide an accurate picture. For reliability to be high, there needs to be consistency and precision evident in the way assessments measure the desired objective.

It is often said that the online environment has made possible the assessment of new skills and capabilities such as Web-based research, design and development, as well as text-based discussion and collaboration between remote and diverse students (McLoughlin & Luca, 2000). To ensure validity, these affordances of online assessment must be used with discretion. It is not acceptable or valid to require group submissions of assignments simply to reduce one's marking load. Is it appropriate to form students into teams so that they jointly complete one assessment task? Are the skills of collaboration and information sharing relevant to the core objectives of the subject or the program of study?

Generally speaking, if your online assessments are authentic and holistic, with a high level of application to learners' own worlds, then validity is likely to be high. On the other hand, authentic assessments pose problems with reliability, as there are higher levels of learner self-direction and a variety of ways in which learners may choose to demonstrate their learning. Even where content and assessment are relatively fixed and finite, there are often surprisingly high levels of disagreement between markers as to the relative merits of a piece of student work (Newstead & Dennis, 1994).

As authentic assessment entails complex tasks with potentially many variables, judgment of student achievement is more likely to be open to interpretation than smaller, discrete tasks. For example, are the criteria for working in teams to produce a single outcome made explicit so that in the case of large classes, there is a marking system that can be reliably applied by a team of markers? Similarly, are the criteria for successful Web-based products clearly espoused to avoid a completely subjective judgement being made in the process of grading. The challenge is for teachers to articulate marking criteria that are sufficiently broad to cover most situations, yet detailed enough that they are useful and provide guidance to students.

Another strategy is to work closely with other markers with occasional double-blind marking to ensure agreed standards and consistency of approach. With the advent of digital archives arising from online interactions, questions of what occurred within the dynamics of interaction and the quality of students' contributions is a relatively simple matter to confirm.

Certifiable as Students' Own Work

While it has always been important to confirm that the enrolled student is indeed the one who has completed the prescribed assessment tasks or attended the exam and is therefore the one who deserves to be awarded the grade, clearly this question of certifying the students' identity has become of special significance with the adoption of electronic forms of assignment completion and submission. As Carroll and Appleton (2001) reported, in this context, in which technology has exacerbated the problem, it is tempting to search for technological solutions. However, we would agree with the authors and advise the adoption of both pedagogical and policy approaches to minimising plagiarism and cheating in online assessment tasks rather than pursuing convoluted and potentially expensive technical solutions. If we wait for technological solutions (e.g., fingerprint and key stroke pattern recognition, optical scans) we may miss the opportunity before us to remember who are our learners, what are they aiming for, and how can we support their learning through engaging and applied processes of learning wherever possible.

In some cases, students are simply not aware of their breach of the rules for plagiarism. It has been reported that some students have mistakenly believed that copying the work of other authors is an appropriate demonstration of their knowledge of the literature in the field (Lok Lee & Vitartas, 2001). In this case, it is important to spend some time educating students regarding academic honesty and the policies within your institution concerning intellectual property and copyright.

Consideration of the assessment process itself can also help to deter cheating and plagiarism. Benson (1996) offered some suggestions for this:

- Clearly link assignments so that each builds on the former.
- Individualise topics to students' own workplace–learning context.
- Use self-directed forms of assessment such as learning contracts.
- Use work-based mentors, supervisors, assessors, and so forth to authenticate students' work.
- Adapt and change assessment tasks regularly.

All these suggestions are easily applicable in the online environment, and we might add attracting commitment through student-generated assessment criteria and the use of technology to appropriately support original tasks such as collaborative projects, Web-site design, and students building on the work of previous cohorts of students through their own analysis and interpretation.

When informing students of the correct approach to collaborative writing and attribution, it is best done within the disciplinary domain using worked examples and with the optimistic view that students are fundamentally enrolled with us to learn, that they are prepared to accept responsibility for exercising self direction and progressing towards achieving the objectives to which they are committed. The trick is to design mutually useful and desirable processes that stimulate commitment to the goals of learning and assessment!

If, however, you do find an incidence of plagiarism or cheating, it is vital to deal with it appropriately and openly. Determining an appropriate course of action may in the first instance be a case-by-case approach, only being forwarded to an examination board if initial actions are not sufficient. A full discussion and list of recommendations can be found in Carroll (2002).

Subject of Continuous Improvement via Evaluation and Quality Enhancement

Though of prime importance to students, assessment processes as summative events are limited in what they might affect in terms of institutional change to teaching and learning (American Association of Higher Education, n.d.). The importance of evaluating the integrity of your syllabus, drawing upon your reflections on assessment practices, employing insights from reflections and feedback, and quality enhancement initiatives cannot be underestimated when moving assessment online.

Does your institution provide assistance with a calibrated tool for obtaining student feedback? Is such feedback available to you when you are revising your subject or transforming it into a more flexible learning experience and reconceptualising assessment to take account of the online context?

How have you been able to benchmark your approaches to online teaching and learning with other institutions teaching the same subjects? With such a volume of subject specific resources available online, have you considered evaluating your teaching, learning, and assessment strategies against other examples you can find on the Internet? Perhaps you could consider a process of (blind) peer review where you circulate your materials and a description of your assessment scheme (or provide access into your online subject site) to a number of your disciplinary colleagues who provide anonymous feedback according to a prescribed set of criteria. This would generate some valuable perspectives on the syllabus and processes of teaching as well as providing you with some benchmarks against which to assess your own professional teaching and assessment activities.

Does your institution or department provide any guidelines for designing, developing, and implementing assessment strategies? Are you able to work with educational design support and Web-design support to actualise your pedagogical ideas within the online context? If you have not been able to find support and assistance for these benchmarking and quality enhancement activities till now, it is probably only a matter of time before the process of monitoring and quality assurance finds a place in your institutional context. We recommend a proactive approach to continuous improvement by starting now to gather information from a range of sources—students, colleagues, disciplinary peers and vocational or professional specialists, to begin developing a profile of your approaches to assessment and their effectiveness in supporting student learning.

Further, to the collection of feedback and implementation of improvements, bringing your knowledge and experience to a wider audience through publication is one of the best ways to share your insights on assessment practice. In this way you will find yourself among a community of practice where you can avail yourself of the most potent insights of others.

Conclusion

These 10 key qualities are not specific to the online environment—they apply equally to any assessment encounter whether it be in face-to-face or distance settings. Good assessment practice remains essentially the same, irrespective of the mode of delivery. However, we have sought to highlight the particular issues that arise in online settings and the implications for the design and management of assessment schemes. We have also sought to highlight issues that need to be considered when there is a transition of assessment practices from traditional modes of classroom delivery to the online environment. As we have argued, the shift to online learning and assessment has sometimes unwittingly promoted old thinking and old practices, which no longer readily apply in this newer medium. We need a clear understanding of the online environment, including its strengths, weaknesses and new opportunities, as well as a reflective core to our practice to be able to effectively make this transition.

References

Alexander, S., & McKenzie, J. (1998). *An evaluation of information technology projects for university learning.* Committee for University Teaching and Staff Development Canberra, Australia: AGPS

American Association of Higher Education. (n.d.). Assessment forum: 9 principles of good practice for assessing student learning. Retrieved from *http://www.aahe.org/assessment/principl.htm*

Benson, R. (1996). *Assessing open and distance learners: A staff handbook.* Churchill Centre for Distance Learning, Monash University Victoria, Australia

Brockett, R. G., & Hiemstra, R. (1991). *Self-direction in adult learning: Perspectives of theory, research and practice.* Routledge: London

Brown, S., Race, P., & Smith, B. (1996). *500 tips on assessment.* London: Kogan Page.

Candy, P. (1991). *Self-direction for lifelong learning.* San Francisco: Jossey-Bass.

Carroll, J. (2002). *A handbook for deterring plagiarism in higher education.* Oxford, England: Oxford Centre for Staff and Learning Development.

Carroll, J., & Appleton, J. (2001). *Plagiarism: A good practice guide.* Oxford, England: Oxford Brookes University Press.

Chambers, E. A. (1992). Workload and the quality of student learning. *Studies in Higher Education, 17*(2), 141-152.

Chambers, E. A. (1994). Assessing learners workload. In F. Lockwood (Ed.), *Materials production in open and distance learning.* London: Chapman,

Gibbs, G. (1992). *Improving the quality of student learning.* Bristol, England: Technical & Educational Services.

Gibbs, G. (1995). *Assessing student centred courses.* Oxford, England: Oxford Centre for Staff Development, Oxford Brooks University.

Hogan, K., & Pressley, M. (Eds.). (1998). *Scaffolding student learning: Instructional approaches and issues.* New York: University of Albany, State University of New York.

Kerka, S., Wonacott, M., Grossman, & Wagner, J. (2000) *Assessing learners online.* Retrieved June 5, 2003, from *http://ericacve.org/docs/pfile03.htm/principles*

Lok Lee, Y., & Vitartas, P. (2001, November 15). *Teaching and learning in Asia.* Unpublished Seminar hosted by Teaching and Learning Centre, Southern Cross University.

McLoughlin, C., & Luca, J. (2000). Assessment methodologies in transition: Changing practices in Web-based learning. *ASET-HERDSA, 5,* 16-526.

Morgan, C., Dunn, L., Parry, S., & O'Reilly, M. (2004). *The student assessment handbook—New directions in traditional and online assessment.* London: Routledge-Falmer.

Morgan, C., & O'Reilly, M. (1999). *Assessing open and distance learners* London: Kogan Page.

Newstead, S. E., & Dennis, I. (1994, May). The reliability of exam marking in psychology: Examiners examined. *The Psychologist,* 216-19.

Prosser, M., & Trigwell, K. (1999). *Understanding learning and teaching: The experience in higher education.* London: SEDA, Kogan Page.

Ramsden, P. (1997). The context of learning in academic departments. In *The experience of learning* (2nd ed.). Edinburgh, Scotland: Scottish Academic Press.

Rimmington, G. M., O'Reilly, M., Gibson, K. L., & Gordon, D. (2003, June 23-28). Assessment strategies for global learning: I theory. *Proceedings of the Ed-Media 2003 World Conference on Educational Multimedia,*

Hypermedia & Telecommunications. Association for the Advancement of Computing in Education, Honolulu, HI.

Rowe, S. (2003). *A virtual classroom: What you CAN do to enrich the learning experience.* Paper presented at the NAWeb, New Brunswick, Canada.

Rowntree, D. (1977). *Assessing students: How shall we know them?* London: Kogan Page.

Chapter V

Factors to Consider in the Design of Inclusive Online Assessments*

Sandra J. Thompson, University of Minnesota, USA

Rachel F. Quenemoen, University of Minnesota, USA

Martha L. Thurlow, University of Minnesota, USA

Abstract

This chapter presents factors to consider in the design of online assessments for all students, including students with disabilities and English-language learners. It presents a process and considerations for the initial transformation of paper-and-pencil assessments to inclusive online assessments, focusing on features of universal design, the use of assistive technology, and an examination of the use of individual accommodations in light of the content tested. The authors hope to convey the importance of implementing a process for addressing these considerations from the beginning of online assessment design. Retrofitting completed assessments can result in concerns about validity.

Introduction

Learners today are a diverse group. In the context of laws such as the Individuals with Disabilities Education Act (IDEA) for children in public preschools and K-12 schools, and the Americans with Disabilities Act (ADA) for individuals who have left school, nearly 12% of the population of learners have been identified as having disabilities of various types. The United States also has many learners for whom English is not their first language. In the past decade there has been a 72% increase in the number of English-language learners in the K-12 public schools, up to nearly 4 million students. With this diversity in learners today, we can no longer afford to think of online learning and online assessment without making sure that these are inclusive of all individuals, including those with disabilities and those who are English-language learners.

For the full benefits of online assessments to be realized, a thoughtful and systematic process must occur to examine the transformation of existing paper-and-pencil assessments. It is not enough to simply transfer test items from paper to screen. Not only will poor design elements that exist on the paper version of the test transfer to the screen, but additional challenges may be introduced that reduce the validity of inferences from the assessment results and possibly exclude some groups of students from assessment participation.

This chapter presents factors to consider in the design of online assessments for all students, including students with disabilities and English-language learners. The focus of this chapter is on operational large-scale assessments with common items that measure grade-level content. We do not address the complex issues involved in computerized adaptive testing.

The chapter begins with opportunities and challenges presented by this approach to assessment and then explores research about effective, universally designed assessments and technology-based accommodations, relating this research to online assessment design features. Finally, we present a process and considerations for the initial transformation of paper-and-pencil assessments to inclusive online assessments.

Opportunities and Challenges of Online Assessments for Diverse Learners

Online technology has been touted as a tool that can be used to empower diverse learners, including students with disabilities and English-language learners (Goldberg & O'Neill, 2000). Specifically, online assessments have been viewed as a vehicle to increase the participation of diverse students in assessment programs. For example, computer operating systems now support a great variety of adaptive devices (e.g., screen readers, Braille displays, screen magnification, self-voicing Web browsers). The National Research Council (NRC, 2001) found online assessments to be effective for students who perform better visually than with text, who are not native English speakers, or who are insecure about their capabilities. According to NRC, "technology is already being used to assess students with physical disabilities and other learners whose special needs preclude representative performance using traditional media for measurement" (p. 286).

Standardization of accommodated administration can be facilitated by online assessments. According to Brown-Chidsey and Boscardin (1999), "using a computer to present a test orally controls for standardization of administration and allows each student to complete the assessment at his/her own pace" (p. 2). Brown and Augustine (2001) cited educator appreciation of a computer's ability to present items over and over, in both written and verbal form, without the need for a nonstandard (and sometimes impatient) human reader. Several studies have shown the positive effects of providing a reader for math tests (see Calhoon, Fuchs & Hamlett, 2000; Fuchs, Fuchs, Eaton, Hamlett, & Karns, 2000; Tindal, Heath, Hollenbeck, Almond, & Harniss, 1998).

With audio and video technology built into online assessments, specialized testing equipment such as audiocassette recorders and VCRs could become obsolete (Bennett et al., 1999). According to Bennett (1995),

test directions and help functions would be redundantly encoded as text, audio, video, and Braille, with the choice of representation left to the examinee. The digital audio would allow for spoken directions, whereas the video could present instruction in sign language or speech-readable form. Among other things, these standardized presentations should reduce the noncomparability associated with the uneven quality of human readers and sign-language interpreters. (p. 10)

Finally, just as the use of accommodations on paper-and-pencil tests has increased awareness and use of accommodations in the classroom, so can opportunities to use the built-in accommodation features of online assessments encourage and increase the use of those features in classroom and other environments. For example, Williams (2002) believed that "it is possible that new developments in speech recognition technology could increase opportunities for individual reading practice with feedback, as well as collecting assessment data to inform instructional decision making" (p. 41). In addition, most online assessments have built-in tutorials and practice tests. These tutorials provide students with both opportunities for familiarizing themselves with the software and immediate feedback.

Despite the potential advantages offered by online assessments, there remain several challenges, especially in the transition from paper-and-pencil assessments. First of all, the use of technology cannot take the place of content mastery. No matter how well a test is designed or what media are used for administration, students who have not had an opportunity to learn the material tested will perform poorly. Students need access to the information tested to have a fair chance at performing well. Hollenbeck, Tindal, Harniss, and Almond (1999) strongly cautioned that the use of a computer, in and of itself, does not improve the overall quality of student writing. They, and other researchers, continue to find significantly lower mean test scores for English-language learners and students with disabilities than for their native English speaking peers without disabilities.

Concerns continue to exist in the area of equity, where questions are asked about whether online assessments put some students at a disadvantage because of lack of access, use, or familiarity (Trotter, 2001). These concerns include unfamiliarity with answering standardized test questions on a computer screen, using buttons to search for specific items, and indecision about whether to use traditional tools (e.g., hand-held calculator) versus computer-based tools. According to Wissick and Gardner (2000), "students will not take advantage of help options or use navigation guides if they require more personal processing energy than they can evoke" (p. 38). The gap in access to technology—sometimes referred to as the "digital divide"—is continuing to grow. According to Bolt and Crawford (2000), "in the context of the overall racial digital divide, a low-income European-American child is three times more likely to have Internet access than his or her African-American counterpart, and four times as likely as a Latino family in the same socioeconomic category" (p. 98).

Some research raises the question of whether the medium of test presentation affects the comparability of the tasks students are being asked to complete. Here are some findings that show added difficulty for some students.

- Online assessments place more demands on certain skills, such as typing, using multiple screens to recall a passage, mouse navigation, and the use of key combinations (Bennett, 1999; Ommerborn & Schuemer, 2001).

- Some people become more fatigued when reading text on a computer screen than on paper (Mourant, Lakshmanan, & Chantadisai, 1981).

- Long passages may be more difficult to read on computer screen (Haas & Hayes, 1986).

- The inability to see an entire problem on screen at one time is challenging because some items require scrolling horizontally and vertically to get an entire graphic on the page (Hollenbeck et al., 1999).

- Few teachers use computers in math instruction, or spreadsheets, so students do not know how to "think on the monitor" (Trotter, 2001).

- Graphic user surfaces present considerable obstacles to students with visual impairments (Ommerborn & Schuemer, 2001).

An additional challenge is the ongoing entry of new Web browsers and new versions of existing browsers. In addition, Web tools such as HTML and document converters are constantly being developed and modified. Unfortunately, several features may not be universally accessible, and advancements in assistive technology are usually several steps behind new Internet components and tools. For example, using an eye-pointing device may increase the time needed to position each eye-pointing frame, leading to increased fatigue, boredom, and inattention by the test-taker (Haaf, Duncan, Skarakis-Doyle, Carew, & Kapitan, 1999). As online assessments become a reality across states and districts, it is important to ensure that the new technology either improves accessibility or is compatible with existing assistive computer technology.

According to Web Accessibility in Mind, an initiative of the Center for Persons with Disabilities at Utah State University (WebAIM, 2001), there are 27.3 million people with disabilities who are limited in the ways they can use the Internet:

> *The saddest aspect of this fact is that the know-how and the technology to overcome these limitations already exist, but they are greatly under-utilized, mostly because Web developers simply do not know enough about the issue to design pages that are accessible to people with disabilities. Unfortunately, even some of the more informed Web developers minimize the importance of the issue, or even ignore the problem altogether.* (p. 1)

Universally Designed
Online Assessments

Universal design is defined by the Center for Universal Design (1997) as "the design of products and environments to be usable by all people, to the greatest extent possible, without the need for adaptation or specialized design." The Assistive Technology Act of 1998 (P.L. 105-394) addressed universal design through this definition:

> *The term "universal design" means a concept or philosophy for designing and delivering products and services that are usable by people with the widest possible range of functional capabilities, which include products and services that are directly usable (without requiring assistive technologies) and products and services that are made usable with assistive technologies.* (§ 3 (a)(17))

A recent article on the application of universal design to large-scale assessments (Thompson, Thurlow, & Malouf, 2004) found that good basic design, whether on paper or technology-based, increases access for everyone, and poor design can have detrimental effects for nearly everyone. According to WebAIM (2001), "everyone benefits from well-designed Web sites, regardless of cognitive capabilities. In this context, 'well-designed' can be defined as having a simple

Table 1. Elements of universally designed assessments

Element	Explanation
Inclusive Assessment Population	Tests designed for state, district, or school accountability must include every student except those in the alternate assessment, and this is reflected in assessment design and field testing procedures.
Precisely Defined Constructs	The specific constructs tested must be clearly defined so that all construct irrelevant cognitive, sensory, emotional, and physical barriers can be removed.
Accessible, Nonbiased Items	Accessibility is built into items from the beginning, and bias review procedures ensure that quality is retained in all items.
Amenable to Accommodations	The test design facilitates the use of needed accommodations (e.g., all items can be Brailled).
Simple, Clear, and Intuitive Instructions and Procedures	All instructions and procedures are simple, clear, and presented in understandable language.
Maximum Readability and Comprehensibility	A variety of readability and plain language guidelines are followed (e.g., sentence length and number of difficult words are kept to a minimum) to produce readable and comprehensible text.
Maximum Legibility	Characteristics that ensure easy decipherability are applied to text, to tables, figures, and illustrations, and to response formats.

*Note. From **Universal Design Applied to Large-Scale Assessment** (Synthesis Report 44), by S. J. Thompson, C. J. Johnstone, and M. L. Thurlow, 2002, Minneapolis, MN: University of Minnesota, National Center on Educational Outcomes.*

and intuitive interface, clearly worded text, and a consistent navigational scheme between pages" (p. 8). The National Center on Educational Outcomes has conducted an extensive review of research relevant to the assessment development process and the principles of universal design (Thompson, Johnstone, & Thurlow, 2002). This review produced a set of seven elements of universal design that apply to assessments (see Table 1).

Universally designed assessments do not result in tests with "simpler" content; the content has been carefully specified, and assessments are designed to provide equitable opportunity for all students by avoiding extraneous content or complexity. Thompson and Thurlow (2002) stated that

> *universal design is based on the same ethics of equity and inclusiveness that are expected for people with disabilities and others in schools, communities, and on the job – an ethic that values differences in age, ability, culture, and lifestyle. Testing conditions should not be affected by disability, gender, race, English language ability, or levels of anxiety about tests. On the other hand, it is important to remember that universal design does not address deficiencies in instruction. Students who have not had an opportunity to learn the material tested will be disadvantaged during testing no matter how universal the design of the assessment.* (p. 2)

Assistive Technology

Even though items on universally designed assessments will be accessible for most students, there will still be some students who continue to need accommodations, including assistive technology. Assessments may be biased if they do not allow for adaptation for use with assistive technology that is needed to facilitate use of a student's primary means of communication. Online assessments need to be accessible for a variety of forms of assistive technology (e.g., key guards, specialized keyboards, trackballs, screen readers, screen enlargers) for students with physical or sensory disabilities. Bowe (2000) stated, "If a product or service is not usable by some individual, it is the responsibility of its developers to find ways to make it usable, or, at minimum, to arrange for it to be used together with assistive technologies of the user's choice" (p. 27).

It is important to note that making online assessments amenable to assistive technology does not mean that students will automatically know what to do. Educators need to be competent in technology knowledge and use. According to

Lahm and Nickels (1999), "educators must become proactive in their technology-related professional development because teacher education programs have only recently begun addressing the technology skills of their students" (p. 56). The Knowledge and Skills Subcommittee of the Council for Exceptional Children's (CEC) Professional Standards and Practice Standing Committee has developed a set of 51 competencies for assistive technology that cross eight categories, along with knowledge and skills statements for each category (see Lahm & Nickels, 1999).

There are several resources available to increase the accessibility of online assessments for students with disabilities. Chishold, Vanderheiden, and Jacobs (1999) offer guidelines on how to make Web content accessible to people with disabilities. They are quick to point out that following these guidelines can also make Web content more available to all users, including those who use voice browsers, mobile phones, automobile-based personal computers, and other technology.

Process for Developing Inclusive Online Assessments

The transformation of traditional paper-and-pencil tests to inclusive online assessments takes careful and thorough work that includes the collaborative expertise of many people. Here are some steps to follow in addressing the transformation process.

- **Step 1. Assemble a group of experts to guide the transformation.** To be effective, this group needs to include experts in assessment design, accessible Web design, universal design, and assistive technology, along with state and local assessment and special education personnel.

- **Step 2. Decide how each accommodation will be incorporated into the online assessment.** Examine each possible accommodation in light of online administration. Some of the traditional paper-and-pencil accommodations will no longer be needed (e.g., marking responses on test form rather than on answer sheet), while others will become built-in features that are available to every test-taker. Some accommodations will be more difficult to incorporate than others, requiring careful work by test designers and technology specialists.

- **Step 3. Consider each accommodation or assessment feature in light of the content tested.** For example, what are the implications of the use of a screen reader when the content being measured includes reading skills and knowledge ranging from decoding and phonemic awareness to comprehension and narrative structure, or the use of a spell-checker when achievement in spelling is being measured as part of the writing process? As the use of speech recognition technology permeates the corporate world, writing on paper without the use of a dictionary or spell-checker may become essentially obsolete and need to be reconsidered.

- **Step 4. Consider the feasibility of incorporating the accommodation into online assessments.** Questions about the feasibility of an accommodation may require review by technical advisors, or members of a policy/ budget committee, or may require short-term solutions along with long term planning. Construct a specific plan for building in features that are not immediately available, in order to keep them in the purview of test developers. Extensive pilot testing needs to be conducted with a variety of equipment scenarios and accessibility features.

- **Step 5. Consider training implications for staff and students.** The best technology will be useless if students or staff do not know how to use it. Careful design of local training and implementation needs to be part of the planning process. Special consideration needs to be given to the computer literacy of students and their experience using features such as screen readers. Information about the features available on online assessments needs to be marketed to schools and available to Individuallized Education Program teams to use in planning a student's instruction and in preparation for the most accessible assessments possible. Practice tests that include these features need to be available to all schools year around. This availability presents an excellent opportunity for students whose schools have previously been unaware of or balked at the use of assistive technology.

Accommodations Considerations for Online Assessments

Most states have a list of possible or common assessment accommodations for students with disabilities within the categories of presentation, response, timing and scheduling, and setting (Thurlow, Lazarus, & Thompson, 2002). Some states also list linguistic accommodations specifically designed for students with limited English proficiency (Rivera, Stansfield, Scialdone, & Sharkey, 2000). The accommodations described below were generated to address the needs of students with a variety of accommodation needs—including students with disabilities, students with limited English proficiency, students with both disabilities and limited English proficiency, and students who do not receive special services but have a variety of unique learning and response styles and needs. Further descriptions of these considerations have been developed by Thompson, Thurlow, Quenemoen, and Lehr (2002).

Presentation Accommodations

Presentation accommodations allow students to access assessments in ways that do not require them to visually read standard English print. These alternate modes of access are visual, tactile, auditory or multisensory. Examples of presentation accommodations include large print, Braille, magnification, instructions simplified or clarified, audio presentation of instructions and test items, instructions and test items presented in sign language or in a language other than English. Here are some considerations for the use of presentation accommodations for online assessments (Thompson, Thurlow, & Moore, 2003):

- Capacity for any student to self-select print size or magnification
- Graphics and text-based user interfaces have different challenges
- Scrolling issues
- Variations in screen size
- Effects of magnification on graphics and tables
- Capacity for any student to self-select audio (screen reader), alternate language, or signed versions of instructions and test items (all students wear earphones or headphones)
- Capacity to have instructions repeated as often as student chooses
- Variable audio speed and quality of audio presentation
- Capacity for pop-up translation
- Use of screen reader that converts text into synthesized speech or Braille
- Alternative text or "alt tags" for images
- Avoidance of complex backgrounds that interfere with readability of overlying text
- Tactile graphics or three-dimensional models may be needed for some images
- Capacity for multiple screen and text colors

Response Accommodations

Response accommodations allow students to complete assessments in different ways or to solve or organize problems using some type of assistive device or organizer. Examples of response accommodations include write in test booklet, scribe, Brailler, tape recorder, paper-and-pencil response, spell-check, calculator, English or bilingual dictionary or glossary. Here are some considerations for the use of response accommodations for online assessments (Thompson, Thurlow, & Moore, 2003):

- Capacity for multiple options for selecting response—mouse click, keyboard, touch screen, speech recognition, assistive devices to access the keyboard (e.g., mouth stick or head wand)
- Option for paper-and-pencil in addition to computer (e.g., scratch paper for solving problems, drafting ideas)
- Option for paper-and-pencil in place of computer (e.g., extended response items)
- Capacity for any student to self-select spell-check option
- Capacity to disable spell-check option when spelling achievement is being measured
- Spelling implications when using speech recognition software
- Capacity for any student to select calculator or dictionary option

Timing and Scheduling Accommodations

Timing and scheduling accommodations increase the allowable length of time to complete a test or assignment and may also change the way the time is organized. Examples of timing and scheduling accommodations include extended time, time of day beneficial to student, breaks, multiple sessions, possibly over multiple days, and order of subtest administration. Here are some considerations for the use of timing and scheduling accommodations for onlne assessments (Thompson et al., 2003):

- Availability of location of computers and peripherals
- Flexible, individualized timing
- Capacity of network system
- Maintaining place and saving completed responses during breaks

- Capacity to turn off monitor or blank screen temporarily
- Test security
- Capacity for self-selection of subtest order

Setting Accommodations

Setting accommodations change the location in which a test is given or the conditions of the assessment setting. Examples of setting accommodations include individual or small-group administration, preferential seating, special lighting, adaptive or special furniture, hospital–home–nonschool administration. Here are some considerations for the use of setting accommodations for online assessments (Thompson et al., 2003):

- Grouping arrangements
- Use of earphones or headphones
- Use of individual setting, if response method distracts other students
- Availability–comparability–location of computers and peripherals
- Glare from windows or overhead lights
- Adaptive furniture
- Test security

Summary

With the reauthorization of Title I, nearly all states are in the process of designing new assessments. As part of this process, several states are considering the use of online assessments, because this is the mode in which many students are already learning. Several states have already begun designing and implementing online assessments. According to a report to the National Governors' Association (2002), "testing by computer presents an unprecedented opportunity to customize assessment and instruction to more effectively meet students' needs" (p. 8).

Careful attention to ongoing research on the challenges cited in this chapter will be critical to the development and inclusive implementation of online assessments. Because many accessibility features can be built into online assessments, the validity of test results can be increased for many students, including students with disabilities and English-language learners, without the addition of special accommodations. However, even though items can be accessible for most students, there will still be some specialized accommodations, and online assessments need to be amenable to these accommodations. Students with disabilities will be at a great disadvantage if paper-and-pencil tests are simply copied on screen without any flexibility.

In conclusion, a report to the Nationals Governors' Association (2002) sums up what we need to remember as the use of online assessments evolve across the United States and throughout the world:

> *Do not forget why electronic assessment is desired. Electronic assessment will enable states to get test results to schools faster and, eventually, cheaper. It will help ensure assessment keeps pace with the tools that students are using for learning and with the ones that adults are increasingly using at work. The technology will also help schools improve and better prepare students for the next grade, for postsecondary learning, and for the workforce.* (p. 9)

References

Bennett, R. E. (1995). Computer-based testing for examinees with disabilities: On the road to generalized accommodation. In S. Messick (Ed.), *Assessment in higher education: Issues of access, student development, and public policy*. Hillsdale, NJ: Erlbaum.

Bennett, R. E. (1999). Using new technology to improve assessment. *Educational Measurement Issues and Practice, 18*(3), 5-12.

Bennett, R. E., Goodman, J., Hessinger, J., Ligget, J., Marshall, G., Kahn, H., et al. (1999). Using multimedia in large-scale computer-based testing programs. *Computers in Human Behavior 15*, 283-294.

Bolt, D., & Crawford, R. (2000). *Digital divide: Computers and our children's future*. New York: TV Books.

Bowe, F. (2000). *Universal design in education: Teaching nontraditional students*. Westport, CT: Bergin & Garvey.

Brown, P. J., & Augustine, A. (2001, April). *Screen reading software as an assessment accommodation: Implications for instruction and student performance*. Paper presented at the American Education Research Association Annual Meeting, Seattle, WA.

Brown-Chidsey, R., & Boscardin, M. L. (1999). *Computers as accessibility tools for students with and without learning disabilities*. Amherst, MA: University of Massachusetts.

Calhoon, M. B., Fuchs, L. S., & Hamlett, C. L. (2000). Effects of computer-based test accommodations on mathematics performance assessments for secondary students with learning disabilities. *Learning Disability Quarterly, 23*, 271-282.

Center for Universal Design. (1997). *What is universal design?* Retrieved from North Carolina State University, Center for Universal Design Web site: *www.design.ncsu.edu/cud/univ_design/ud.htm*

Chishold, W., Vanderheiden, G., & Jacobs, I. (1999). *Web content accessibility guidelines*. Retrieved from University of Wisconsin, Trace R & D Center Web site: *http://www.w3.org/TR/1999/WAI-WEBCONTENT-19990505*

Fuchs, L. S., Fuchs, D., Eaton, S., Hamlett, C. L., & Karns, K. (2000). Supplementing teacher judgments of mathematics test accommodations with objective data sources. *School Psyhology Review, 29*, 65-85.

Goldberg, L., & O'Neill, L. M. (2000, July). Computer technology can empower students with learning disabilities. *Exceptional Parent Magazine*, 72-74.

Haaf, R., Duncan, B., Skarakis-Doyle, E., Carew, M., & Kapitan, P. (1999). Computer-based language assessment software: The effects of presentation and response format. *Language, Speech, and Hearing Services in Schools, 30*, 68-74.

Haas, C., & Hayes, J.R. (1986). What did I just say? Reading problems in writing with the machine. *Research in the Teaching of English, 20*(1), 22-35.

Hollenbeck, K., Tindal, G., Harniss, M., & Almond, P. (1999). Reliability and decision consistency: An analysis of writing mode at two times on a statewide test. *Educational Assessment, 6*(1), 23-40.

Lahm, E. A., & Nickels, B. L. (1999). Assistive technology competencies for special educators. *Teaching Exceptional Children, 32*(1), 566-63.

Mourant, R. R., Lakshmanan, R., & Chantadisai, R. (1981). Visual fatigue and cathode ray tube display factors. *Human Factors, 23*(5), 529-546.

National Governors' Association. (2002). *Using electronic assessment to measure student performance*. Education Policy Studies Division: National Governors' Association.

National Research Council. (2001). *Knowing what students know: The science and design of educational assessments*. Washington, DC: Board on Testing and Assessment, Center for Education. Division of Behavioral and Social Sciences and Education, National Academy Press.

Ommerborn, R., & Schuemer, R. (2001). *Using computers in distance study: Results of a survey amongst disabled distance students.* Retrieved from FernUniversität—Gesamthochschule in Hagen Web site: *http://www.fernuni-hagen.de/ZIFF*

Rivera, C., Stansfield, C. W., Scialdone, L., & Sharkey, M. (2000). *An analysis of state policies for the inclusion and accommodation of English language learners in state assessment programs during 1998-1999.* Arlington, VA: George Washington University Center for Equity and Excellence in Education.

Thompson, S. J., Johnstone, C. J., & Thurlow, M. L. (2002). *Universal design applied to large-scale assessment* (Synthesis Report 44). Minneapolis, MN: University of Minnesota, National Center on Educational Outcomes.

Thompson, S., & Thurlow, M. (2002). *Universally designed assessments: Better tests for everyone!* (Policy Directions No. 14). Minneapolis, MN: University of Minnesota, National Center on Educational Outcomes.

Thompson, S. J., Thurlow, M. L., & Malouf, D. B. (2004). Creating better tests for everyone through universally designed assessments. Retrieved from *http://www.testpublishers.org/atp_journal.htm*

Thompson, S., Thurlow, M., & Moore, M. (2003). *Using computer-based tests with students with disabilities* (Policy Directions No. 15). Minneapolis, MN: University of Minnesota, National Center on Educational Outcomes.

Thompson, S. J., Thurlow, M. L., Quenemoen, R. F., & Lehr, C. A. (2002). *Access to computer-based testing for students with disabilities* (Synthesis Report 45). Minneapolis, MN: University of Minnesota, National Center on Educational Outcomes.

Thurlow, M. L., Lazarus, S., & Thompson, S. J. (2002). *2001 state policies on assessment participation and accommodations* (Synthesis Report 43). Minneapolis, MN: University of Minnesota, National Center on Educational Outcomes.

Tindal, G., Heath, B., Hollenbeck, K., Almond, P., & Harniss, M. (1998). Accommodating students with disabilities on large-scale tests: An experimental study. *Exceptional Children, 64,* 439-450.

Trotter, A. (2001). Testing computerized exams. *Education Week, 20*(37) 30-35.

WebAIM (2001). Introduction to Web accessibility. Retrieved from *www.webaim.org/intro/*

Williams, S. M. (2002). Speech recognition technology and the assessment of beginning readers. In National Research Council, *Technology and assessment: Thinking ahead: Proceedings of a workshop.* Washington, DC:

Board on Testing and Assessment, Center for Education. Division of Behavioral and Social Sciences and Education, National Academy Press.

Wissick, C. A., & Gardner, J. E. (2000). Multimedia or not to multimedia? That is the question for students with learning disabilities. *Teaching Exceptional Children, 32*(4), 34-43.

Endnote

*   This chapter was developed with support from the National Center on Educational Outcomes at the University of Minnesota, and its funding agency the Office of Special Education Programs, US Department of Education (Cooperative Agreement H326G000001), as well as NCS Pearson, Inc. Opinions expressed herein do not necessarily reflect those of OSEP, the US Department of Edcuation, or NCS Pearson.

Chapter VI

Best Practices in the Assessment of Online Discussions

Katrina A. Meyer, University of Memphis, USA

Abstract

This chapter develops the rationale for several best practices in the assessment of online discussions. It provides instructors with an introduction to the differences between face-to-face and online discussions, how to evaluate online discussions, how to perform these assessments, and how to use assessment information to improve future online discussions. These best practices are intended to be an initial guide to the novice online instructor. The increasing use of online discussions in both traditional and distance classes will likely generate new forms of assessment, new rubrics, and new insights, and the instructor will need to stay informed of these developments.

Introduction

The enormous growth in use of the Web to enhance campus-based courses or to deliver entire courses has required that instructors adapt assessment methods to new learning activities or existing activities that occur in new ways, such as online discussions. With the advent of course management systems such as WebCT and Blackboard, an instructor can use three different methods for holding an online discussion. E-mail can be used to conduct a conversation, although its major drawbacks are its public nature and its occasional unreliability. A chat allows students to discuss course content in a synchronous mode (at the same time) and in a way that allows the conversation to unfold visibly in nearly real time. And a threaded discussion or discussion board allows students to contribute to a discussion asynchronously, or whenever they can log into the course, read others' contributions, and formulate their own thoughts or ideas to contribute to the group's discussion.

In every case, these discussions leave a written or printable record, and although students and faculty have been communicating as part of the teaching and learning enterprise for centuries, this is the first time that a fleeting and temporary communication can be frozen in text, removed from the immediate demands of a face-to-face encounter, and evaluated as many times as necessary to learn what occurred and determine what to do differently. Other than videotaping and transcribing classroom activities, there has never been an opportunity to see and evaluate what occurs in these communications between and among students and instructors. Though various forms of discussion are not entirely new to academe, the form of online discussion is finally open to careful evaluation.

Online discussions represent several new opportunities and challenges for instructors. Instructors need to learn how to design effective online discussions and continually improve discussion assignments for students to encourage greater and deeper learning. They need to evaluate students' contributions to the discussion for assessment purposes and use these results to determine students' grades, help students learn and evaluate their own role in the discussion, and learn how best to guide a discussion without stifling students' willingness to participate. In each of these cases, the instructor faces several challenges related to assessment. First, instructors need to understand how online discussions compare to the more familiar face-to-face or classroom discussions. Second, they need to decide how online discussions should be evaluated and what criteria might be used. Third, they must learn how to use traditional or computer-based methods to conduct these assessments. Fourth, they need to practice using assessments of online discussions to improve the design and conduct of future online discussions. Finally, instructors need to keep informed of the best practices for the assessment of online discussions.

How Do Online Discussions Compare to Face-to-Face Discussions?

It is not surprising that many instructors who begin to implement online discussions, either to enhance a traditional class or in an online course, wonder if the online discussion is the same as face-to-face discussions held in class. The classroom discussion is what they are familiar with, and this comparison is reasonable (Meyer, 2004b). The natural first step for the instructor who is new to online learning is to specifically compare the face-to-face and online discussion modes from the points of view of the student and the instructor. Comparing these two modes of discussion is an important way for instructors to teach themselves about the differences and similarities between them and to test different assessment methods for the online discussion. They will likely come to understand that each mode of discussion has its advantages and disadvantages, and each may encourage different types of learning or be advantageous to students with different learning styles. The face-to-face discussion has a certain energy and benefits students who are quick thinkers and comfortable speaking up in class. The online discussion allows students to reflect on the previous posting and prepare a more thoughtful response and, therefore, benefits students who prefer reflection and are more inhibited (Carnevale, 2003; Meyer, 2003). Understanding these differences in discussion mode and how each discussion mode benefits different students is important for the instructor, who must rapidly learn how to best use and evaluate online discussions to support the instructional objectives of the course.

This period of comparing the two discussion modes will likely lead the instructor to experiment with different discussion settings (online or face-to-face) for different topics or to provide a variety of settings to benefit a wider range of student learning preferences or styles of processing information. There is no clear and consistent indication of which mode works best in different settings, so this is obviously an important area for further experimentation and evaluation. As experience is gained with conducting online discussions and evaluating student learning as a result of participating in online discussions, instructors will likely discover that some topics (perhaps one that is controversial or highly charged) work better in one setting than the other, or that some students are so dependent on facial clues that they are unable to participate in the online discussion.

How Might Online Discussions be Evaluated, and What Criteria Might be Used?

At the current time, there are three main tools for evaluating online discussions: content analysis, rubrics, and frameworks. This section should provide the reader with a basic understanding of these methods or tools, the types of learning they evaluate, and their uses. Online discussions can be evaluated for qualities of the student or the group using a variation of content analysis. For example, a discussion transcript can be reviewed for instances or expressions of support, such as personal comments or answering a question (Fahy, 2003). The instructor can also look for instances of "social climate" or the amount of non-class-related or social interaction among class members (Oren, Mioduser, & Nachmias, 2002), or the instructor may want to look for evidence of "active interaction" among students or the frequency of messages posted by students per week (Rovai & Barnum, 2003). The instructor may want to pay attention to students' different use of "linguistic qualifiers and intensifiers" as a way to understand how the different genders combine social interchanges with online discussions (Fahy, 2002). An instructor may be interested in these qualities because they help contribute to the formation of learning communities or simply to better understand what is going on in their class-related online discussions.

There have been several studies of critical thinking that have used content analysis. For example, Garrison, Anderson, and Archer (2001) developed four stages of critical thinking and used these to identify and evaluate the level of a student's posting to an online discussion. Newman, Webb, and Cochrane (1995), Garrison et al. (2001), and Jeong (2003) also used variations of content analysis to evaluate critical thinking in online discussions. For example, Garrison et al. (2001) used a four-stage model of critical thinking (moving from "triggers" to "exploration" to "integration" to "resolution"), which Meyer (2003, 2004) used to evaluate students' contributions to online discussions and change subsequent discussion assignments.

The method of content analysis depends on determining the qualities of interest and then identifying the cues, words, phrases, or characteristics that indicate the presence of the quality. The instructor reviews the transcript of the discussion for these indicators or may ask others to code the discussions, in which case interrater reliability will be important to ensure. Once coded, the online discussions can be analyzed based on certain qualities' frequency of occurrence, patterns of qualities, proximity of qualities to other occurrences, and which students have the quality of interest.

A rubric is an "authentic assessment tool which is particularly useful in assessing criteria which are complex and subjective" (Pickett, n.d., ¶1). Rubrics improve the objectivity and consistency of assessment and require faculty to clarify criteria beforehand and to do so in very specific terms. Three excellent examples of rubrics are assessments of (a) the effectiveness of student participation in online discussions during an entire course (Edelstein & Edwards, 2002), (b) the interactive qualities of distance learning courses (Roblyer & Ekhaml, 2000), and (c), the presence of critical thinking and critical engagement in courses (Brown, n.d.a, n.d.b).

The value of rubrics is that they allow the evaluator to assess the strength or weakness of the student's contribution to an online conversation, based on a continuum for a particular quality, such as pertinence to the topic, new ideas brought into the discussion, or any other quality that the instructor wishes to assess. For example, the rubric might evaluate the student's new ideas from "no new ideas offered" to "an occasional new idea contributed" to "often contributes new ideas to discussion." Another rubric might then evaluate the value of these new ideas, whether other students built upon them in their own postings, and whether students evaluated ideas in a balanced manner.

In fact, there is a need for many more rubrics to be developed. A rubric for social presence (or the ability of students to contribute in such a way as to make their personalities come to life in their online contributions) would be helpful. Social presence is a quality attributed to written communications that has been tied to student satisfaction and effective learning (Gunawardena & Zittle, 1997; Richardson & Swan, 2003) and may be a quality that is useful to develop in students who are new to online discussions. A close variation is "teaching presence," which includes such functions as the ability to facilitate discussions and provide direct instruction and which has also been tied to student satisfaction and learning (Shea, Pickett, & Pelz, 2003). A rubric to assess teaching presence might be useful in evaluating the instructor's contributions to the online discussion and helping instructors who are new to online instruction improve their skills. Such a rubric might help identify online teacher behaviors that stimulated discussion, others that had a more deleterious effect, and messages that students interpreted in positive or counterproductive ways.

Another important rubric that is needed is one for interactivity. The theoretical underpinnings of much of online learning is based on three interactions: with the content, with other students, and with faculty and interaction has been deemed of vital importance for student involvement and learning (Meyer, 2002; M. G. Moore, 1993). Thus, a rubric that assessed the amount and quality of interaction of all three types would be a necessary addition to the rubrics available for assessing online discussions.

Additional rubrics are needed to assess the individual's contribution to the social construction of knowledge or the development of an online learning community. These skills are important for students and essential to the development of effective online discussions and classes. Another future rubric might emphasize what is currently known about what matters most for student learning in the online setting and what we know about good practice in any setting. An example may be a modification of the "Seven Principles of Good Practice" (Chickering & Gamson, 1987), which have been applied fruitfully to the online learning environment (Chickering & Ehrmann, 2000; Hutchins, 2003). The Seven Principles recognize good practice as:

1.   encouraging contact between students and faculty,
2.   developing reciprocity and cooperation among students,
3.   encouraging active learning,
4.   giving prompt feedback,
5.   emphasizing time on task,
6.   communicating high expectations, and
7.   respecting diverse talents and ways of learning.

For example, a rubric might help identify instances wherein feedback was provided to students (from the instructor or other students), how prompt it was (occurring within the hour, day, or week), and also describe the nature of the feedback (from being supportive or asking questions or providing additional information or critiquing the student's response).

In addition to rubrics, the online discussion can be evaluated through a variety of frameworks, whose development may have preceded the onset of online modes of learning or may be developed specifically for online learning. These frameworks can be used to evaluate the student's individual contribution to the discussion (in other words, his or her "posting"), the student's overall contribution to the discussion (across several postings), the level of the student's development or thinking process, or the process used by the group to discuss the topic. King and Kitchener's (1994) reflective judgment model and Perry's (1999) stages of intellectual development may also be useful means for determining the stage at which a student is operating (Meyer, 2004a), even though these frameworks were developed prior to online coursework. These models offer instructors an approach to determining the level of student thinking and ethical development, and perhaps designing future discussions that develop student thinking further.

Bloom's taxonomy (Anderson & Krathwohl, 2001; Bloom & Krathwohl, 1956) can also be used in this fashion to determine the level of student's postings, the level at which a student consistently contributes, or the progress of an entire online discussion (Meyer, 2004a, 2005). Other online assessments might include an evaluation of how closely the discussion corresponded to good models of problem solving (see National Research Council, 2001, for additional suggestions). In fact, other earlier frameworks for classifying, understanding, or critiquing student thinking or his or her processing of information may be usefully applied to online discussions in order to evaluate the student, his or her thinking, and the group's overall effort at discussion.

How Can the Instructor Perform these Assessments?

There are at least three means of performing these various assessments. The first and most rudimentary technique may be simply printing off the online discussion and coding by hand each posting, elements within the posting, or the discussion as one would in qualitative research. This is a time-consuming and laborious task and depends upon instructors' familiarity with the evaluation tool (i.e., rubric or framework) and their ability to distance themselves from the discussion being evaluated. If, especially, the discussion includes instructors' own contributions, they must be able to assess their own postings with honesty and objectivity, in case there are lessons to be learned about how, when, and why instructor comments close, stall, or redirect an ongoing discussion (G. Brown, personal communication, November 4, 2003).

The second technique involves saving the online discussion to a word processing document or other software package and either coding contributions manually, asking Word to search for phrases of interest, or by using a software program such as Ethnograph (Qualis Research, n.d.), coding software that is commonly used in qualitative research. No software, however, alleviates the need for the instructor to determine what is important in the discussion to assess and what it means.

The last technique would be a fully online assessment. In this case, the instructor might be able to code the posting directly into the discussion or a software product could be programmed to do the coding. Because manual coding is such a time-intensive procedure, a product for coding online discussions might be a welcome and useful addition to instructors' evaluation tools or to the developers of course management systems such as Blackboard, WebCT, or Desire2Learn. Moving the technique to a more automatic or preprogrammed mode may be one way to lessen the amount of time it currently takes for instructors to assess online discussions, estimated by Lazarus (2003) to be over 200 minutes per week on average. Furthermore, students could also use such software to identify immediately the nature of their contribution to the discussion so that they could assess their own performance.

How Can These Assessments Improve Future Online Discussions?

These evaluations have great potential for improving online learning, both in terms of improving instructor practice but also increasing student learning. The printed documentation of what occurred during the online discussion allows the instructor to look for a variety of problems and successes, take the time to analyze what occurred, and prepare alternative approaches that can be tested in another discussion or subsequent class. For example, discussions that are short, end prematurely, or do not go very deeply into the topic are an opportunity for the instructor to evaluate whether the problem was the topic, student preparation for the discussion, its introduction or presentation to the group, instructor role in the discussion, or a lack of guidance by the instructor to the group on what was expected. To solve the latter problem, many guides suggest assessing points to students for participation in online discussions (Berner, 2003) or providing clearer directions in the syllabus about what the instructor is looking for or wants to occur in the discussion (W. S. Moore & Rousso, n.d.). Schulte (2003) provided an example of an instructor who modified her choices for the learning environment based on experiences in previous courses and especially on an assessment of students' interactivity during online discussions. Berner (2003) also provided a thoughtful revision of online course requirements based on earlier assessments of student participation and learning.

W. S. Moore and Rousso (n.d.) drew upon an existing knowledge base of good assessment practices to develop principles for designing and evaluating quality online assessments. Many of the principles—having a clear purpose, providing specific instructions for students, including clear evaluation criteria—apply as well to online discussions as any other educational activity. Such aids to designing appropriate assessments for online discussion can be a useful tool for instructors desiring to improve their own assessment practices and learn from discussions that are not optimal in terms of student learning. In any case, the goal of assessment is to improve the educational value of the online discussion through careful reflection and analysis of the transcripts of the discussion.

What Might be Some Best Practices For Assessing Online Discussions?

The assessment of online discussions is a field that is experiencing rapid development but is still in its infancy. The following suggestions should be considered tentative guidelines for instructors desiring to evaluate online discussions.

1. **It matters how the instructor sets up the discussion.** Especially for students who are new to online discussions or need direction or encouragement to participate, the instructor needs to provide an understandable context for the discussion, a purpose or goals, and guidelines for participation (e.g., how many times a student should post to the discussion; Meyer, 2005; W.S. Moore & Rousso, n.d.).

2. **It matters what the purpose of the evaluation is.** Online discussions can be evaluated for several purposes or qualities (e.g., participation, quality of contributions, existence of course learning outcomes, final exams or course grade), and therefore the instructor can look for a variety of outcomes. It is good instructional practice to indicate to students how the discussion will be evaluated, what evaluation criteria will be used, and how the evaluation will be used (e.g., student grades; W. S. Moore & Rousso, n.d.).

3. **It matters how and at what level an online discussion is initiated.** Evidence exists that subsequent student postings are more likely to occur at the same level of the original posting. Thus, the instructor may wish to initiate a discussion at the level desired or have students initiate the discussion using questions or triggers that are at a specified level (Meyer, 2004a, 2005; Williams & Murphy, 2002).

4. **It matters how the instructor interacts in the discussion.** With some students, the instructor is considered the expert or authority. Once the instructor contributes to the online discussion, discussion ends. With other students, the instructor is considered another member of the discussion with equal authority. This may be as much a reflection of the student–instructor relationship, the student's enjoyment of discussion (Berner, 2003), or the prevailing instructional philosophy that might stress democratic values (Finkel, 2000) and student responsibility. In any case, an instructor should carefully assess how and whether and when they should choose to participate in the discussion.

5. **It matters what rubric or framework is used.** The choice of rubric or framework tends to focus the evaluation on one quality and ignore other qualities that may be as valuable to the evaluation. Either choose several rubrics, choose a framework that will specifically look for the qualities desired in the discussion, or vary evaluation tools so that multiple lenses may be used to assess discussions in the course (Meyer, 2004a; W. S. Moore & Rousso, n.d.).

6. **More rubrics or frameworks are needed.** Future evaluators have ample opportunity to develop new rubrics or frameworks or apply frameworks that have been developed for other purposes.

Conclusion

This is an exciting time for instructors who are designing, participating in, and evaluating online discussions. New rubrics and frameworks useful for evaluation will likely appear in each issue of the growing number of online journals devoted to online learning. The software tools that allow online discussions to occur will also continue to develop and offer new ways for instructors to use existing software to evaluate their students' contributions. New ideas for using online discussions in the learning process will also be developed, as more instructors go online and design learning experiences that are imaginative, creative, and use the technology in ways not currently possible or conceivable.

In the meantime, it will remain important to follow the best practices for assessment of student learning developed prior to the advent of online discussions. And given the number and volume of new research studies on online learning as well as online discussions, it is likely that further research will revise or change dramatically the best practices mentioned earlier. In other words, the best practices for assessing online discussions will likely be a moving target for some time.

References

Anderson, L. W., & Krathwohl, D. R. (Eds). (2001). *A taxonomy for learning, teaching, and assessing: A revision of Bloom's taxonomy of educational objectives*. New York: Longman.

Berner, R. T. (September/October, 2003). The benefits of bulletin board discussion in a literature of journalism course. *The Technology Source*. Retrieved from *http://ts.mivu.org/default.asp?show=article&id=1036*

Bloom, B. S., & Krathwohl, D. R. (1956). *Taxonomy of educational objectives: The classification of educational goals*. New York: Longmans, Green.

Brown, G. (n.d.a). Guide to rating critical thinking. Pullman, WA: Washington State University. Retrieved from *http://wsuctproject.ctlt.wsu.edu/ctr.htm*

Brown, G. (n.d.b). Guide to rating critical engagement. Pullman, WA: Washington State University. Retrieved from *http://www.ctlt.wsu.edu/Critical_Engagement.doc*

Carnevale, D. (December 12, 2003). Introverts do well in online chats. *The Chronicle of Higher Education, 50*(16), A29.

Chickering, A. W., & Gamson, Z. (1987, May). Seven principles of good practice. *AAHE Bulletin*, 3-7.

Chickering, A. W., & Ehrmann, S. C. (1999). Implementing the seven principles: Technology as lever. Retrieved from *http://www.aahe.org/technology/ehrmann.htm*

Edelstein, S., & Edwards, J. (2002). If you build it, they will come: Building learning communities through threaded discussions. *The Online Journal of Distance Learning Administration, 5*(1). Retrieved from *http://www.westga.edu/~distance/ojdla/spring51/edelstein51.html*

Fahy, P. J. (2002). Use of linguistic qualifiers and intensifiers in a computer conference. *The American Journal of Distance Education, 16*(1), 5-22.

Fahy, P. J. (2003). Indicators of support in online interaction. *International Review of Research in Open and Distance Learning, 4*(1). Retrieved from *http://www.irrodl.org/content/v4.1/fahy.html*

Finkel, D. L. (2000). *Teaching with your mouth shut*. Portsmouth, NH: Boynton/Cook.

Garrison, D. R., Anderson, T., & Archer, W. (2001). Critical thinking, cognitive presence, and computer conferencing in distance education. *The American Journal of Distance Education, 15*(1), 7-23.

Gunawardena, C. N., & Zittle, F. J. (1997). Social presence as a predictor of satisfaction within a computer-mediated conferencing environment. *The American Journal of Distance Education, 11*(3), 6-26.

Hutchins, H. M. (2003). Instructional immediacy and the seven principles: Strategies for facilitating online courses. *The Online Journal of Distance Learning Administration, 6*(3). Retrieved from *http://www.westga.edu/~distance/ojdla/fall63/hutchins63.html*

Jeong, A. C. (2003). The sequential analysis of group interaction and critical thinking in online threaded discussions. *The American Journal of Distance Education, 17*(1), 25-43.

King, P. M., & Kitchener, K. S. (1994). *Developing reflective judgment*. San Francisco: Jossey-Bass.

Lazarus, B. D. (2003). Teaching courses online: How much time does it take? *JALN, 7*(3), 47-54. Retrieved from *http://www.aln.org/publications/jaln/v7n3/pdf/v7n3_lazarus.pdf*

Meyer, K. A. (2005). The ebb and flow of online discussions: What Bloom can tell us about our students' conversations. *JALN, 9*(1), 53-63. Retrieved from *http://www.alm.org/publications/JALN/v9.1/pdf/v9:1_meyer.pdf*

Meyer, K. A. (2002). Quality in distance education: Focus on on-line learning. *ASHE-ERIC Higher Education Report Series, 29*(4).

Meyer, K. A. (2004a). Evaluating online discussions: Four different frames of analysis. *JALN, 8*(2). Retrieved from *http://www.aln.org/publications/jaln/v8n2/pdf/v8n2_meyer.pdf*

Meyer, K. A. (2003). Face-to-face versus threaded discussions: The role of time and higher-order thinking. *JALN, 7*(3). Retrieved from *http://www.aln.org/publications/jaln/v7n3/pdf/v7n3_meyer.pdf*

Meyer, K. A. (2004b). Putting the distance learning comparison study in perspective: Its role as personal journey research. *The Online Journal of Distance Learning Administration, 7*(1). Retrieved from *http://www.westga.edu/%7Edistance/ojdla/spring71/meyer71.html*

Moore, M. G. (1993). Three types of interaction. In K. Harry, M. John, and D. Keegan (Eds.), *Distance education: New perspectives* (pp. 19-24). New York: Routledge.

Moore, W. S., & Rousso, E. (n.d.). *Principles for designing and evaluating the quality of online assignments/assessments.* Unpublished manuscript.

National Research Council. (2001). *Knowing what students know: The science and design of educational assessment.* Washington, DC: National Academy Press.

Newman, D. R., Webb, B., & Cochrane, C. (1995). A content analysis method to measure critical thinking in face-to-face and computer supported group learning. *Interpersonal Computing and Technology, 3*(2), 56-77.

Oren, A., Mioduser, D., & Nachmias, R. (2002). The development of social climate in virtual learning discussion groups. *International Review of Research in Open and Distance Learning, 3*(1). Retrieved from *http://www.irrodl.org/content/v3.1/mioduser.html*

Perry, W. G., Jr. (1999). *Forms of ethical and intellectual development in the college years: A scheme.* San Francisco: Jossey-Bass.

Pickett, N. (n.d.). Rubrics for web lessons. Retrieved from *http://edweb.sdsu.edu/triton/july/rubrics/Rubrics_for_Web_Lessons.html*

Qualis Research. (n.d.). The Ethnograph. Retrieved from *http://ww.QualisResearch.com*

Richardson, J. C., & Swan, K. (2003). Examining social presence in online courses in relation to students' perceived learning and satisfaction. *JALN, 7*(1), 68-88. Retrieved from *http://www.aln.org/publications/jaln/v7n1/pdf/v7n1_richardson.pdf*

Roblyer, M. D., & Ekhaml, L. (2000). How interactive are YOUR distance courses? A rubric for assessing interaction in distance learning. *The Online Journal of Distance Learning Administration, 3*(2). Retrieved from *http://www.westga.edu/~distance/roblyer32.html*

Rovai, A. P., & Barnum, K. T. (2003). On-line course effectiveness: An analysis of student interactions and perceptions of learning. *Journal of Distance Education, 18*(1), 57-73.

Schulte, A. (2003). Discussions in cyberspace: Promoting interactivity in an asynchronous sociology course. *Innovative Higher Education, 28*(2), 107-118.

Shea, P. J., Pickett, A. M., & Pelz, W. E. (2003). A follow-up investigation of "teaching presence" in the SUNY learning network. *JALN, 7*(2), 61-80. Retrieved from *http://www.aln.org/publications/jaln/v7n2/pdf/v7n2_Shea.pdf*

Williams, C. B., & Murphy, T. (2002). Electronic discussion groups: How initial parameters influence classroom performance. *Educause Quarterly, 25*(4), 21-29. Retrieved from *http://www.educause.edu/ir/library/pdf/eqm0244.pdf*

Section III

Challenges in Online Assessment and Measurement

Chapter VII

Challenges in the Design, Development, and Delivery of Online Assessment and Evaluation

Clark J. Hickman, University of Missouri-St. Louis, USA

Cheryl Bielema, University of Missouri-St. Louis, USA

Margaret Gunderson, University of Missouri-Columbia, USA

Abstract

This chapter introduces online assessment and evaluation and explores strategies to overcome challenges in their design, development, and delivery in online settings. It argues that both assessment and evaluation are process-driven, involving many stakeholders including faculty, administrators, and students. This process includes needs assessment, strategies for asking good questions, knowing where to find information, and how to communicate findings in persuasive ways. By developing a comprehensive system, one based on reliable research principles, faculty and administrators can harvest considerable data that enable stakeholders to refine educational programs and maximize learning.

Introduction

What types of data do you collect to report outcomes and the impact of teaching, learning, and technology programs on your campus? How do you justify new programs or make decisions about improving or terminating less effective ones? Are you and others frustrated with the course-evaluation process at your institution? Are administrators pushing for online course evaluations, and do you know what opportunities and problems accompany the implementation of electronic evaluation systems?

These are just a few issues and concerns that confront educators in our digital world. There are many forms and purposes for assessment and evaluation, and each can provide clues in the mystery of how we can improve. The key is to carefully plan and design the different forms of evaluation or assessment so as to obtain the maximum amount of information—information that may generate new knowledge or verify already known knowledge. This chapter proposes to address the challenges of assessment and evaluation and suggests ideas for meeting those challenges. While some educators view the words *assessment* and *evaluation* as interchangeable, we differentiate them in this chapter by the following operational definitions:

- **Assessment:** The process of gathering and discussing information from multiple and diverse sources to develop a deep understanding of what students know, understand, and can do with their knowledge as a result of their educational experiences; the process culminates when assessment results are used to improve subsequent learning (Huba & Freed, p. 8).

- **Evaluation:** The determination of a thing's value; in education, it is the formal determination of the quality, effectiveness, or value of a program, product, project, process, objective or curriculum (Worthen & Sanders, p. 23).

Student Assessment

Online instructors often are interested in how to assess students' preparation, how to evaluate their work and effort, how to maintain efficiency with an overloaded schedule, how to grade, and so on. Although instructors are often ultimately responsible for assessing students, they are not the only source. For example, students may assess themselves. Self-assessment will reveal to both instructors and students how much the students have become self-directed learners. Another helpful source is from a student's peers. This is particularly valuable for courses emphasizing frequent discussion or collaborative work. In distance education courses, the community may also become involved in student assessment. For unique situations instructors and students may make arrangements to use community facilities or preceptors to supervise students. And finally, the instructor's own evaluation of students' performance within the course should be broader than examinations. Within an online course, evaluation may also take into account student performance on assignments, individual or group projects, and participation in online discussions.

Course and Program Evaluation

For departments investing in online courses or programs, ongoing evaluation is necessary. Different stakeholders will be interested in various issues. Thus, having more information from a variety of sources will help to determine the course or program's effectiveness as well as needed areas of improvement. Sources for evaluation include students, faculty, the sponsoring department, student support staff and faculty support staff, as well as faculty or administrators with content expertise and those with experience in effective online education delivery. As with assessment, these evaluations are most revealing when they are multi-faceted. However, it takes planning to successfully design and bring together the needed information from so many various sources.

Designing an Online Assessment or Evaluation Process

Huba and Freed stated that "gathering feedback from others and spending time reflecting on it are critical in helping us understand our practice as teachers" (2000). In addition, assessment and evaluation of programs and services are important for administrators and service providers to conduct and critically review. As we examine the various best practices or examples of online assessment and evaluation, a common thread can be seen throughout: that of taking the time and effort to plan. Careful planning allows planners to maintain a learner-centered focus throughout the process. This process may be viewed at multiple levels for different purposes and utilize multiple methods—all contributing toward helping to document what works well and how we can improve it.

Plan a Systematic Approach

The first step in an effective process is to plan a systematic approach, which implies the need to examine the elements that are related to the functions of evaluations and how planning their intended purpose is an important sequence. Reeves and Hedburg (2002) offered one of the most comprehensive models for a systematic approach to evaluation. Their functions for assessment and evaluation include six functions that can also be arranged as parts of the different stages of project development.

Creation and Design

- **Review:** for project, course, or program conceptualization
- **Needs assessment: identify the critical needs that the proposed project is intended to meet:** identify issues that will impact the selection of content, delivery system, or authoring systems; for initial design of the project, course, or program

Development and Refinement: Formative Evaluation

- **Formative evaluation through the development phase:** conduct evaluation early and often; check the project's functionality, usability, appeal, and effectiveness; begins with the planning stage and continues through the assignment, course, or program's completion
- **Effectiveness evaluation during implementation:** Some instructors view this as a type of continuous formative evaluation; that is, to shift focus if the course is not proceeding according to plan. Ideally, formative evaluation should occur at least once before the midpoint of a course. Students are asked to provide feedback about how they are experiencing the course, the mode of instruction, and the online environment. The instructor should provide guides or rubrics for student performance (i.e., are e-mail discussion postings of the quantity and quality desired) to let students know whether they are performing satisfactorily on course assignments.

Formative evaluation can help everyone involved in the teaching and learning exchange make improvements or reinforce learning outcomes. *Learners* will be involved in specific and more frequent assessments of their ongoing gains in knowledge and skills. Formal feedback systems, such as midsemester course evaluations, can help faculty make better decisions based on their student's experiences, early in the semester. Students appreciate the opportunity to offer course feedback because the probability exists that their professors will be attentive to how they are learning as the semester progresses. According to Margaret Cohen, Director of the Center for Teaching and Learning, University of Missouri–St. Louis, "motivation and learning are enhanced when faculty are aware of the specific concerns and learning needs of their students" (M. Cohen, personal communication, March 1, 2004).

Instructors benefit by using classroom assessment techniques to assess what students know and what they do not understand or cannot do. Classroom assessment techniques are small-scale assessments conducted continually in classrooms to determine what and how students are learning. Instructors can better tailor lectures and learning activities once they are aware of the gaps and gains in their students' comprehension. Additionally, *planners* can make improvements to new and existing events and programs by involving participants in data collection and decision making. The processes are similar to total quality management (TQM) or continuous quality improvement techniques.

Completion and Summative Evaluation

- Complete an impact evaluation as the course or program becomes institutionalized, that is, determine whether the knowledge, skills, and attitudes learned transfer to the intended context of use (i.e., the workplace, the next course in a program series). Because impact evaluation presents so many long-term challenges, it is not often implemented.

- Complete a maintenance evaluation to help make decisions about the course or program's future, that is, address the different variables impacting students; address perceptions of the gestalt of the course or program (i.e., the overall sense of how all the parts fit together).

Whether data are collected for the evaluation of an individual's performance or the evaluation of a department, funded program, or institution, summative data can be collected to make decisions about the merit or worth of these individuals, programs, or institutions. The purposes for summative evaluation often include decisions about continuation of programs, accountability for dollars expended, or documenting successful completion of a course of study.

The second step in successful planning is to identify and select the appropriate method or tool and collect data. There are a variety of different methods and techniques for student, course, and program assessment and evaluation. Questionnaires, minute papers, and peer and self-assessments can be used independently or together to inform administrators, instructors, and students. A systematic plan that will provide the greatest impact will be one that utilizes a variety of methods or techniques but will also avoid over-evaluating participants and thus raising their frustration level.

Analyze and Report Results

The final step in the planning process is to determine how to analyze and report results for optimal utilization. Patton described the processes of assessment and evaluation as supporting change "by getting people engaged in reality testing, that is, helping them think empirically, with attention to specificity and clarity, and teaching them the methods and utility of data-based decision making" (Patton, 1997, p. 103). He advocated for a team approach in planning, implementing, and making the evaluation outcomes relevant to continuing program efforts. Consulting is a necessary skill for evaluation specialists. Feedback is the communication of data and preliminary judgments back to decision makers and others responsible for making changes in programs. Joint development of implications, recommendations, and action planning is the logical and essential step before final evaluation reports are distributed to key stakeholders.

Utilizing several means to investigate the course or program will help validate what really happened to the individual, the program, and the organization. Look for unintended consequences, get in contact with nonparticipants, and probe both successful and failed approaches to help supply missing pieces of the puzzle. *Why* and *how* questions are as important as *who, what, where,* and *when* questions.

It is crucial to involve the primary users of the evaluation in analyzing data, "because, in the end, they are the ones who must translate data into decisions and actions" (Patton, 1997, p. 302). Analysis should lead to a concise, balanced, and easily understood report that provides (a) description and analysis of patterns or themes, with possible comparisons to relevant benchmarks, ideals, or competing programs; (b) interpretation; (c) judgment of merit or worth; and (d) recommendations that follow and are supported by the findings.

In summary, the scope of planning entails taking the time to collaborate with the other stakeholders in order to identify the purpose and needs for assessment or evaluation, define and plan the sequence of a systematic approach, identify and select the appropriate methods or tools, plan how data will be collected and analyzed, and how the results will be reported and interpreted.

Challenges

Challenges will be addressed as global concerns as well as the specific constraints and opportunities of the online environment. The remainder of the chapter focuses on four special challenges for the online educator:

1.   Picking the right evaluation for the purpose
2.   Designing an effective assessment strategy
3.   Asking the right questions
4.   Communicating recommendations and building in follow-through

Addressing these details in advance affects time and resources and offers stakeholders greater confidence in results and recommendations.

Challenge 1: Picking the Right Evaluation for Your Purpose

Picking the right evaluation necessarily depends on the primary purposes of the evaluation and the type of course being evaluated. In this section, we will discuss the evaluation of students' acquisition of skills and knowledge in both noncredit and professional development courses as well as evaluation of the course. We will also explore ways to evaluate students earning academic credit at a distance.

Assessing Participants in Distance Education: Noncredit Courses

People engaging in noncredit courses tend to do so for either personal interest or professional development. As such, evaluation in these courses usually takes one or more of the following forms:

*   **Needs assessment:** Doing a needs assessment early can avoid inappropriate content and disappointed reactions among participants. There are two basic questions that drive needs assessments: What are the specific needs of the intended audience? What do they know versus what they would like to, or need to, know? Depending on the topic, this can be easily assessed by either of two methods: giving participants a cafeteria-style checklist to indicate areas of interest, or by specific skills-based tests that determine, quickly and accurately, what skills are needed to meet a

competency-standard. With on-the-job training, this is frequently done in concert with management's expectations of optimum skills and abilities needed to do required tasks, compared to assessments of employees' abilities to do them.

- **Self-assessments:** Adult learners are more easily attracted to learning situations if they perceive knowledge or skill deficiencies in themselves. Self-assessment instruments are good ways to inform learners of learning needs. A thoughtful needs assessment instrument, whose results are shared with the participant, can serve a dual purpose of both needs assessment and self-assessment instrument.

- **Action plan for further instructional needs:** Unless a course is designed to be comprehensive in terms of knowledge or abilities, there is usually subsequent instruction that could be taken for advanced learning(e.g., Part 2, Advanced). An assessment toward the end of one class can inform participants as to their readiness for advanced work.

- **Assessment for skill transfer:** Sometimes an instructor will need to check that his or her material has been understood and that participants have adequately acquired the desired skills or knowledge to better perform their jobs. In this instance, it may be desirable to demonstrate to an employer that their employees' time was not wasted and that the program achieved its objectives.

Common evaluation systems include testing for knowledge; assessing how the participants will apply the knowledge change or improve behaviors; and an assessment of the participants' views of the course—what they liked best, what they liked least, areas to improve, and how they plan to apply their new knowledge or skills. Strategies for crafting the proper evaluation for various situations will be discussed next.

Assessing Participants in Distance Education: Credit Courses

Courses offered for academic credit, or any assessment situation that requires stricter verification of results, presents different challenges in online and other distance education settings. The obvious problem is, how do we know those persons submitting assignments or, especially, tests are the people they purport to be? They are not physically present, and they are filing material electronically or sitting for exams with distant proctors who may not personally know them. Instructional Technology personnel will warn you that whatever safeguard you think is "failsafe", a determined cheater will find a way around it. Inevitably, a determined cheater will find a way around failsafe technology. The following are methods to assess learning in a less secure, online distance environment.

Authentic assessment as a form of testing. It is usually better to think of student evaluation not in terms of online, or distant, tests; rather, strive to access actual learning through more creative means—means that demonstrate acquisition of intended learning objectives or intended skills. Not only can an authentic assessment get rid of paper-and-pencil tests that tend to assess rote memory, it can assess at a deeper, more realistic, level of knowledge acquisition and skill transfer by analyzing a student's ability to critically analyze knowledge and integrate concepts to transfer skills from the classroom to a real-life situation (Huba & Freed, 2000).

Here are some ways to do this:

1. Ask participants to write an essay or paper demonstrating their acquired knowledge. This can take the form of a critical analysis of a topic or an essay on how this program will be integrated into their job (e.g., a teacher or trainer can produce a curriculum unit; a graduate student can critically analyze the topic of the course relative to literature, to practice, or both; an extension agent can provide action plans on changing some aspect of the job function).

2. Ask participants to create a detailed action plan for changing some behavior or practicing some skill, depending on the nature of the program. After a trial period, ask the participant to evaluate the success of the newly acquired behaviors and skills and assess whether further modifications in strategy are warranted, then retest the skills and reevaluate. Although this strategy may be time-consuming and may not be appropriate for all programs, the benefits are obvious: It is a concerted trial-and-error method that allows skill practice and refinement under the direction of the instructor. It is an effective means to practice skill refinement under safe and controlled conditions. This is especially effective for participants trying to change behaviors such as managerial leadership, handling difficult people, assertiveness training, conducting effective meetings, public speaking, and mastering the art of negotiation, among others.

3. Some instructors abandon formal assessments of students through tests, papers, and essays. Instead, they assess student learning through carefully constructed questions or case studies conducted through the course's group discussion boards (chat rooms). Such questions or cases probe students for mastery of content and their ability to integrate that content into real-life situations. If an instructor carefully monitors the responses, he or she will be able to determine who is contributing, what they are contributing, and how others are responding to them. An observant instructor will also

recognize the students are themselves continuing to learn from each other in this process. Students who are not contributing properly (i.e., being silent or contributing in a way that signifies confusion) can be contacted individually for clarification or remediation of content. While this approach may be time-consuming for instructors, its primary advantage is that it allows assessment to occur unobtrusively and, some may say, more authentically, in that students are behaving naturally as they would in more real-life situations (e.g., see Brown, 2000).

Traditional testing through distance education. If a test in the form of a physical document needs to be produced for whatever reason (e.g., requirements of the department, an agency, a funder), there are two primary ways to secure this form of assessment from fraud:

1. If possible, design the test to be "open book or open note." In this design, the test is not so much based on rote-memory facts but on a more sophisticated integration of the topics of the course into a real or imagined setting. Hence, it does not matter if books or notes are used, because this application, or integration, of material is not explicitly stated in these documents. Moreover, someone who has not been through the course is unlikely to be able to "sit in" for a test taker and satisfactorily complete the test. Such written tests can be designed in both objective (e.g., multiple choice) or subjective (short answer or essay) formats.

2. If open book or open note is not acceptable or desirable, then an alternative method of evaluation may be taking a proctored exam in a public computer lab. Public libraries, universities, and extension centers are three common venues for performing this service. Arrangements are made with a proctor, to whom the rules of the test are explained (e.g., time limits, breaks, ability to log in or out, time and date, whether notes or books can be used). The proctor will require a photo ID to verify student's identity. It is extremely important for the instructor to supply a means of contact during administration of the exam, in the event of computer problems or other administration questions.

Examining your personal philosophy of assessment. Too often, the assessment of students or evaluation of courses is done out of habit and with little reflection. We test because we always have or because it is expedient. Sometimes, it is a management expectation. We go through course evaluations for the same reasons. With some guided reflection, however, we can think through our own personal philosophy of assessment and, as a result, create evaluation and assessment systems that are more appropriate for the course or the students.

Here are some questions instructors may ask themselves to reflect on their philosophy of assessment:

1. Why is it important for students to know the content of this course? Do they *want* to know the content, *need* to know the content, or both?

2. How important is it to the students and other course stakeholders (e.g., employers, school administrators) that the course conveys skills that can be transferred to an outside-the-classroom situation?

3. How can students demonstrate that they have mastered the content? What constitutes sufficient proof?

4. Do I ask students to critically analyze information and integrate it within complex concepts?

5. Do I trust students not to cheat on tests or bluff on personal assessments?

6. How can I improve this course? Are there any constructive criticisms that seem to recur? Is it possible or beneficial to respond affirmatively to these criticisms?

7. Do I do a reality check (e.g., "How's it going, folks?") during my courses to make sure participants are not struggling, frustrated, falling behind, or failing to master material?

Challenge 2: Designing an Effective Assessment Strategy

The success or failure of an online evaluation system will rest, primarily, on the evaluation instrument. Toward this end, it is important to remember the KISS principle: Keep It Super Simple. Cumbersome forms that rival a 1040 tax form will invite errors and nonresponses. Common errors made by evaluators include designing forms that are too long, too short, or have unclear questions or phrases.

Design is contingent on purpose. For formative evaluations, it is especially critical to keep the forms and, indeed, the whole process, simple. In formative evaluations, one frequently conducts several probes at various intervals throughout a course or program. The evaluation should be brief and focused on evaluating very specific areas or evaluating within specific timeframes. For summative evaluations, however, the process can be somewhat longer and can assess the experience in total. The following suggestions help to answer the question, how do I design the best instrument?

Determine Your Purpose by Carefully Analyzing What it is You Want to Know, and Who is the Best Person to Give this Feedback.

The lack of forethought in evaluating courses or programs is a common error. This is particularly evident in organizations in which there is a general template evaluation form that is used for all courses or programs, everywhere. These forms (often called "happiness indexes") may seem convenient, but they rarely provide specific information about individual activities. So what kind of course or program evaluation do you do: quick glance and feel or a detailed inspection? More important, what should you do with different programs? The type of program evaluation required obviously depends on what it is you want to know. Common deciding factors include how new the program is, how certain you are of the outcomes, and the particular needs of the stakeholders involved in sponsoring the course. The following variables necessarily influence the proper approach taken:

- *Course or program is new or under major revision.* This type requires extensive evaluation, and needs to be done in concert with all major stakeholders.

- *Course is established but evaluation results have been mixed or ambiguous.* These situations require careful scrutiny. Program managers want to know who is complaining and who is complimenting. What does each person in the group have in common? Anticipating the likely causes for the feedback gives clues as to what to assess in further evaluations to get at the underlying cause of mixed evaluations.

- *Course is established and evaluations are consistently positive.* In these situations, simple happiness index evaluations usually suffice. However, should the pattern of responses change, so should the evaluation mechanism to more accurately identify the source of problems. Benchmarking successful programs can be helpful when adding courses or future programs. However, it also can be used to improve outcomes for those courses that may not reach the intended benchmarks.

The most effective way to approach any type of evaluation is to think it through with all the major stakeholders. You will find that each stakeholder values different aspects, and all needs can be easily accommodated in one system. Figure 1 identifies common interests among major stakeholders. Notice that, although each stakeholder has a different perspective on the educational activity, the lines can blur in their interests.

Figure 1. Specific interests by stakeholder

Stakeholder	Evaluation Interest
Instructor	What did the students learn? Did the students have trouble with the format of the course or the logistics of the course (e.g., time, online issues, technology, confirmations, and technical assistance)? Did the content meet the students' needs? Was content missing that student needed to proceed to the next stage of learning? How could the content be improved in the future?
Instructional Designer	How did the format work for this content? For students? For this instructor? For this organization? Was the instructor adept at this medium? Is the medium a good format for this course? How can the instructional design be improved from the instructor's and student's perspective?
Administrator	Were students pleased, overall, with the format of the course and the quality of the content? Was the course financially profitable? Was the marketing effective? Should this enterprise be continued with this or further courses?
Students	Did they have an opportunity to assess the course accurately? What is it they would like to be able to comment on? Would they be interested in seeing the results?

Once the Purpose has been Determined, Identify the Most Appropriate Methods of Tools to Find Out What You What to Know

The online environment offers an array of different tools for evaluators. Do not look on evaluation as necessarily a one-time probe that utilizes only one type or format. It is helpful to use a mixed method approach for different groups or for different stages in the process. It does not mean, however, that evaluators should use multiple methods for all users and all stages of the evaluation process. Evaluations are time-consuming, both for the organization and for the participants. Never evaluate as a formality or because you feel you should. If you over evaluate, participants will be annoyed and either skew the results in anger or stop responding. Be sure you are clear with yourself and with the participants as to why you are evaluating, and do no more than is necessary to get accurate results.

Purpose: Assessing Student Performance

Faculty often have many reasons for wanting to assess their students, including (a) to determine what a student knows, (b) to provide data for the revision of curriculum for the classroom teacher and the system, (c) to compare what a student can do before and after instruction (with or without instruction), (d) to anticipate educational needs of students as they progress, and (e) to determine if instructional objectives are being met (Chapin & Messick, 1999).

To accomplish these goals, there are increasing opportunities and tools for online instructors to use. Many online course management systems include quiz or survey functions as well as grade books and grading or portfolio systems.

Publishers often are able to provide digitized test banks and answer keys, and many also provide content modules designed for downloading into course management systems. However, these opportunities also come with a variety of concerns; for example, proctoring and other security issues and technical problems in test taking. Addressing these concerns often starts with an instructor's own assessment philosophy and a willingness to adjust prerequisites or assignment structure to take advantage of online methods.

The use of rubrics, or scoring tools, is a way of describing evaluation criteria, grading standards, based on the expected outcomes and performances of students. Typically, rubrics are used in scoring or grading written assignments or oral presentations; however, they may be used to score any form of student performance. Each rubric consists of a set of scoring criteria and point values associated with these criteria. In most rubrics the criteria are grouped into categories so the instructor and the student can discriminate among the categories by level of performance. Rubrics for online instruction have the added benefit of helping students understand the importance of using the technology effectively as a learning environment.

Another helpful technique is that of peer review—students assessing students. Although utilized mostly for individual or independent work, peer review of assignments also may be viewed as a group method, depending upon how much interaction is incorporated into the review process. For an effective peer review process, the course should include multiple assignments that are complex enough to require substantive revision for most students. Students should submit a draft or other preliminary work, consider responses from peers (as well as the instructor), revise, and finally edit. The final version of the assignment should be the result of a work in progress. It is very helpful for writer and reviewer alike to have a rubric or guidelines of expectations as well as information about the review process.

When considering student assessment, many faculty think of quizzes or examinations, first. However, there are other types of assessment techniques that may be especially helpful for online faculty. Classroom assessment techniques have been defined as "small-scale assessments conducted continually in college classrooms by discipline-based teachers to determine what students are learning in that class" (Cross & Steadman, as cited in Huba & Freed, 2000, p. 124). The following classroom assessment techniques are especially suited to distance education or online course delivery:

- **E-mail minute:** The minute paper is adapted to e-mail. Students are asked to answer briefly two to four questions. Examples of questions include, how was the pace of the class, what 1-2 key ideas do you recall, were examples clear, were topics presented sufficiently, and what specific questions do you have. Encourage sending additional comments privately through e-mail. Return a summary of class responses via e-mail.

- **One-sentence summary:** Students are asked to form a long summary statement, answering who does what to whom, when, where, how and why. This will help instructors assess student's ability to identify critical features of a particular topic or lesson.

- **Direct paraphrasing:** Students are asked to paraphrase a topic, concept, lecture, or article for a specific audience. This helps assess how well students understand and can explain a topic from an individual's perspective. It is a useful technique for those fields in which practitioners are expected to explain or to instruct others.

- **Two-way fast feedback:** Instructors give feedback on the students' feedback. Questions are answered on a Likert scale, with a comment box following. Questions relate to aspects of the lecture and notes' clarity and on student's preparation beforehand. The essential step in this process is the reverse feedback, according to Huba and Freed (2000). Faculty will be able to "fix" the assignment or clarify content at the time it becomes a problem to students.

- **Plus/delta feedback tool:** Students are instructed to answer questions on the left side, top, and bottom with "plus" comments and the right side, top, and bottom with "delta" comments or changes they recommend. Questions on each side are related: left side – 1) the teacher/course – what's working, 2) the student – what's working; right side – 1) the teacher/course – what needs changing, 2) the student – what needs changing.

Another form of student assessment is termed performance-based or authentic assessment. Performance-based assessment techniques offer several advantages to the instructor. They can represent the natural flow of units or modules, where teachers have ongoing opportunities to gather data about student progress. A student's actual work is the focus. Examples of authentic assessments include essays, open-ended problems, portfolios, observation of students solving hands-on problems, sample of tasks, or student self-assessment. One more unique way of assessing knowledge or progress in understanding concepts and relationships is concept mapping. Computing software makes it easier to correct, manipulate, and adapt a concept map. Evaluation of students' maps could include the number of concepts, focal concepts, organization and validity of linkages, number of linkages, horizontal versus vertical flow, categories of links, and completeness of interrelationships.

Faculty may need to evaluate small-group activities. This is particularly complicated if the end result is a product, a presentation, or both. Some instructors choose to have the group project submitted only to the instructor. However, others use the project results as a learning activity for the entire class. Groups present their projects to their peers. Questions and answers from the students may follow this, with the instructor moderating. In assessing the student's work, the rubric provided by the instructor needs to address how much of the grade will focus upon collaboration, how much on individual contribution, and how much on the final product.

Purpose: Evaluation of Courses or Programs

Evaluations of online courses or programs often emerge as a two-edged sword. Palloff and Pratt (1999) explain the additional constraints that the online classroom presents. Instructors are able to easily establish a sense of presence in face-to-face classes. However, in many distance education classes, the students and instructor may never meet, and the "presence" is known by the number, length, and quality of e-mail posts. Although this creates a difficult evaluation process, it also serves, on some level, to make the feedback received from students more valuable. This is because student feedback relates directly to their experience of the course and the material they have studied, rather than reflecting on the personality of the instructor.

To obtain that valuable feedback, there are a variety of tools available for online courses and programs. One such tool is the survey format—mailed questionnaires, telephone surveys, or individual and group interviews. The data collection and interpretations from surveys will help answer evaluation questions, assess needs and set goals, and can establish baseline data against which future comparisons are made. Selecting the appropriate survey tool or method will vary with the parameters of each situation. Figure 2, adapted from Isaac and Michael (1990), overviews various survey methods' advantages and disadvantages.

The development of a questionnaire is perhaps the most underestimated step in survey methods. To begin, Worthen and Sanders (1987) recommended listing the "big" questions that you and other stakeholders wish to have answered. By reviewing relevant literature, including similar assessments or evaluations which have been conducted, questions that will yield the most useful information can be determined. At some point, you and the members of your survey team will probably have to hone the survey to the most important questions, for the sake of focus, brevity, and the respondents' goodwill.

The main difference between using questionnaires and doing interviews (individual or group) is the ability to ask additional questions in response to comments. You can clarify and probe the individual responses to get at deeper feelings and meanings. Interviewers will ideally be trained in the techniques they will use to

Figure 2. Advantages and disadvantages of various survey methods

Survey Methods	Advantages	Disadvantages
Questionnaire	Inexpensive Simple to construct Self-administering	Low response without follow-up Little opportunity to clarify responses
Telephone Interview	Less costly than face-to-face interview Increased comfort level with home environment Increased candor	Impossible to reach those without telephones May be viewed as intrusive
Individual Interview	In-depth responses and probing possible Nonverbal cues can add to understanding	Expensive and time-consuming Biases and influences of interviewer Difficult to summarize findings
Group Interview	More efficient than individual interviews Group behavior, interaction and consensus possible Synergy of group process	Intimidating for some people Encourages conformity May polarize disparate opinions Biases and influences of interviewer
Focus Group	Useful for determining perceptions and feelings about products and services Triangulation of quantitative assessment data	Not intended for reaching consensus in planning or decision making Moderation skills important to outcomes

engage and record, and will use a well-planned script to standardize the questions asked.

An effective observation technique that evaluators frequently use during individual interviews for formative evaluation is the think-aloud technique. The evaluator asks the participant to read or use the materials being evaluated, following all the directions as if they were a student. While participants proceed, they are also asked to voice aloud their thoughts as they go through the materials or activities. Initially, the evaluator may need to prompt the participant by asking "What are you thinking now?" or "Tell me your thoughts as you work through this part." As the participant thinks out loud at each point, voicing questions, confusion, understanding, excitement, and so forth, the evaluator is taking notes or recording the interview.

Focus groups provide a group-oriented interview approach to obtaining perceptions and feelings; however, they are not intended to develop consensus or make decisions (Krueger, 1994). Interviews are usually scheduled for 1 to 2 hours, and groups are typically composed of 6 to 10 relatively homogeneous people. Participants are asked to reflect on the open-ended questions asked by the moderator while an assistant moderator records observations and verbal responses. The questions will be set in a context that is understandable and logical to the members. The topics of discussion are carefully predetermined and sequenced, based on a prior analysis of the circumstances surrounding such an in-depth discussion.

One of the most significant concerns of online educators is how to increase response rates. Sorenson and Johnson (2003) indicated that "researchers feel the lower response rate may be due the perceived lack of anonymity of responses, lack of compulsion to complete ratings online, student apathy, inconvenience, technical problems, and required time for completing the ratings." The following are ideas for increasing response rates in the online environment:

- Target audiences for specific survey information
- Legitimize the survey with an invitation and statement of purpose by an administrator or stakeholder, known by the respondents
- Provide an incentive (e.g., cash, drawing for one or more prizes, copy of survey results)
- Incorporate the evaluation into a time normally dedicated to the course (i.e., assignment, chat time); timing is everything (try to avoid end-of-semester crunch)
- Plan follow-up with nonrespondents, (e.g., reminders and second copies of surveys mailed)
- Identify a percentage of nonrespondents to reach via telephone interview

After Determining the Purpose and Suitable Methods, Investigate Commercial Packages that Might Assist You in Creating an Online Assessment or Evaluation

Many colleges and universities have purchased course management systems that enable multiple functions needed to conduct online courses and evaluation procedures. Such functions include discussion boards, uploading links and documents, group and individual email capabilities, announcement boards, assessment (quizzes) and survey capabilities. While WebCT, Blackboard, and eCollege are among the most common packages, there are many others that also provide assessment and evaluation features. *Edu-Tools* provides a helpful comparison (see *http://www.edutools.info/course/help/howto.jsp#compare*). When considering assessment and evaluations in course management systems, an additional helpful software is Respondus. This software package allows creating and managing exams that can be printed to paper or published directly to Blackboard, eCollege, WebCT, as well as other systems. Thus the same surveys, quizzes, or exams could be utilized in multiple course management systems.

For online evaluation surveys, program developers have several choices. One could conceivably use programs imbedded in some of the platforms above, but it may be easier to use software programs that were specifically designed for surveys. One such program is Flashlight Online, which is readily available to educational institutions and is specifically designed for public surveys. This program guides you through the construction of survey items, which you can then just use—or you can save these items into a "bank" for future use on other surveys. After you finish constructing your survey, it is posted on the Flashlight Web site where you can then direct participants to go and answer it. As they submit their responses, the Flashlight program records the responses and can give you a running total of how many have responded as well as a running total on how each item is being answered. This type of program not only saves you considerable time in tallying each item as it is returned (along with not making any human errors in the process), but it also gets responses back to you quickly. Reminders to respondents can be more timely, too.

As this is being written, there are several dozen commercially available online survey services; most can be tried the first time, at no cost. A few follow:

- Advanced Online Surveys: *http://www.advancedsurvey.com*
- Flashlight Online: *http://www.tltgroup.org/programs/flashlightonline.html*
- Hosted Surveys: *http://www.hostedsurvey.com/Pricing.htm*
- Question Mark: *http://www.questionmark.com*
- Web Surveyor: *http://www.websurveyor.com/web-survey-tour.asp*
- Zoomerang: *http://www.zoomerang.com/Login/index.zgi*

Challenge 3: Asking the Right Questions

What are you asking? And, how are you asking it? As previously stated, the key to effective evaluation is asking the right questions of the right stakeholders. We have discussed the stakeholders and their roles in the process, we have reviewed possible methods and online tools, and now we focus on some strategies to help in the design of an actual evaluation instrument.

Evaluation instruments can focus on many different parts of the educational activity, but they typically revolve around these global areas:

1. Satisfaction with instructor
2. Satisfaction with content

3. To what extent does the participant feel learning has occurred and how will behavior or life change because of it?

4. Satisfaction with environmental issues and logistics (e.g., creature comforts, registration, confirmation, time of day/day of week, quality of food or refreshments)

5. What did the participant like best and least?

6. What other courses would the participant like to have offered?

7. Would the participant recommend this program to a friend/colleague? Why or why not?

8. How did they find out about the program?

9. Is the format of the course (distance, online, etc.) conducive to the format and are they comfortable with that format as learners?

10. How valuable were the texts and supplementary material, chat rooms, and so forth?

Taken minimally, these 10 areas could constitute a basic, simple evaluation form for almost any program. But, as you will also notice, using these 10 questions as a guide in a cookie-cutter template evaluation will also give you only limited feedback on course after course after course. Pretty soon, with enough experience, this type of feedback is so general that you could practically fill it out yourself without bothering the students.

More thorough evaluations take these 10 areas of questioning, plus others that are unique to particular courses, and craft pointed questions that probe areas of particular concern. Again, this is best done in concert with the entire stakeholder team to make sure that all areas are probed adequately. At the same time, remember the old analogy about what a camel is: A camel is a horse designed by committee. You don't want a camel evaluation! You want something that is thorough enough to give you accurate data, yet simple and straightforward enough so that people actually take the time to fill it out!

Despite what you might read in journals and texts, there are no firm rules of thumb regarding the number of questions to ask. The number should be determined by the need. At the University of Missouri–St. Louis, we conduct evaluations with a few as 10 and others with as many as 65 questions. Generally, most of our evaluations are between 20 and 35 questions.

Here are some tips in writing questions that will help you remain focused and help you keep it simple:

1. For each question, ask yourself, why do I want to know this? Ask it of the other stakeholders, as they contribute questions. For example, say an administrator wants you to ask "What is your household income?" This question appears on many evaluation forms. Why do they want to know this? Are they curious, or is there a valid reason to know this? Another is, "What courses would you like to see offered?" Are you genuinely interested in knowing this for future program development, or do you have a clear five-year plan of development and will not implement outside ideas anyway?

2. Ask a blend of open-ended as well as fixed-response (yes/no, multiple choice, or continuum) type questions. Give people a chance to insert feelings and opinions in open-ended formats, by asking "Any additional comments" sections at the end or adding this option as part of other questions. However, a whole evaluation of open-ended questions is tedious to fill out and the quality of written responses will probably be disappointing.

3. Be careful about issues of reliability and validity. *Reliability*, simply stated, refers to how consistent responses are. This consistency can be in terms of how consistently one person answers the questions throughout the form, or it can be in terms of how consistently each person responding answers a particular question. *Validity*, simply stated, refers to the extent to which you are measuring what you purport to measure. When you state that you are measuring *satisfaction* with items 5, 6, and 7, are you really measuring *satisfaction,* or are you inadvertently measuring *popularity of instructor, reputation of institution,* or even *lack of complaints.*

So, how do you achieve reliability and validity? To accurately assess this would require statistical treatments of data you receive. Some basic statistical programs will give you a reliability measure—that is easily done on programs such as Statistical Protocols for the Social Sciences (SPSS) program. Any faculty member in an assessment or research department can easily do this for you. What you end up with is an alpha number, which signifies reliability. Any number greater than .65 usually indicates acceptable reliability. If you do not want to subject your instrument to this type of test, you can overall increase your reliability by paying close attention to items you ask and how you ask them, as demonstrated below.

For validity, true validity tests are expensive and time-consuming and usually not performed on evaluation instruments. However, we do pay attention to "face validity." In other words, you ask yourself, "Does this question belong?" "Is this question appropriate to this course?" The best way to achieve face validity is to have all stakeholders examine the form to ask these very questions. Once all are satisfied that questions that need to be asked are asked, and irrelevant or redundant questions are taken out, you can assume the instrument has face validity.

Your biggest task is to get respondents to focus in on what you want to know—and for each to focus in the same way. Here is a common question that could be asked on an evaluation form:

When I had a question about the material, or became confused, I contacted the instructor:

A.   rarely or never

B.   sometimes

C.   often

D.   always

In this particular example, we have to assume each participant defines *rarely, sometimes, often,* and *always* the same way. We use these terms often in our everyday life, and of course everyone knows that they mean! As a matter of fact, researchers from Pepperdine University (McCall & Walters, 1998) did some investigating among graduate students and faculty as to their "common understanding" of such terms. Here are some examples of his findings when he asked respondents, on a scale of 0-100 to define:

- *A few* Response range of 0 to 80
- *A couple* Response range of 2 to 100
- *Almost all* Response range of 5 to 100
- *Not very many* Response range of 0 to 90
- *Virtually all* Response range from 0 to 100
- *Almost none* Response range from 0 to 90

As you can see, people are capable of internally-assigning any value to your seemingly straightforward phrase-choices. A stronger way of asking the above evaluation question might be:

When I had a question about the material, or became confused, I contacted the instructor:

A.   Rarely or never (0 to 25% of the time)

B.   Sometimes (26 to 50% of the time)

C.   Often (51-75% of the time)

D.   Almost always (76 to 100% of the time)

Although this question, like all questions, is not perfect, it does focus the respondent on what you mean by *rarely, often,* and so on.

Here is another example of an innocent question that can lead to reliability problems. Say you want to ask an open-ended question about offline work habits:

I find that I become frustrated and stop working on class assignments when:

Now, the purpose of this question was meant to solicit feedback about topics or assignments that were problematic to the point of frustration or shutting down. So, anticipated responses would follow the scheme of "when the material doesn't follow the text;" or "when I can't reach the instructor for clarification;" or "when I have not done the background reading and am not prepared."

Instead, what we got mixed in were comments, such as,

It's Friday night and I've had a busy week; or When the baby is sick and needs a lot of attention; and, When my husband [wife] is on my case about something.

The problem with this question is the innocent word *when.* It was being interpreted two different ways by respondents and, thus, was a threat to the instrument's reliability.

An improved question might be,

Where there any times in the course when you felt frustrated and stopped working on class assignments, or felt like giving up? What was it about that course that created this frustration and how did you cope with it? Is there anything that the instructor or sponsors could have done to help you?

Again, while not perfect, this does focus the respondent on what it is you really want to know, yet does not focus them too much with choice answers. You will get their original thoughts in their own words.

Be careful of other common mistakes in constructing evaluation items, such as giving people an unrealistic continuum on which to answer a question:

On a scale of 1 to 10, how would you rate this instructor:

0	1	2	3	4	5	6	7	8	9	10
Awful				*Good*				*Exceptional*		

The vague terms *awful, good,* and *exceptional* notwithstanding, the real problem with this type of question is the large 0-10 continuum. For example, try to explain to someone the difference between a ranking of a 3 versus 4; or 7 versus 8. You probably cannot. Neither can the respondents. Thus, two people feeling basically the same way about the instructor could easily circle two different numbers. This, then, is another threat to reliability.

Avoid asking double-questions, such as,

> *I found this course to be at a convenient time and price.*
> *A. True   B. False*

It could be, of course, that either the time or the price was convenient, but a true response does not let you know which. Or, it could be both were. A false response could mean neither or just one were. You do not know.

Avoid presuming questions, such as,

> *Did we make adequate provisions to assist you in using the technology needed for this course?*

On the surface, this seems like a straightforward and innocent question. However, it really is a presuming one. The presumption lies in that the fact that it is believed that the university *should* be providing such assistance.

Say that no provisions to assist in using technology were provided, and I did not need provisions or did not feel it was anyone's responsibility but my own. In such a case, I would answer yes in that none were provided and none were needed. Or, I could answer no because I knew none were provided, even if I did not need them, anyway.

Instead of presuming, this question could be strengthened by dividing it into two questions: (a) Do you feel you needed assistance in using the technology before the class? and (b) If so, was the assistance available sufficient?

Avoid leading questions. In general, any question that begins with

Do you not agree that.... or Would you agree that...

There are countless other examples of well-intended questions that failed to achieve the task of asking what we really wanted to know. Testing evaluation instruments on a live class and refining it is the best way to achieve an optimum form. Having colleagues and course stakeholders look at it also helps insure comprehensiveness and straightforward grammar. Finally, examine your evaluation instrument and system against your assessment and evaluation philosophy. Do they match? Are you asking what you really want to know—and are you really willing to respect the feedback you are given?

Challenge 4: Communicating Recommendations and Building in Follow-Through

Assessment of Students

Assessment of student's knowledge or behavior change can be facilitated by access to online testing tools. Addressing the technical requirements and support needs for both instructors and students tops the list of needs. Institutions investing in course management systems will likely have support staff and training programs in place. Ideally, individual faculty will have oriented their students to online course requirements, including expectations for feedback.

Communicating feedback online is complicated by several inherent constraints: A *text-only* context, especially in abbreviated or brusque e-mails, may lead to misunderstandings between the student and instructor. This can be problematic when both the online learning environment is new and the course content is complex and difficult for students.

The lack of opportunities for a real-time conversation with back-and-forth exchanges and nonverbal cues, as in face-to-face classroom settings, may also prevent a speedy resolution of these misunderstandings. New communication tools combining audio and video transmissions have begun replicating the face-to-face, instantaneous exchanges and move us beyond text-based messages. Interested parties are invited to search for Web conferencing tools. The following Web conferencing tools have been used by the authors for a variety of instructional purposes: Centra, *www.centra.com*, Elluminate, *www.elluminate.com*, and Netmeeting, *www.microsoft.com/windows/ netmeeting.*

Figure 3. The communication exchange

Steps in Asynchronous Communication Between Instructor and Student
Step 1: Assessments completed
Step 2: Assessment message e-mailed or posted by instructor
Step 3: Student accesses and reads message
Step 4: Student decides to respond
Step 5: Student drafts response
Step 6: Student sends the message
Step 7: Instructor opens and reads the e-mail
Step 8: Instructor drafts a response
Step 9: Instructor sends the message
Step 10: Student reads reply and decides no further communication is needed

Second, there is a persistent time lag inherent in electronic communications, which can slow the process. We increasingly expect quick responses to our e-messages. To illustrate the potential time gaps, notations of the critical stages in the process of assessment reporting are charted in Figure 3.

Figure 3 includes two communication loops; notice the nine stages wherein competing activities might intervene and extend the time of this asynchronous exchange. Preparing students for what to expect regarding instructor's feedback with scoring rubrics and time frame will help prevent much anxiety and frustration.

Finally, anticipating the problems students may have (i.e., what is required for an assignment or following steps of a process) and developing Frequently Asked Questions can be good insurance against misunderstandings and need for extended communications. One of the authors, an instructor in adult education, routinely uploads *exemplars*—samples of papers—to give her students a head start on major projects. She has also created more detailed assignment descriptions, along with examples, to guide her students iteratively. Finally, there is one discussion board located on the course Web site for answering course-related questions. She believes these strategies have resulted in improved communications with her students.

Pros and Cons of Communicating Issues of Student Assessment Online

Those of us teaching online courses must prepare for assessing and communicating results electronically. We summarize by asking our readers, "What are the benefits and drawbacks of trying to communicate assessment-type issues with students through an online medium?" Benefits include instantaneous feedback, expedience for both instructor and student, and maintaining a record of what was said.

Electronic communication is also two-dimensional, with limited opportunity for dialogue and potential for delayed feedback. We have described these limitations and have suggested several proactive techniques. Further, we speculate that instructors and program leaders will eventually want to offer multiple communication channels for sending and receiving assessment messages. For example, for the student in the same town as the instructor, could a face-to-face meeting be scheduled? Or, could Web conferencing, virtual chat, *and* telephone service better serve students? Or, should a combination of feedback strategies be implemented?

Program Evaluation

Evaluation of a course or program is a means for improvement as well as a decision-making mechanism. An evaluation's impact on decision-making varies greatly, however. In a review of 20 national health program evaluations, Patton (1986) concluded that impact tended to be modest, at best. He found that "none of the impacts described was the type in which new findings from an evaluation led directly and immediately to the making of major, concrete program decisions. The more typical impact was one in which the evaluation findings provided additional pieces of information in the difficult puzzle of program action, permitting some reduction in the uncertainty within which any federal decision maker inevitably operates" (p. 34).

According to Carter (1994), there are five ways to increase the use of evaluation results: (1) develop realistic recommendations that focus on program improvements, (2) explore multiple uses of study data, (3) constantly remind decision makers of findings and recommendations, (4) build expectation for stakeholder discussions and planning based on findings, and (5) assign evaluation staff to assist in implementing recommendations. Following this short list of maximizing evaluation results, the authors specify how to organize data and deduce recommendations.

1. **Develop Realistic Recommendations:** Carter (1994) suggested a focused analysis of the first 10% of findings in the study. He maintained that recommendations for changes to the course or program should grow from this analysis.

2. **Explore Multiple Uses of Data:** Sharing data with other campus groups or organizations will create goodwill and can result in unintended program changes. A longitudinal study report, "Blackboard Use by Faculty and Students at the University of Missouri–St. Louis," has been made available in PDF via the course management system login page. Faculty and

administrators have reported using the data for their own grant proposal development and presentations.

3. **Remind Decision Makers of Findings and Recommendations**: Agency newsletter articles and presentations for key stakeholders and managers likely to implement recommendations are all means to remind decision makers. When new managers join the organization, reports can help educate them to its history and current initiatives.

4. **Share Findings and Recommendations:** Broad dissemination is key to getting the right information to the right people for their use. Consider whether different types of reports must be created for the various stakeholders involved.

5. **Assign Evaluation Staff in Implementing Recommendations:** A productive team will include evaluation and program staff to oversee implementation. Each brings expertise to the table that will keep suggested changes in the foreground and will identify possible stumbling blocks and resistance.

How to Organize Data

Simplicity of presentation is the guiding principle. Although social scientists advocate for more or greater sophistication in data analysis, most decision makers, according to Patton, relate better to simple frequencies and charts than to detailed statistical tests and descriptions.

You may have used the more sophisticated tools and techniques to analyze the data, but think creatively about how best to present the findings to key stakeholders and decision makers. Sounding like a Zen master, Patton explained, "Simplicity as a virtue means that we are rewarded not for how much we confuse or impress, but for how much we enlighten" (Patton, 1997, p. 310). He continued, "The point is to make complex matters understandable without distortion" (p. 312).

There are several additional guiding principles for determining how best to represent the findings.

Ideas for Balancing the Data

First, the presentation of data should be balanced, making sure that the findings:

- are categorized in more than one way;
- include absolute numbers and percentages;
- include the mean, range, and standard deviations; and
- list both positive and negative quotes.

Second, understanding the data will be facilitated by comparisons to relevant benchmarks, ideals, or competing programs. Statistics will be more meaningful to the stakeholders if they are compared to relevant comparative data. The range of possible comparisons is detailed in a Menu of Program Comparisons (Patton, 1997, p. 314). First, outcomes compared to like programs will yield a more balanced view. Reports of faculty using a course management system at one institution can be compared to similar use by another group of faculty and may provoke needed changes in support to increase adoption rates. Additionally, year-to-year trends will help identify specific areas experiencing change, decline, or growth.

In a third type of comparison among representative programs, clients, or students, the outcomes of various samples or groups can support or contrast individual results. For example, comparison of data from two different studies concerning students' connectivity indicated the same percentages (37%), stating they had DSL or cable high-speed connections at the University of Missouri–St. Louis, thereby strengthening the reliability of that statistic for both studies.

Comparisons between stated program goals or participant's goals and actual outcomes can help describe the relative success of program efforts and identify improvement areas. The classic example is to ask participants to list their learning goals in a specific course or training program. At the conclusion, they will be asked to rate the degree of accomplishment of personal goals.

Other comparisons listed by Patton (1997) include comparisons to external standards set by a professional organization, accreditation, or licensing organization as well as the ideals established for program functioning. He advised evaluators to engage stakeholders in determining which comparisons are "appropriate and relevant" for the specific evaluative need.

Guides for Making Recommendations

Making recommendations is considered the natural conclusion of data analysis and interpretation by some evaluators, while others consider this phase to be inaccurate and inappropriate.

The objection goes away when an evaluator involves the stakeholders in a collaborative process of discussing implications, however. Patton offers the following 10 guidelines for making recommendations:

1. Ideally, the nature and content of the evaluation report should be negotiated with stakeholders before data are collected.

2. "Recommendations should clearly follow from and be supported by the evaluation findings" (p. 315).

3. There are different kinds of recommendations. Distinguish those dealing with central questions from peripheral issues; also, pose recommendations according to a time line (e.g., those to be implemented short term and longer term); and suggest recommendations for various groups or those that require action versus those for further discussion.

4. List recommendations along with their pros and cons.

5. Do a cost-benefit analysis for the recommended changes to a program.

6. "Focus on actions within the control of intended users" (p. 325).

7. Be sensitive to the political ramifications of the recommendations.

8. Be careful and precise in the wording.

9. Let time help percolate recommendations in collaborative discussions with stakeholders.

10. Facilitate a meeting to consider the range of recommendations.

In conclusion, it is important to focus on *learning and improvement*, rather than *absolute judgment*, when findings are reviewed. That is, place the emphasis on organizational learning and collaborative approaches to get the best outcomes and usefulness from the evaluative data you collect.

Pulling It All Together

This chapter has explored several options in assessing both students and courses. We should never forget, though, that student assessment and course assessment are frequently interrelated. It comes as little surprise that those who do well in a course also think more highly of it; but the reverse is also true—those who enjoy a course and for whom the course is designed, tend to put forth the requisite effort to excel in mastering the content. Student assessment—whether through tests or projects—also informs us as to the strengths and weaknesses of our curriculum and our delivery method. Did we cover every section in depth? Did we provide enough examples? Was there continuity in the course design and content? Problems in these areas can easily surface in formative assessments, if the instructor is careful to analyze the mastery trends of whole class. Thus, we recommend using your student assessment mechanisms to validate your course evaluation mechanisms—and vice versa. Key questions to ask: Are the two sources of evaluation congruent? What have you learned about the strengths and weaknesses of your course? What do you plan to change or retain in your future online classes?

One of the themes embedded in this chapter has been "collaboration." Evaluation systems are rarely designed or conducted in isolation. The best evaluation systems bring all stakeholders together to craft a system (not just an instrument) that meets the needs of each holder. Common evaluation committees include the faculty member, an instructional designer (this is especially true for distance learning courses), an administrator, and perhaps even a former student. In courses offered aggressively to the public outside of degree programs (e.g., continuing education professional development type courses), it is helpful to also include a marketing representative. The goal of this group is to develop a *system*, as well as an instrument. When will evaluations be performed—will they be formative or summative? What is the purpose of the evaluation? What information is required? How will students be assessed? How will the results be compiled and who will do it? How will it be shared? Are there specific plans for this committee to reconvene and analyze the results? Does each member promise to look at the results objectively and formulate plans to correct any weaknesses?

The authors of this chapter recently had the opportunity to put theory and practice into action as we collaborated and developed a Web-based, train-the-trainer program that was part of a U.S. Department of Agriculture grant in 2003-2004. Our topic was "Assessment and Evaluation of Distance Education and Online Programs." We relied heavily on our own professional experience, but also couched the development of the module in terms of systematic or meta-assessment processes (Huba & Freed, 2000; Reeves & Hedburg, 2002) and the concepts of utilization evaluation (Patton, 1997). The overall evaluation for this program reinforced the importance of addressing the variety of stakeholders and their specialized interests.

Through a variety of assessment methods, the authors got acquainted with participants before the start of the program. It was learned that the group was more diverse, with more faculty members, than was anticipated. The participants said they valued interactions with one another, although later the participants did not get involved to the extent the authors had anticipated. The participants indicated they would be using office hours to complete the modules, and a majority reported they progressed in brief snatches of time during their work-days. They also expressed that they wanted the activities to directly relate to the job and not be busywork.

Assignments were designed to engage participants in the specific content of each destination while encouraging practice and application to individual learning needs. For example, each assignment in the assessment and evaluation module led to designing an online survey for a specific institutional purpose (i.e., the Micro Project). Discussion forums attempted to draw on professional experiences and expertise, with topics ranging from "Reflections on motivating adult learners," to "What evaluation needs are present in our organization?"

As one can see by the above descriptions and examples, what begins to emerge from this process is an overall institutional philosophy of assessment. Each organization and each academic culture has its own philosophy of need, methods, and frequency of evaluation. Instructional designers, administrators, and faculty have much influence on how this philosophy develops and is enforced. What is important is not the fine details of how an evaluation system operates at one school versus another. What matters is that one is in place that is consistent with the overall goals of the organization and one that leads to systematic improvements in instruction, and therefore, learning.

Finally, we cannot forget to assess the assessment. Have we met the four different challenges successfully? Are the assessments we conduct giving us the information we need in an accurate and timely manner? Are the correct people involved? Is our system practical and efficient? Are the data from our course evaluation instruments consistent with student assessments and consistent with our own observations and informal conversations with students and other stakeholders? The extent to which we can answer yes to these questions is the extent to which we are confident that we are doing the best possible job in providing superior learning opportunities for our students.

References

Brown, G. (2000). Learning and the Web: Reflections on assessment. In D.G. Hanna & Associates (Eds.), *Higher education in an era of digital competition: Choices and challenges.* Madison, WI: Atwood.

Carter, R. (1994). Maximizing the use of evaluation results. In J. S. Wholey, H. P. Hatry, & K. E. Newcomer (Eds.), *Handbook of practical program evaluation* (pp. 575-589). San Francisco: Jossey-Bass.

Chapin, J. P., & Messick, R. G. (1999). *Elementary social studies: A practical guide* (4th ed.). London: Longman.

Huba, M. E., & Freed, J. E. (2000). *Learner-centered assessment on college campuses: Shifting the focus from teaching to learning.* Boston: Allyn & Bacon.

Isaac, S. & Michael, W.B. (1990). *Handbook in research and evaluation* (2nd edition). San Diego: EdITS Publisher.

Krueger, R. A. (1994). *Focus groups: A practical guide for applied research* (2nd ed.). Newbury Park, CA: Sage.

McCall, C.H., & Walters, L.E. (1998). Words and their value to the survey researcher. Paper presented at the *Annual Conference of the American Educaitonal Research Association,* San Diego.

Palloff, R.M., & Pratt, K. (1999). *Building learning communities in cyberspace: Effective strategies for the online classroom.* San Francisco: Jossey-Bass.

Patton, M. Q. (1986). *Utilization-focused evaluation* (2nd ed.). Newbury Park, CA: Sage.

Patton, M. Q. (1997). *Utilization-focused evaluation* (3rd ed.). Thousand Oaks, CA: Sage.

Reeves, T., & Hedburg, J. (2002). *Interactive learning systems evaluations.* Englewood Cliffs, NJ: Educational Technology Publications.

Sorenson, D. L., & Johnson, T. D. (Eds.). (2003). *Online student ratings of instruction: New directions for teaching and learning, no. 96.* San Franscisco: Jossey-Bass.

Worthen, B. R., & Sanders, J. R. (1987). *Educational evaluation: Alternative approaches and practical guidelines.* White Plains, NY: Longman.

Chapter VIII

Creating a Unified System of Assessment

Richard Schuttler, University of Phoenix, USA

Jake Burdick, University of Phoenix, USA

Abstract

The approach to curriculum design and development are essential elements to consider when creating a unified system of learning assessment. Although there are differences between facilitating curriculum online and in the classroom, the modalities do not need to be treated as separate paradigms when establishing assessments. By creating an approach not governed by delivery, education can be made more effective in all modalities. When academic programs are designed in a sound and systematic approach, learning assessment is a natural function of curriculum deployment. Assessing learning without assessing its associated curriculum design is essentially the act of looking at an effect without mentioning its causes. This chapter suggests that the goal of any institution in developing curriculum should be to create an assessment methodology that is integrated throughout its curriculum in a conscious manner, not specific to a modality.

Introduction

The effectiveness of online education is still under debate. However, acceptance of this learning modality is beginning to evolve, as the breadth of programs expands and online education becomes ubiquitous. More significant, online learning's market segment, which was once overwhelmingly composed of adult learners, is beginning to include younger students as well. With this influx of students and, consequently, learning styles, educators and administrators have questioned whether an online class can provide an outcome-based education wherein students never see their teacher.

This line of inquiry assumes that the physical distance between student and faculty dictates the learning process and that modality is a crucial factor in designing curriculum and assessments. While there are clear differences between online and classroom settings, the modalities do not need to be treated as separate paradigms. Potentially, by creating an approach not governed by delivery, education can be made more effective in all modalities, and the perception of distance can be diminished.

Dilemma of Modality-Based Assessment

Over the past 15 years, the appropriateness of online modalities, specifically the delivery of information and the assessment of intended outcomes, has been debated. Conversely, many online, nontraditional universities have experienced success in terms of student growth and retention, and traditional universities are now considering, or already adding, online education to their offerings. The potential of the medium and its ability to reach a broad base of students has seemingly outweighed several of its detractors. This shift has not stymied the discussion of discrepancy between modalities; rather, the addition of new voices brings even more questions regarding its viability in comparison to *traditional* education and its differences in approach, especially in the conceptualization of assessment. The true issue of this debate is, perhaps, the perception that online and face-to-face delivery and assessment are mutually exclusive. Part of the debate appears to suggest that, because online is a different modality of content delivery, there should also be a *different* approach to assessing learning.

Division has seemingly grown between the two modalities of learning, as evidenced in the sheer amount of journals, conferences, and other media dedicated specifically to online education. The discussion is hardly limited to market position, as a great deal of attention to modality differences and exclusivity exists in current educational theory and practice. In fact, some of the literature in support of online delivery and assessment has taken on a moderately defensive posture, essentially reacting to the perceived differences and ensuing criticism. Perhaps this position is nothing more than resistance to change, but nonetheless, it has a presence that is felt throughout academia. The discussion suggests that online educational delivery systems need to be designed in a consciously different manner to reach the same outcomes as traditional methods.

Further critique of online assessment focuses on practical issues, such as the potential for plagiarism and cheating and the lack of specific cues that relate comprehension (Rittschof & Griffin, 2003). These concerns, while clearly valid, might also be addressed with a conscious curriculum design strategy, one that incorporates assessment methodologies that invoke the affective, or personal, elements of learning. In either modality, these assessments, if designed appropriately and revised on a consistent basis, are superior to rote assessment methodologies, such as multiple-choice testing or essays that require mere summarization of content. Clearly, quantitative testing has its place in certain content areas, such as mathematics and the natural sciences, but even in those instances, it can be supported with other, more interactive classroom practices and assessments. Furthermore, as cheating and plagiarism are not solely endemic to the online environment; face-to-face educational settings can equally benefit from the addition of these practices.

Tension between the two modalities has been a prevalent, yet downplayed, theme in the literature and has created an artificial gap between modalities, fragmenting educational thought and practice. Bridging this gap may only be a matter of considering curriculum and assessment design independent of modality. If curriculum is developed with focus on content, rather than delivery method, there should be no difference in the assessments created for either modality. This practice is currently modeled by University of Phoenix.

Curriculum Across Modalities

University of Phoenix, regularly acknowledged as the leader in online education with over 100,000 online students, does not incorporate separate assessment and measurement systems for online and on-ground delivery of education (University of Phoenix, 2004). Rather, a common curriculum is created for online, classroom, and mixed-modality course delivery. This consistent means of analysis stems from a systematic curriculum development model constructed with the University's dispersed locations and online modalities in mind. By recognizing the needs of students and faculty across all modalities, the university crafts curricula that can be applied in any educational delivery system, effectively eliminating the perception that an online education is intrinsically different.

Background

In the early 1970s, Dr. John Sperling, a Cambridge-educated economist and professor (later to become the founder of University of Phoenix), realized that a large number of adults were attending universities that utilized education designed for younger (18-22 years old) students. These universities were applying the same pedagogical approach to providing education to all segments of the student population. This approach works well within a traditionally based institution composed of younger students, but according to the research and writing of Malcolm Knowles (1973), adults prefer a different mode of learning. Much of that difference stems from the level of experience these adult students bring to the classroom, experience that their younger counterparts will likely not possess. Furthermore, the existing pedagogical model did not allow for the difficulty many adults encountered when conforming to a university's schedule and location. Sperling, through field-based research in adult education, recognized the need for change and began conceiving, researching, and applying an andragogical model, based on the research and writings of Knowles and David Kolb (1984).

The outcome of Sperling's research, combined with over 27 years of application, produced a learning content management system that contributed to the creation of a standardized, transmodality curriculum for use in the classroom, online, or in a blended format. This design utilizes a cascading system of assessment that can be communicated easily across a dispersed faculty and student population. While consistently being refined, the model has served the University well, and academic consistency has been maintained despite the University's unprecedented geographical expansion and student growth.

Curriculum Design Structure

As previously stated, the University's consistency in delivery and assessment is rooted in its curriculum design process, a key component of Sperling's model. A central development body, led by an academic dean, oversees curriculum development at University of Phoenix. This group begins program design with high-level encompassing goals, which will eventually drive the program's more granular materials, such as assignment prompts, questions for class discussion, and faculty materials. Furthermore, while faculty are afforded a great deal of academic freedom in their teaching, the University clearly communicates that course objectives need to be adequately represented and thoroughly assessed within a given course. Thus, the University is able to map a student's learning outcomes directly to the greater vision and intent of that student's program, regardless of delivery modality.

Curriculum at University of Phoenix is designed to engender a personal and practical learning experience. This focus is represented through the concepts of active learning and collaboration, both of which are integral to each program's design. Active learning integrates affective learning with cognitive learning to ensure that students are capable of translating course content to their personal and professional lives (Krathwohl, Bloom, & Masia, 1999; Bloom, 1984). The learning model of many traditional universities, with content delivered via lectures and reading and assessed through highly quantitative tests, does not afford students the same intimate relationship with knowledge as the active and affective learning paradigm. When taking tests, students are merely prompted to repeat the concepts they have heard or read. This assessment only involves the lower tiers of Bloom's (1984) cognitive hierarchy and is ultimately less meaningful for students who are not engaged in the learning opportunity. Curriculum designed with a focus on purposeful and personal interaction with knowledge and supported by self-reflective, application-based assessment tools can aid students in developing increased levels of meaning and retention in their studies.

Student collaboration at University of Phoenix occurs in formal Learning Teams in which adult students interact with, teach, and mentor one another. The structure of these teams mimics the team format of real-world practices, and teams are encouraged to combine their talents to achieve beyond the abilities of their individual members. Teams assist in bridging the perceptual *distance* in online education, as students are consistently interacting and mentoring their peers in a communal academic setting. Lovitts (2001) suggested that the lack of a "community atmosphere" is a major contributor to doctoral students leaving their degree programs before completion, and the same may be true of a student in any academic program that lacks a sense of human interactivity (p. 108).

The Curriculum Development Process

Academic programs are created in a collaborative effort between the university and subject matter experts from fields or backgrounds that correspond to course content. The following describes the curriculum development process used in the creation of the Doctor of Business Administration program. While this program is used as the unit of analysis in this chapter, a similar process has been used to create all University of Phoenix academic programs.

In the initial phases of planning for the Doctor of Business Administration program, a steering council, composed of four senior-level executives who were not University employees or faculty, created thematic and topical guidelines for the program's curriculum—overarching concepts that would weave throughout coursework and be realized fully at the completion of the degree in capstone assessment practices. This general framework established the context and concepts of the program and provided a vision for future developers to follow.

Upon the conclusion of the steering council's sessions, the dean for the program convened another group of highly skilled academics and practitioners. These individuals either held DBA degrees or possessed backgrounds that could help to inform the creation of the program. This group, called the Academic Program Council, was solicited from the University's faculty pool in order to represent the unique needs of that group. The dean serves as the chair of this council, and the group is charged with the development of the domains, subdomains, and competencies of the degree program. The Academic Program Council strengthens the role of faculty in the oversight of curriculum as well as impacts decisions regarding curriculum design, development, and revision.

Domains

In the instructional structure created by the Academic Program Council, domains represent the broadest areas of content within a degree program. The Doctor of Business Administration degree program's domains include strategy, global business, and research. Domains designate a topical area around which other individual outcome statements can be grouped. Domains are described as nouns and designed with a brief explanatory paragraph to provide a contextual framework for developers.

Subdomains

A program's domains are broken down into more manageable and actionable content areas, subdomains. An example of a subdomain from the DBA domain *strategy* is *environmental scanning*, a concept encompassed by the domain, but more manageably assessed across a curriculum. Subdomains are cascaded into related competencies that state the measurable expectations of the subdomain. When students meet the expectations of an area's subdomains, they should possess the requisite knowledge to satisfy the domain as it is described in the program.

Competencies

From the subdomains, competencies, or statements of student ability, are created. These competencies are usually written in the form of an exceedingly broad objective, and the creation of these competencies leads to identification of course content, topics, objectives, and assignments. These statements establish, by design, a vehicle for developing a system of assessment. Competencies are written in verb form, much like objectives; however, competencies speak to a much broader scope of knowledge than a typical course objective. Figure 1 details the cascading effect of the university's curriculum design structure for the domain of *strategy* in the Doctor of Business Administration degree program:

After the Academic Program Council creates this framework for the program's content and assessment, individual courses, including course topics and objectives, are mapped directly to programmatic competencies. In this manner,

Figure 1. The Doctor of Business Administration domain of strategy cascading into the subdomain of environmental scanning and analysis and its corresponding competencies

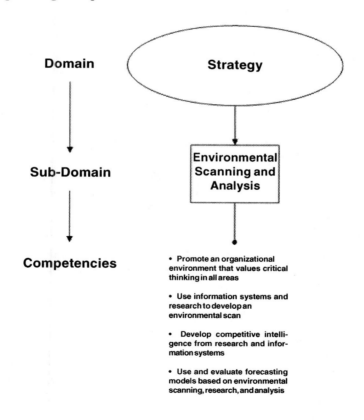

curriculum is organically derived from the identified topical areas. Each course within the DBA program is not directly created by the council; however, at least one member of the council is typically included in each course creation session. This involvement allows for a consistent and involved approach to curriculum design, and it ensures that the program's design is systematic and that assessment methodologies can be standardized and accurately mapped to the program's intent.

The other members selected to create each course are subject matter experts who are academically prepared and working in the field. For example, in creating a finance course, a team might consist of a chief financial officer from a large Fortune 500 company as well as a comparable officer from a nonprofit organization. These diverse perspectives of course content allow for both breadth and depth of knowledge during the curriculum development process.

Assessing Learning

Learning at University of Phoenix is measured by the student's satisfactory completion and understanding of course objectives, represented by their earned grade. Thus, each objective is mapped directly to a course assignment. For example, one objective in a first-year DBA leadership course is to "analyze leadership through doctoral-level research." This objective is measured by a written assignment asking students to locate peer-reviewed research articles related to the historical foundations of leadership and leadership research after 1990. The students must then briefly outline the research methodologies used in the articles and discuss the lessons learned from the articles as they apply to the topics of class discussion. Furthermore, to invoke an affective response to content, students are often asked to relate course topics to their current workplace environment or to one within which they have previously worked. This associative process brings additional emotional resonance and meaning to the academic event.

All objectives written for University of Phoenix courses are assessed directly and without emphasis on modality. Essentially, all faculty are given the same tools for delivering education—classroom or online—with the only difference being the means by which they guide students to achieving intended outcomes. To truly actualize this process, a much greater investment in learning and teaching systems must be made. In many traditional delivery and assessment models, resources are limited, and the paradigm resembles the *one to the many* lecture hall approach to teaching. However, assessing learning—not recapitulation—requires active involvement by a facilitator, not a lecturer.

In either modality, it is suggested that higher order learning is only gained via active participation of students and faculty alike. As such, "The teacher moves from acting as an authority figure to become. . . the guide, the pointer-out, who also participates in learning in proportion to the vitality and relevance of his [or her] facts and experiences" (Brookfield, as cited in Mezirow, 1991). Learning is difficult to gauge from the results of a rote quantitative examination; this practice is seemingly better suited to assessing memorization. Thus, course and programmatic curriculum should include highly participative and engaging activities and assessments that assist the faculty member in facilitating active learning practices.

Highly participative curriculum involves a redefinition of teachers as facilitators and students as learners. A thorough understanding of the terminology as it applies to online and other new approaches to learning may make understanding online education easier and, thus, more acceptable in its assessment relationship to the classroom setting. In the lecture hall of the traditional brick and mortar university, a teacher, who may not even have related practical experience, discusses theorists and his or her interpretation of their work. In most cases, very few students have any background in the content being conveyed and leave the lecture hall with the opinion that the teacher's view of the content is the only one and, often, the *right* one.

In these lecture halls, students are—theoretically—listening, but they may not be learning. The learning assessment is normally given in the same large lecture hall several weeks later when the students are asked to write a paper or to answer a set of related questions on a test. If those students were asked several weeks, months, or years later what they learned from that test, they may not even remember the test, let alone the content. This model may have served a well-defined purpose at one time; however, a reassessment of this approach might be appropriate for the contemporary educational environment.

Several colleges and universities across the world teach under this philosophy. However, the alleged success of the traditional method and modality of learning stated above does not account for the growth of online education. Online education is evolving quickly, with noted results in not only the retention of those seeking the education but of those who remain in degree programs and graduate.

In an online academic program, the instructor is a facilitator of content. University of Phoenix applies this same sound and systematic approach—the andragogical—to both the face-to-face and online learning environments. The successful online classroom is not centered on lecturing; rather, an abbreviated lecture is provided, with an emphasis on the applied nature of course content. In the lecture hall, a student attends class once or twice a week to receive content, hoping their questions are answered during the lecture. In the online classroom setting used at University of Phoenix, successful students participate in their electronic classrooms at least five out of seven days a week. This continual participation and attendance supports an active approach for students to learn and gain the model course objectives.

In the online classroom, a facilitator submits a written lecture in the beginning of the week, much as a facilitator would in a classroom setting. This lecture might be 1000 words that explore the facilitator's personal experience with the theory discussed in that week. This lecture concludes with several discussion questions directed toward the students to aid them in synthesizing the theory they addressed in the textbook with the application of that theory in the facilitator's lecture. Students respond to the discussion questions with an account of their experiences as they relate to the topics. Once one student responds to a given discussion question, other students respond to the same discussion question and participate by adding on to their peers' answers. Thus, a dynamic and interactive learning environment is established.

With observable effectiveness, the implication of facilitators versus teachers and learners versus students exceeds mere terminology. Online learners are encouraged to respond to one another's discussion question answers. The facilitator adds comments and further questions to bring the learning to a higher level of critical thought. The dynamic atmosphere between one facilitator and 14 students offers a learning experience that often supercedes that of a lecture hall environment, or even a small classroom setting, in the brick-and-mortar classrooms of most traditional higher educational institutions.

Emerging Trends

Assessing learning online will, in many ways, mirror the trends that are occurring in traditional learning environments. Learning assessment will always be a point of concern as long as learning institutions create programs for the greater public. The formal and well researched learning assessment systems at the lower levels of education (K-12) do not effectively translate to the higher education classroom. Thus, higher education institutions—especially publicly funded universities—are challenged not only to devise means of assessing the learning that has taken place in the classroom but also to predict the success of their students after graduation.

Three emerging trends challenge the basic assumptions, beliefs, and attitudes of learning assessment. These trends include a greater emphasis on the affective domain of learning (Krathwohl, Bloom, & Masia, 1999); the formalization of programmatic rubrics mapped directly to domains, subdomains, and competencies; and a collaborative approach to assessing learning via team-based case studies.

Affective Assessment

Learning in any environment must be a transformative process, one in which students actively participate and are assessed in two distinct domains—the cognitive and the affective. Mezirow (1991) alluded to the need for this level of learning, stating "transformative learning involves reflectively transforming the beliefs, attitudes, opinions, and emotional reactions that constitute our meaning schemes or transforming or meaning perspectives (sets of related meaning schemes)" (p. 223). Such a close psychological and emotional *proximity* to programmatic content might decrease sensations of isolation in online learning and allow for assessment practices that transcend rote repetition and measure the impact of learning in students.

Curriculum developed only from a cognitive context does not convey mastery of a subject and cannot completely engage a learner, regardless of the medium by which curriculum is delivered. Cognitive curricula ultimately promote what Kegan (2003) called "instrumental competence," a task-based understanding of a subject (p. 10). The goal of affective learning is the development of value systems through personal response to course content. As such, curriculum should be developed with a blended approach to knowledge, enabling both the cognitive and affective domains within courses and their corresponding assessments. This methodology would allow students to become receptive to the wealth of knowledge presented, creating greater capacity for learning and an opportunity for self-synthesis of that learning. Accordingly, assessment could be designed to address both cognitive and affective dimensions of learning in either modality. Regardless of the medium used, a dual focus in assessment methodology would promote more substantial, personal, and lasting learning, effectively combating the perceived *distance* of an online program.

These approaches, which include Jack Mezirow's (1990, 1991, 2003) concept of transformational learning and Robert Kegan's (1982, 1994) adult psychological development models, can be integrated with the standard cognitive approach to address deeper effects and outcomes. The framework for assessing these concepts has centered on personal response assignments that build on one another over the course of a curriculum, requiring consistent reevaluation of a student's assumptions. Other strategies include self- and team-based reflective measurements of personal growth at stages within the programs. By combining affective-oriented, personal educational practices with cognitive strategies, students can gain knowledge of content *and* a contextual framework into which that content fits. This integration of learning strategy may begin to shift the paradigms of student and faculty collaboration in educational performance, undoubtedly opening new venues to assess learning and new challenges in its acceptance—much like the trials of online learning itself.

The need for an integrated model of curriculum and assessment development is apparent in the very design of Bloom's (1984) cognitive taxonomy, one of the most predominant tools of curriculum and assessment development. Bloom's model is often represented as an ascending scale of lower to higher level cognitive processes, with higher order processes subsuming lower order processes. Problematically, this model does not completely account for the acquisition of higher order knowledge, the role of critical thinking, or the need for students to internalize learning on deeper psychological and emotional levels, all of which are important components of a sound evaluation process. The term *evaluate* suggests that students must have an internal value system to use in their judgment of academic concepts. The development of these value systems, however, is not incorporated into Bloom's cognitive hierarchy.

These assertions are not a rejection of the concepts inherent within Bloom's cognitive (Bloom, 1984; Krathwohl et al., 1991) framework and the instructional design principles it has fostered. Rather, these authors' intention is to explore the ways that the cognitive domain intersects and informs affective learning and outcomes. The relationship between cognitive and affective notions was illustrated by Piaget (as cited in Kegan, 1982), in that "there are not two developments, one cognitive and the other affective, two separate psychic functions, nor are there two kinds of objects: all objects are simultaneously cognitive and affective" (p. 83). This interrelationship needs greater articulation within both online and face-to-face modalities, as in many ways, the knowledge that is most important to student success is gained by understanding ideas in terms of content and context.

Another manner of understanding the relationship between cognitive and affective learning is illuminated in the recent revision to Bloom's (1984) cognitive taxonomy, which establishes a metacognitive domain that involves several tasks significant to education, specifically, "self-awareness, self-reflection, and self-regulation" (Pintrich, 2002, p. 220). These tasks require students to achieve distance from their enculturated identities and question the underlying assumptions of those identities. Students should experience deep change throughout the program through cycles of personal reflection, evaluation, assessment, and revision. The programs' curriculum should encourage, not merely accommodate, this process.

Additionally, by incorporating an affective domain of learning, as originally described by Krathwohl et al. (1999), universities enable students to create meaning that is specific to their personal and professional needs and styles. While technological limitations may negate the nonverbal cues that aid in classroom-based assessment and interactivity, assignments that require personal interaction with knowledge can help instructors in gauging students' comprehension of course material. Further, affective processes often integrate directly with cognitive ones, especially those deemed as higher order in Bloom's (1984) taxonomy. In order to craft a personal response to a subject, a student must be able to analyze the material and his or her reception of it, evaluate for dissonant or congruent beliefs, and synthesize a new understanding.

In the aforementioned Doctor of Business Administration program, both the cognitive and affective attributes of the program's domains are illuminated and assessed. While business is often stereotyped as being mechanical and emotionless, current trends in business and leadership literature call attention to the soft skills needed for success in contemporary organizations. As the program's developers became uncomfortable with the limitations of applying only Bloom's (1984) cognitive approach to assessment, other assessment technologies that are inclusive of the affective taxonomy (Krathwohl et al., 1999) were suggested for crafting a more enriching process of postsecondary educational development. This distinctly humanist approach to curriculum aids in bridging the perceived gap between online and classroom assessment practices. Students, regardless of modality, experience challenges to their assumptions and values, ideally undergoing deep change in the process of learning and becoming intentional and active learners (Mezirow, 1991).

Rubrics

One method of assessing learning without differentiating modalities is the formation of programmatic rubrics. Rubrics allow for a criteria-based assessment based on certain levels of content acquisition by students. With the aforementioned approach to creating curriculum via content domains, subdomains, and competencies, one can easily apply a rubric to assess the extent competencies were learned. Figure 1 displayed one content domain, one of its subdomains, and the related competencies from the Doctor of Business Administration program; Table 1 displays the associated rubric for assessing learning in regards to these competencies.

A rubric designed in this manner gives faculty a baseline by which to assess their students and provides curricular consistency, despite the potential size and dissipation of a university.

Table 1. Doctor of Business Administration rubric

Competency	Unsatisfactory	Basic	Proficient	Advanced
Strategy: Environmental Scanning and Analysis				
Promote an organizational environment that values critical thinking in all areas	Fails to demonstrate cognizance of critical thinking principles in organizational model and values design	Inconsistently or inadequately addresses critical thinking principles in organizational model and values design	Integrates assessments, training, initiatives, and communication related to critical thinking in organizational model and values design	Develops strategies to integrate critical thinking into organizational models that extend or transcend current research and/or practice
Use information systems and research to develop an environmental scan	Fails to integrate research into environmental scanning practices and suggestions	Inconsistently or inadequately integrates research into environmental scanning practices and suggestions	Uses research effectively in environmental scanning practices and suggestions	Critically evaluates and uses research to optimize and iteratively improve environmental scanning practices
Develop competitive intelligence from research and information systems	Fails to gather effective competitive intelligence data	Inconsistently or inadequately integrates research into the development of competitive intelligence	Uses research effectively to improve an organization's competitive advantage	Critically evaluates and uses research to optimize and iteratively improve an organization's competitive advantage
Use and evaluate forecasting models based on environmental scanning, research, and analysis	Fails to implement forecasting models or utilize effective methodologies in proposing a forecasting model	Inconsistently or inadequately implements forecasting models or utilizes effective methodologies in proposing a forecasting model	Uses and evaluates forecasting models based on a sound research foundation	Critically evaluates and uses forecasting models based on a comprehensive research foundation

Collaborative Case Studies

In the traditional approach to learning assessment in higher education, the final comprehensive examination and the thesis or dissertation have been standard components for more than 100 years. Although these approaches to learning assessment have apparent value, neither evidences how students will succeed in their chosen career. The final comprehensive examination can be a reflection of cognitive learning or, in some cases, mere memorization. It serves as *snapshot* of a certain field at a particular moment in time, but it offers no evidence that the student will be able to apply their knowledge. The scholarly document—a master's thesis or a doctoral dissertation—often focuses on an area that the student may never work with again. As academic institutions change to meet the needs of their students, capstone learning assessment may need to take on new practices to predict success adequately.

Collaborative case studies, while more frequently encountered in business-related programs, may find application as capstone assessments in other content areas. The collaborative effort can assess a student's knowledge, skills, and abilities as well as the student's ability to work with diverse perspectives to demonstrate competency. Because of its practical focus, this assessment might offer a better predictor of student capability beyond the academic environment.

A collaborative case study designed to assess learning could incorporate all aspects of the curriculum in a program. To that extent, the rubric that is derived from a program's domains, subdomains, and competencies would be parallel to the rubric used to assess student performance on the case study. Small teams of three to six students could respond to case study questions and present their findings to both faculty and peers in a common setting. This methodology allows for individual application of knowledge, the synthesis of diverse viewpoints surrounding theories and practice, a dialogic model of presentation, and the ability to engage in an approach that mirrors the workplace.

The emerging trend of viewing curriculum design and learning assessment in a holistic manner and of applying course content to pragmatic issues drives the use of a comprehensive case study as a primary means of assessing programmatic learning. When curriculum and assessment are designed in a holistic manner and without an overwhelming emphasis on delivery methodology, they strengthen universities' ability to provide integrative programs that withstand the scrutiny of auditors, evaluators, critics, and skeptics.

Conclusion

Curriculum development is a key point in the assessment of learning. Learning content management systems need not be considered separate functions or approaches to classroom, online, or any other curriculum delivery format. If the academic program is designed in a sound and systematic approach, learning assessment is a natural function of curriculum development and deployment. Assessing learning without assessing its associated curriculum design is essentially the act of looking at an effect without mention of its causes.

The misnomer of *distance* education still looms as one of the most influential detractors to its acceptance in all spheres of academia. By creating an environment that promotes peer interaction, developing a faculty model that promotes a facilitative relationship, and devising learning strategies and assessment methodologies that stimulate students' construction of meaning in an academic environment, universities can enable a sense of *closeness* that supercedes proximity. Thus, online students and faculty, through their relationship with knowledge, can develop the same level of interactivity and engagement as those in a traditional classroom setting.

Although the aforementioned approach was designed for an andragogical model, it remains relevant to other educational environments. The need is apparent for academic institutions to focus on the content of the program without labeling it as either *online* or *classroom*, as doing so creates an arbitrary division between the two modalities. The goal of any institution in developing curriculum should be to create an assessment methodology that is integrated throughout their curriculum in a conscious manner, not specific to a modality.

References

Anderson, L. W. (2002, Autumn). Curricular alignment: A re-examination. *Theory Into Practice, 41*(4), 255.

Bloom, B. S. (1984). *Taxonomy of educational objectives, handbook 1: Cognitive domain.* Boston: Addison-Wesley.

Kegan, R. (1982). *The evolving self.* Cambridge, MA: Harvard University Press.

Kegan, R. (1994). *In over our heads: The mental demands of modern life.* Cambridge, MA: Harvard University Press.

Kegan, R. (2003). What form transforms? A constructive-developmental approach to transformative learning. In J. Mezirow, & Associates (Eds.), *Learning as transformation: Critical perspectives on a theory in progress* (pp. 35-70). San Francisco: Jossey-Bass.

Kolb, D. A. (1984). *Experiential learning.* Englewood Cliffs, NJ: Prentice Hall.

Knowles, M. S. (1973). *The adult learner: A neglected species* (4th ed.). Houston, TX: Gulf.

Krathwohl, D. R., Bloom, B. S., & Masia, B. B. (1999). *Taxonomy of educational objectives, handbook 2: Affective domain.* Boston: Addison-Wesley.

Lovitts, B. (2001). *Leaving the ivory tower: The causes and consequences of departure from doctoral study.* New York: Rowman & Littlefield.

Mezirow, J. (1990). *Fostering critical reflection in adulthood: A guide to transformative and emancipatory learning.* San Francisco: Jossey-Bass.

Mezirow, J. (1991). *Transformative dimensions of adult learning.* San Francisco: Jossey-Bass.

Mezirow, J. (Ed.). (2003). *Learning as transformation: Critical perspectives on a theory in progress*. San Francisco: Jossey-Bass.

Pintrich, P. R. (2002, Autumn). The role metacognitive knowledge in learning, teaching, and assessing. *Theory into Practice, 41*(4), 220.

Rittschof, K. A., & Griffin, B. W. (2003). Confronting limitations of cyberspace college courses: Part I—identifying and describing issues. *International Journal of Instructional Media, 30*(2), 127.

University of Phoenix. (2004). *University of Phoenix fact book*. Phoenix, AZ: Author.

Chapter IX

Legal Implications of Online Assessment:
Issues for Educators

Bryan D. Bradley, Brigham Young University, USA

Abstract

This chapter provides a survey of basic legal issues that online assessment developers and users need to be aware of and account for in their assessment design, development, implementation, and interpretation activities. High-stakes assessments such as professional certification and college admissions exams are particularly vulnerable to legal challenge when there is evidence of problems with respect to validity, reliability, testing fraud, or unfair bias. The chapter provides straightforward suggestions that will help assessment developers and users avoid legal problems with their exams.

Introduction

Assessment of knowledge and skills has been a part of every documented society throughout history. Individuals desiring to belong to a societal group or institution such as a professional association are often required to demonstrate their competence within the relevant knowledge and skill domain of that organization. Examples of such organizations include companies that provide product or skills-based certifications, licensing bodies (such as the American Medical Association), and government agencies. Each of these organizations defines the criteria for the admission and acceptance of licensed or certified professionals and then designs various assessment instruments accordingly. This practice is becoming even more prevalent in the early years of the new millennium.

As individuals pursue new levels of knowledge, professional qualification, college admission, and limited scholarships, they usually submit to assessments that serve a "gate-keeping" function that determine the potential of candidates to pursue further professional growth opportunities. Online assessment has become a very effective tool in this endeavor with its capabilities to not only deliver "traditional" item types, such as multiple choice, true–false, and short answer, but with its powerful capacity to deliver many other item types such as drag-and-drop, point-and-click, and very complicated task simulations. With the advent of U.S. government mandates such as the No Child Left Behind Act, state education organizations are becoming increasingly interested in online assessment for the high-stakes testing programs in the public schools (Electronic Education Report, 2002). This movement of increased online assessment coupled with legislative and regulatory interest is potentially confusing in both the educational and industrial sectors of society. Harris (1999) stated that the "assessment landscape is dotted with legislative, regulatory, and legal crabgrass that at times distracts test publishers from their principle business of providing quality tests and assessment services to test users and test takers" (p. 1).

The stakes are high—for test publishers and users and the test takers. Each group has rights and expectations that need to be legitimized and protected by law. Test developers and test users look to protect and ensure their rights with respect to issues such as financial, copyright, validity, and reliability. Additionally, developers and users need to protect the intellectual and technological innovations of the assessments in order to ensure that the results and interpretations maintain integrity and value within the target knowledge and skill domain that they assess.

Test takers want to have confidence that their scores on a given assessment closely reflect their true abilities. They have rights of equal access to assessment materials, fairness in comparisons with other test takers, and proper interpretations and judgments of their scores. When any of these rights are compromised, the purposes of the assessment process and the value of the resulting outcomes are thwarted. Consider the following two scenarios—one from a professional setting and the other from academia.

Mario and the Certification Exam

Mario is a qualified computer systems consultant who has several years of experience in designing, installing, and fine tuning sophisticated enterprise communications and data systems around the world. As his employer's business has grown, the executives of the company created the policy that all computer systems consultants must be certified in the products that they recommend and install. Mario has invested time and money in studying the latest technologies and has prepared to take the appropriate online certification exam for his discipline.

Upon taking the exam, Mario is frustrated to discover that the assessment items primarily test his ability to remember terms, procedure steps, and system part specifications. In Mario's work environment, he is accustomed to looking up such information in a readily accessible database table on his PDA as he performs much more complicated tasks in serving his clients. Still upset with the content of the exam, Mario clicks on the final item, reviews his scores on the screen and discovers that he did not pass. Without passing the exam, he will not receive the certification and his employment is now in jeopardy. What are the legal implications in this case?

Kristina and the Graduate School Admissions Exam

Kristina is trying to get into a prestigious graduate program and needs to achieve a score in the top 10% of all examinees on an online, standardized exam in order to be considered by the school's admissions committee. On her first attempt, she scores in the top 25% category—a very poor score for the goal she is hoping to achieve. She feels that she just needs more preparation to do well on the exam.

Still determined, Kristina purchases a set of self-study materials offered by a company that guarantees that she will raise her scores on the exam after completing that course. After completing the self-study exercises, she retakes the exam and finds that many of the items on the exam are worded the same as the practice items in the self-study package. The items and options are so closely identical to the study-guide items that she hardly needs to consider the incorrect options before selecting the correct answers.

Kristina excitedly proceeds through the test, completing it in less time than most students. Upon receiving her scores, Kristina learns that her scores are well above the cutoff point and that she is eligible for a scholarship that she otherwise would not have qualified for. What are the legal implications in this case?

As we consider both Mario's and Kristina's cases, we see that both test takers were disadvantaged by their experience with the online assessments. Because

Mario received a low score on an invalid online exam, he was unfairly hindered in his career path. In Kristina's case, she appears to have benefited from using pirated test items. However, she is entering a graduate program that is arguably challenging beyond her abilities to succeed. If she had achieved a score that was closer to her true score, she probably would not have passed and would have turned her attention to better preparation or to pursue a career path more closely suited to her talents and abilities. Kristina's exam results have also resulted in an award of opportunity and scholarship resources that should have been given to a more deserving and qualified applicant.

How can test developers and publishers protect and defend their rights of ownership, administration, and interpretation of their exams? How can test takers understand and protect their rights with respect to online assessments and the interpretation of their results? As we examine these questions, how should we take into account legal issues such as Voluntary National Tests, No Child Left Behind Act, and the Americans with Disabilities Act?

This chapter will survey the issues regarding legal and institutional policies surrounding online assessment. The core concerns addressed in this chapter include the various criteria for legally defensible exams, such as ensuring and safeguarding assessment validity, reliability, security, and the elimination of unfair bias against individuals based on race, nationality, culture, and sex. In addressing the psychometric topics of validity and reliability, this chapter will not delineate statistical models from a mathematical point of view. Indeed, volumes of articles and books are available on the theory behind those topics. Nor will this chapter explore differences between alternative psychometric approaches such as classical theory or item response theory. It is sufficient to note that each approach has its strengths and weaknesses in producing evidence regarding the design, outcome, and benefits of individual assessments. The purpose for this chapter is to encourage online assessment developers and users to consider these topics as legally acceptable criteria for judging the issues surrounding online assessments.

Validity

Validity is arguably the most important legal issue in assessment (Messick, 1993). According to the American Educational Research Association (AERA), the American Psychological Association (APA), and the National Council on Measurement in Education (NCME; 1999), validity is defined as "the degree to which evidence and theory support the interpretations of test scores entailed by proposed uses of tests" (p. 9). Messick defines validity further, stating that it "is an integrated evaluative judgment of the degree to which empirical evidence and theoretical rationales support the *adequacy* and *appropriateness* of *inferences* and *actions* based on test scores or other modes of assessment" (p. 13, italics included). In other words, *validity* answers such questions as:

- What is the purpose of the assessment?

- What are the relevant characteristics of the target audience or group of individuals who will be taking this assessment and benefiting from its results and interpretations?

- Are there any groups of individuals (e.g., minorities or other groups) that may be disadvantaged by the format, language, or other characteristics of this assessment?

- What are the knowledge and skill domains that the assessment will cover?

- How are the various knowledge and skill components weighted on this assessment? Is one component more important or critical than another? How are the weightings determined?

- Are the items on the assessment congruent with the established purposes, goals, and objectives of the exam?

- Do the items discriminate properly? Do the qualified test takers do well and nonqualified test takers do poorly?

- What evidence do we have that supports our answers to the above questions?

These are complex questions that require much scrutiny on the part of test developers and users. The answers to each question require ample evidence from a variety of sources to determine validity with any legitimate degree of confidence. Nitko (1996) stated that validity is critical in the interpretation of any assessment results. "To validate [the] interpretations and uses of…assessment results, [assessment developers and users] must provide evidence that these interpretations and uses are appropriate" (p. 36).

For decades, experts in psychometric research such as Cronbach and Meehl (1955), Messick (1993), and others have explained that an assessment's validity must be determined and defined both from its design and from the resulting scores. The validity evidence from the design of an assessment stems from the documentation of procedures and decisions as the assessment is envisioned, designed, and implemented. This documentation includes the composition of stakeholder committees and focus groups that participated in any discussion and decisions about the target audience, the domain of the assessment, and how scores will be interpreted. Jones (2000) states that the assessment developer "must have a thorough understanding to the measurement model underlying the testing program, and the resources required to implement the model in an effective and efficient manner" (p. 1).

To understand the measurement models, we need to consider three basic types of validity, namely, content, criterion, and construct validity. Content validity refers to stakeholder decisions about how the assessment content relates to and is representative of the domain it covers. Criterion validity refers to the relationship of scores of selected groups or individuals. For example, assessment developers expect prepared, experienced, and knowledgeable test takers to perform assessment tasks more proficiently than unprepared or inexperienced individuals. Developers also expect that the assessment items will match the performance criterion that is being tested. If the domain being assessed requires cognitive tasks such as analysis or evaluation, the assessment activities or items should prompt those actions from the test takers. With content and criterion validity established, developers next verify construct validity.

The presumption of construct validity, as noted by Messick (1993), "is based on an integration of any evidence that bears on the interpretation or meaning of the test scores" (p. 17). In other words, construct validity is a composite of all evidence that points to whether or not the assessment of the construct or "thing" in question (e.g., job role, academic domain, personality, dexterity) is fair, comprehensive, and technically accurate. Palomba and Banta (1999) stated that "the ability to rule out alternative explanations for measured outcomes is the hallmark of construct validity" (p. 105). Content, criterion, and construct validity combine to establish the core integrity and legal defensibility of any assessment. As test developers and users compile as much evidence as possible from multiple sources, their claims of construct validity will be credible under legal scrutiny.

Evidence of validity is generally accumulated as documentation of the test design, development, and implementation process. This documentation should include the following:

1.   Statement of the assessment's purpose and appropriate use
2.   Specification of data to be collected that, when considered together, will justify claims of validity
3.   Definition of assessment content in terms of performance objectives
4.   Definition of performance-objective weightings and an explanation and justification of those weightings

Statement of the assessment's purpose, appropriate use. This statement is usually crafted through consensus by a representative group of stakeholders.

Stakeholders of online testing include individuals or groups who hold a vested interest in the assessment, its outcomes, and its uses. In an academic setting, stakeholders are usually colleagues who are responsible for curriculum development and implementation. In a professional certification setting such as a particular branch of the sciences, stakeholder groups are often represented by *advisory councils*; a group of selected professionals that define the certification program and set the parameters of any online assessments that are associated with that program. These individuals are keenly aware of the importance of the accuracy of the online examinations and their responsibility to ensure that the exams adequately represent the appropriate mix of content and difficulty that are associated with successful knowledge and skill performance in the respective assessment domains.

Specification of data to be collected that, when considered together, will justify claims of validity. This specification is generally a list of the compiled set of evidence that will support validity claims for the corresponding assessment. Assessment developers should include knowledge and skill statements derived from events such as curriculum planning sessions, task analyses, and knowledge and skill surveys. As developers select specific knowledge and skill components for the assessment, they should also document how they made those selections. For example, developers should indicate the weightings that they applied to each knowledge and skill component and how those weightings were determined (e.g., through survey data, stakeholder mandates).

Definition of assessment content in terms of performance objectives. This definition of assessment content is the actual set of performance objectives gleaned from the aforementioned data sets. These performance objectives should be statements that describe specific and demonstrable cognitive behaviors (Anderson & Krathwohl, 2001; Bloom, 1956). These objectives should also reflect the context and assumptions that relate to the online assessment. For example, if the items on an online assessment are to be primarily multiple choice, drag-and-drop, or other traditional test item types, the test designers should not write performance objectives that describe behaviors that are best suited for a simulation or real-time application.

This congruence between performance objectives and the actual behaviors that examinees do during the assessment is a critical issue in the legal defensibility of an online assessment. If assessment developers produce items that do not match the performance articulated in the objectives, validity will be in question. Test takers who perform poorly on the assessment will have leverage to challenge the validity of the exam and the decisions derived from the resulting scores.

Definition and justification of performance objective weightings. Assessment developers and users rarely have opportunity to test on all performance objectives in a single administration of an assessment. This concern is especially the case with online assessments. With the advent of Computer Adaptive Testing (CAT), and with the resource cost in classrooms and testing centers, most online assessments do not cover the entire domain or construct being investigated. Therefore, assessment developers must ensure that the objectives they test on are important and justified so that the assessment users can be assured that validity is maintained regardless of the length of the assessment or number of items presented in a CAT.

Psychometric experts and standards (see Cohen & Wollac, 2003; Palomba & Banta, 1999; AERA, APA, & NCME, 1999) stress that a test blueprint is effective in selecting the objectives that should be included on an assessment and justifying how extensively each objective should be covered. Assessment developers create blueprints both formally and informally, depending on the purpose and domain they are assessing. For purposes in this chapter, I will discuss the formal establishment of a blueprint so that legal defensibility can be maintained.

To create a formal assessment blueprint, the test developer must execute a series of procedural steps to gather and evaluate data regarding the importance and usefulness of each performance objective. This data can include the consensus of a focus group of representative stakeholders, interviews and surveys among experts in the target domain, or a combination of these strategies. The benefit of gathering this information is that assessment developers are able to document the thoroughness of the effort to establish performance objectives and their respective weightings that accurately represent the domain of the assessment.

With the blueprint in place, assessment developers can write the appropriate number of items according to that specification. Developers can also produce equivalent forms of an assessment based on the blueprint requirements. Rules for CAT item presentation can also be established, based on the composition of the blueprint.

To summarize, validity is the most critical aspect of legal defensibility for online assessment developers and users. Online test takers have a fundamental right of to know that the assessment is valid from a design, development, implementation, and scores interpretation standpoint. Although assessment developers often find that the establishing of validity can be a daunting task, there are basic charac-teristics of validity that can be identified and articulated within coherent documentation. Developers will then rely on this documentation as the basis for establishing and defending the assessment validity.

Online assessment developers and users can adequately defend validity in a legal challenge to the extent that their documentation is thorough and represents stakeholder consensus, accurate performance objectives, and proper weightings of congruent assessment items within the understood content domain. Conversely, if validity is not steadfastly pursued and documented, test takers will have the legal advantage in any challenge of the assessment's composition or interpretations of scores.

Reliability

In the disciplines of psychology, engineering, manufacturing and any other discipline that inherently relies on measurement and statistics, *reliability* is a definitive term that reflects on the data and measurements within one or more observations of an event (such as an online assessment or a test on the quality of a manufactured product). Within the context of online assessments, reliability refers to whether or not the same results will occur over repeated administrations of the same assessment. Developers like to think that their assessments measure what they are intended to measure and that those measurements will be consistent.

Developers also hope that knowledgeable and prepared examinees will score high and that poorly prepared individuals will score low. Assessment developers and users are concerned when poorly performing examinees correctly answer an item and the better performing examinees get that item wrong. Similarly, developers and users are equally concerned when assessment items are either too easy or too hard, thus limiting the assessment's ability to discriminate between the high or low performing examinees. Like validity, reliability needs to be determined from a variety of views or analyses of the assessment. Reliability measurements give us an understanding of the performance of the assessment in meeting its intended purpose. These measurements are based on well-established statistical and psychometric theories and methodologies such as classical reliability theory, generalizability theory, and item response theory.

Data from online assessments give us detailed information about the performance of the test taker with respect to the content and context of the item or assessment activity (e.g., simulation or real-time task within an online application). Feldt and Brennan (1993) stated that this data include aspects or characteristics of test taker "consistency and inconsistency" that need to be quantified. Feldt and Brennan additionally stated that "all measurement must be presumed to contain error" (p. 105).

There are several accepted strategies for measuring and analyzing assessment reliability. To effectively understand reliability for a given assessment, test developers should include a plan for analyzing reliability as a component of the assessment design. Again, while it is out of the scope of this chapter to delve deeply in the statistical formulas, it is important to note that the plan and documentation of assessment reliability should include information regarding critical concepts such as factor analysis, alternate form reliability, and internal consistency.

As assessment developers and users consider legal issues surrounding assessment reliability they should do so within the context of five basic assumptions of an assessment that should be considered legally reliable:

- **Assumption 1:** Assessment scores reflect the examinee's "true" abilities and knowledge within the assessment's scope and domain along with unexplained error that must be accounted for.

- **Assumption 2:** Examinees will score about the same on multiple attempts of the same assessment.

- **Assumption 3:** Assessment developers have taken the necessary steps to minimize the effects of factors that would cause an examinee to score either higher or lower than his or her true ability within the assessment scope and domain. These factors may include "subtle variations in physical and mental efficiency of the test taker, uncontrollable fluctuations in external conditions, variations in the specific tasks required of individual examinees, and inconsistencies ion the part of those who evaluate examinee performance" (Feltd & Brennan, 1993, p. 105).

- **Assumption 4:** Examinees are assessed on defined knowledge and skills that are specified in the assessment purpose and objectives. It is assumed that examinees are not being inadvertently assessed on indistinct or obscure criteria.

- **Assumption 5:** Scoring algorithms, evaluator rubrics, and answer keys are accurate and consistently applied.

As psychometricians and others evaluate online assessments, they determine if the above assumptions were followed. To the extent that these assumptions are deemed true, assessment users and examinees can be reasonably confident in the accuracy of the scores and their interpretations. If any of these assumptions are not true or are questionable for a given assessment, there may be some legal vulnerability associated with the reliability of that assessment.

For example, in Mario's case, which was considered earlier in this chapter, he was exposed to an exam that violated Assumption 1 and, possibly, Assumption 4. Mario's assessment experience included items that assessed whether or not he had memorized information that he was accustomed to looking up on his PDA. This inconsistency could make the reliability of the assessment questionable.

Mario would be well within his rights to legally challenge the results and decisions derived from this assessment—especially if reliability coefficients or other statistics indicate one or more problems.

For another example, Kristina, whose case was also considered earlier, profited by knowing the items and answers prior to taking the assessment. Kristina obviously scored higher than her true score or ability within the domain and also inconsistently with her previous attempt in taking the graduate school admission exam. Therefore, with respect to Kristina's experience, Assumptions 1 and 2 were violated. The fact that many potential test takers (in addition to Kristina) had access to the items through a third party puts the reliability of the assessment in question. From a legal standpoint, the test publisher and users who make decisions based on the scores of examinees cannot determine the actual reliability of the exam. They cannot answer questions regarding which examinees had cheated by memorizing compromised items. In this case, the test publisher and test users should consider their options to close down the online assessment, cancel scores that are suspicious, and work on republishing the exam with new items. Furthermore, the test publisher should pursue an investigation to see if the perpetrators of the item sharing violated any legally binding concerns such as nondisclosure agreements or copyright infringement.

The measurement and interpretation of reliability for a given assessment is dependent on the overall strategy of the assessment. If a test developer plans on creating an adaptive online exam (where the number of items seen by an individual examinee depends on that examinee's response to previous items), then a proven method for analyzing the assessment performance would be the steps in item response theory (Van der Linden & Hambleton, 1996).

If the assessment developer plans on fixed-length, single or multiple forms of an assessment, then that developer would be wise to follow classical reliability or generalizability theory. Although it is out of the scope of this chapter to discuss the details of the statistical models, strategies, and formulas that are employed in evaluating and defending assessment reliability, there are two points that are noteworthy.

First, some error variance is always assumed in the scores due to individual examinee variation, such as health issues, levels of concentration, degree of forgetfulness, psychological readiness, and luck in guessing. Second, assessment items and exercises are representative samplings from the assessment domain and may inadvertently emphasize some topics over others.

To ensure reliability, online assessment developers and users should follow these steps:

1. Start with a pool of valid items that reflect a balance with the assessment blueprint.

2. Beta test the items among an adequate sample of target examinees.

3. Analyze the beta assessment data to select the final pool of items. This analysis should include as much data as possible including item performance, examinee performance, and item options analyses.

Security

As stated previously in this chapter, in order for an assessment to be legally defensible, it must be designed and implemented so that the data will reflect closely to the true score of each examinee. Foster (2003) stated that the advent of online assessments has actually alleviated security problems that are very prevalent with paper-and-pencil assessments. The following is a list of the security strengths of online assessments according to Foster:

• "Complete and equivalent test forms are randomly selected. Test-takers can't anticipate which set of questions they will see.

• Computerized adaptive testing creates individually tailored exams, reducing the exposure level of questions and making it difficult for less competent individuals to succeed by cheating.

• Encryption and password protection provide secure transmission of tests and test results.

• Test development that is entirely computerized prevents the copying and distribution of question pools and tests." (p.1)

Foster (2003) stated that the security problem with many online assessments stems from the long shelf life of those assessments and that many examinees share information about the items, possible answer options, and other information that tend to give future examinees an unfair, unethical, and (perhaps) illegal advantage in taking the exam. As was mentioned earlier in this chapter, there are organizations that set up "brain dumps" of online assessment items under the guises of being study aides. Foster calls the illegal and unethical practice of copying and sharing online assessment items "test piracy" and "test fraud."

Foster (2003) also cautioned assessment users to be vigilant about securing the test environment so that devices such as PDAs, pagers, cell phones with cameras and text messaging capabilities, and other technological equipment, be strictly controlled and prohibited from the site where the assessment is administered. Foster stated that cheating compromises an assessment's effectiveness in meeting its purpose. Scores from examinees who cheat compromise the value and usefulness of the data sets for those assessments in ways that make the resulting analyses meaningless.

In addition to the use of technology and item piracy as means for cheating, there is also the security problem of imposters who take online assessments in the name of other individuals. In the relatively short time period where online assessments have been used in both academia and industry, there have been many cases where substitute individuals have taken high-stakes, online assessments in the name of the actual person whose name is registered for the exam. In this way the substitutes (who know the content being assessed) take the test in the name of other test takers. The test data, including scores (usually passing) and time spent on the assessment is then associated with the person who should have taken the test.

Unfortunately, cheating sometimes is initiated by unwise educators. Some teachers who are under pressure to have their students do well on standardized assessments have given answers to their students. This form of cheating is especially tempting for teachers who on test day have assumed the role of proctor for paper-and-pencil assessments. Computer-based assessments can solve this problem because items are randomly presented to the examinees. Moreover, with online assessments, there are no paper answer sheets that educators could alter after a student has completed the exam.

There are several ways in which we can use security measures to prevent aberrant or erroneous scores on online assessments:

1. Keep items, media, answer keys, and other assessment materials secure. This is also important during the assessment design and development phases.

2. Create and enforce policies that ensure assessment security, such as prohibiting personal cell phones, PDA, and other send-and-receive devices. Make sure that this policy is widely publicized and understood by the assessment target population.

3. Use trained proctors to observe examinees when they take the assessments. These proctors should be very familiar with the equipment in the testing center and be able to help examinees if they encounter any technical problems (e.g., passwords, frozen computer screens, malfunctioning computers, keyboards, navigation through the assessment).

4. Require examinees to present multiple forms of identification to hinder the use of imposter examinees.

5. Use computer adaptive testing or multiple test forms so that each examinee views a different version of the same assessment. In this way test item pirating will be more difficult to achieve.

6. Monitor the test data frequently. If pass rates start to increase after the assessment has been published for a while, that evidence may indicate that people who have taken the assessment are sharing information with future examinees.

7. Monitor online chat groups to see if examinees are discussing their experiences with the assessment.

8. Pilot test new items and rotate them into the test pool to keep the content of the assessment fresh. New items can be tested as "nonscored" items in the current assessment. Once you have enough data to ensure that the item is effective, you can include it as a live item in other forms of the assessment.

In summary, security is vital for effective and legally defensible online assessments. Assessments that have been compromised due to cheating or overexposed items cannot withstand legal scrutiny. Data derived from the assessment administration will not be conclusive as to which examinees are competent in the target domain. Nor will we be able to know which assessment items perform well or poorly. Assessment security must be diligently planned, implemented, and defended.

Elimination of Unfair Bias

The terms *bias* and *discrimination* are generally thought of as having negative connotations in today's language usage. With respect to measurement and assessment, they are descriptive words that are important in working with and communicating about an assessment's purpose and effectiveness. Inherently, a good assessment must be fair and biased in that it focuses on the content of the knowledge and skill domain fairly while discriminating (or discerning) those examinees who are competent within that domain.

This usage of the two terms is completely legal and desirable within the framework of online assessment. What assessment developers must guard against is the *unfair bias* in assessments which disadvantages examinees due to factors which are out of the scope of the assessment (such as sex, age, language skills, ethnicity, and religion).

As discussed earlier in this chapter, assessment users have the right to expect that examinees are being scored on their competence in the assessment domain—not on extraneous and irrelevant traits such as working a computer mouse, reading ability, or understanding humor imbedded in the assessment items. Similarly, examinees have the legal right to expect that they are being judged only on the criteria that has been established as the foundation and justification of the assessment.

The standards for fairness in assessment practices have been well defined by the Joint Committee on Testing Practices (1988) of the APA. According to the APA standards, online assessments are fair and unbiased when all legitimate examinees have access to the assessment, can read, hear, and otherwise understand the task(s) they need to perform on the assessment, and can understand the feedback and scores they receive. The APA stresses that assessment developers and users should look into "the performance of test takers of different races, gender, and ethnic backgrounds when samples of sufficient size are available. Evaluate the extent to which performance differences may have been caused of the test" (Standard #15).

One of the top assessment development organizations in the United States stressed that assessment users should do the following to ensure fairness:

1. "Treat all test takers equally so that differential treatment does not cause unfairness

2. Provide reasonable accommodations for test takers with disabilities

3. Involve diverse groups in the test development process to obtain a variety of perspectives

4. Review questions to ensure they are not elitist, racist, or sexist, controversial, offensive or upsetting

5. Review tests to ensure they reflect the diversity of the test-taking population

6. Use an empirical measure of differential item functioning to determine whether test takers of similar levels of knowledge in different groups are performing at the same level on test questions

7. Tell test users how to use scores fairly and how to avoid common misuses of the scores." (ETS, 2002, p. 6)

If item analyses reveal that specific groups of examinees have "aberrant response patterns," assessment users must be very careful in how they interpret the results of the assessment for those individuals (Bracey& Rudner, 1992). From a legal standpoint, assessment users must accept that such "scores cannot be interpreted in standard ways" (p. 2). If test users accept low scores within specific groups of examinees, they can leave themselves open for very costly legal challenges. In the event that assessment users find aberrant score patterns, they should immediately examine and evaluate the items in the exam and pull all items that are potentially offensive or misleading to the identified examinee population group.

Another aspect of fairness in online assessment is accessibility. This is especially important with respect to the Americans with Disabilities Act. As a general rule, persons with learning disabilities, psychiatric disabilities, physical disabilities or attention-deficit/hyperactivity disorder have the right to access the content of online assessments in such a way that their disabilities will not inhibit their opportunity to understand the assessment task and to perform that task in a way that is fair to those individuals. Assessment developers and users should carefully determine if disabled individuals are a part of their target examinee audience. If that is the case, then the assessment should have alternative forms or technological enhancements that will accommodate those examinees.

Fair bias and fair discrimination is essential to an effective and meaningful online assessment. *Unfair* bias and *unfair* discrimination is legally disastrous for any assessment program. Although we expect a certain amount of error in the response patterns of examinees, we do not expect apparent trends within specific groups such as sex, race, religion, and ethnicity.

Conclusion

This chapter has presented a review of the standard safeguards for legally defensible online assessments, namely, validity, reliability, security, and the elimination of unfair bias and discrimination. The key to legal defensibility within any assessment offering is to document all decisions regarding the assessment including its purpose, goals, content, weightings, target examinee audience, evaluation strategies and models (e.g., generalizability theory or IRT), rules for assessment item tryouts and final item selection, rules for scoring items and interpreting individual scores, and safeguards to reduce risk for unfair bias and discrimination. Additionally, test users should establish plans to ensure exam security so that cheating and item pool compromising will be easily detected and quickly addressed.

As Foster (2003), Harris (1999), and ETS (2002) asserted, government officials and academic institutions are becoming more interested in the legal issues and ramifications surrounding online assessment. Regulations are evolving and online assessment is gaining in its presence throughout the academic and professional training domains. The legal scrutiny of online assessment will only increase as this evolution continues. Examinees will also press for legal fairness as they submit to online assessments for education and professional opportunity. As assessment developers, users, and respective regulators adhere to established standards of design, use, and interpretation of online assessments, the target examinee audience will be well served and the decisions rendered from these assessments will be legally justified.

References

AERA, APA, & NCME (1999). *Standards for educational and psychological testing.*

Anderson, L.W., & Krathwal, D.R. (Eds.) (2001). *A taxonomy for learning, teaching, and assessing: A revision to Bloom's taxonomy of educational objectives.* New York: Longman.

Boom, B.S. (1956). *Taxonomy of educational objectives: The classification of educational goals.* New York: David McKay Company.

Bracey, G., & Rudner, L. M. (1992). *Person-fit statistics: High potential and many unanswered questions. Practical Assessment, Research & Evaluation, 3*(7). Retrieved April 28, 2004, from *http://PAREonline.net/getvn.asp?v=3&n=7*

Cohen, A. S., &. Wollac, J. A. (2003). Helpful tips for creating reliable and valid classroom tests: Getting started—The test blueprint. Retrieved April 29, 2004, from *http://wiscinfo.doit.wisc.edu/teaching-academy/LearningLink/LL403Cohen.htm*

Cronbach, L.J. & Meehl, P.E. (1955). Construct validity in psychological tests. *Psychological Report, 52,* 281-302.

Electronic Education Report. (2002). State interest in online testing blossoms, as new federal testing mandates loom. *Electronic Education Report, 9*(16).

ETS. (2002). Basics of assessment. Retrieved April 29, 2004, from *http://www.ets.org*

Feldt, L. S., & Brennan, R. L. (1993). Reliability. In R. L. Linn (Ed.), *Educational measurement* (3rd ed.). Phoenix, AZ: Oryx Press.

Foster, D. (2003, January). Test piracy: The darker side of certification. Retrieved April 28, 2004, from *http://www.caveon.com/df_article2.htm*

Harris, W. G. (1999, April 20-22). *From my window: A trade association perspective on assessment.* Paper presented at NCME annual meeting, Montreal, Quebec, Canada.

Joint Committee on Testing Practices. (1988). *Code of fair testing practices in education.* Washington, DC: Author.

Jones, J. P. (2000). Promoting stakeholder acceptance of computer-based testing.Retrieved April 2004 from *http://www.testpublishers.org/ journal02.htm*

Messick, S. (1993). Validity. In R. L. Linn (Ed.), *Educational measurement* (3rd ed.). Phoenix, AZ: Oryx Press.

Nitko, A. J. (1996). *Educational assessment of students* (2nd ed.). Englewood Cliffs, NJ: Prentice Hall.

Palomba, C.A., & Banta, T.W. (1999). *Assessment essentials: Planning, implementing, and improving assessment in higher education.* San Francisco: Jossey-Bass Publishers.

Van der Linden, W. V., & Hambleton, R. K. (1996). *Handbook of modern item response theory.* New York: Springer.

Chapter X

Legal Implications of Online Assessment:
Issues for Test and Assessment Owners

Robert R. Hunt, Caveon Test Security, USA

Abstract

At a time when information, including purloined test and assessment content, moves at "Internet speed," test and assessment owners are usually comforted to know that the law provides meaningful protection if threshold test security measures are observed. This chapter explores the question of whether that protection extends to the use of online assessment which promise greater access, convenience and savings. Applying threshold security requirements derived from copyright and trade secret laws, this analysis indicates that in its widely practiced and current form, online assessment would fail to qualify for protections by which disclosure of text content could be swiftly condemned. Online test and assessment owners are cautioned to explore threshold security alternatives and to carefully weigh the importance of their tests, as well as investments in the creation and distribution of a test instrument, against the potential loss of test content.

Introduction

As with most other types of information, use of the Internet for the delivery of assessments can increase access, convenience, even learning, and vastly lower the cost on both sides of the ledger (test and assessment owners and test takers). Faced with such potential, test and assessment owners can hardly afford to bypass consideration of an online test delivery strategy.

Consideration of such a strategy, however, calls for a clear understanding of an assessment's value. Apart from the time-honored necessity to seek valid and reliable scores, most types of tests and assessments also derive value from the measures taken to secret their contents before, during and after administration. "Test security" in short, plays a critical role in the efficacy (as well as the cost) of tests and assessments—a lesson reflected in U.S. copyright law, which has long provided special protections for "secure tests."

In recognition of the relationship of security to the value of certain types of assessments, professor Robert Kriess (1996) noted that copyright law specifically departs from its overarching policy of ensuring creators the rewards of their work in exchange for access. Owners of secure tests, in other words, inherit all of the typical rights and protections of copyright ownership *without* a corresponding requirement of disclosure.

To attain and maintain this enigmatic copyright status, however, courts have reiterated that nondisclosure of test and assessment content should be understood as a privilege as well as an obligation to keep that content secure (*College Board v. Pataki*,1995). To assist in that effort, copyright law allows secure test owners to condemn virtually any unauthorized use of their test or assessment content.

The questions of law and practice explored in this chapter involve whether online assessment, envisioned as a more convenient, flexible and cost-effective alternative to brick-and-mortar testing centers with professional staff and other security infrastructure, satisfy the legal thresholds for security necessary for secure test copyright and trade secret protections.

Online Insecurity?

From a professional certification and licensure perspective, in the recent past the phrase *online assessment* has indicated roughly the opposite of secure, professional test delivery in which test-taker identity is systematically authenticated, and test-taking behavior is closely scrutinized. While skeptical of the relative insecurity of online assessment, "high-stakes" testing programs have also admired its lower cost and flexibility.

Realizing that security is an impediment to entry into the lucrative professional certification and licensure markets, online assessment vendors have furiously explored security alternatives including remote test-taker authentication and monitoring technologies. In the absence of such technologies, however, educators, trainers and corporations have generally reserved online assessment delivery for lower stakes activities, including measurement of training outcomes and course progress. Many in the field of testing agree, however, that even these varieties of tests and assessments are also important and valuable and that the dissemination of their content on the Internet and elsewhere would erode confidence in learning outcomes.

Why Legal Protection of Tests and Assessments is Useful

Absent the familiar hallmarks of test security (test-taker authentication and supervision), under copyright and trade secret law, tests and assessments may be regarded as nonsecure, and therefore, either not protected (trade secret), or subject only to general restrictions applied to the use of other types of publicly accessible material (copyright).

Under copyright law, this analysis means determining whether copying and distribution of test content is a "fair use." Remember the copy center at your college or university? Chapters and excerpts from expensive textbooks being copied and bound by students, or even at the request of professors? Whereas publishers and courts have expressed their disfavor with some of these practices, the underlying reality is that the purpose of copyright is to *promote* access to creative works, including (absent special exceptions) tests and assessments. If no exception is recognized, test and assessment content can be legitimately reproduced for a variety of educational and nonprofit purposes (Nimmer & Nimmer, 2003).

Contrast the treatment of tests and assessments eligible for secure-test copyright and trade secret protections. These laws provide a variety of legal remedies to test and assessment owners, including criminal penalties, injunctions, and damages against not only individuals who misappropriate assessment content, but also third parties who sell or share that content. Short of legal action, the threat of these provisions operate as a powerful deterrent.

Test and assessment owners must ultimately decide whether the spread of content via Internet "braindumps" (i.e., Web sites that sell purloined test content) and similar outlets pose a threat to their pedagogical objectives; the purpose of this chapter is to help them do so in an informed and reasoned manner. An awareness of the utility and limits of secure test copyright and trade secret protections can assist assessment owners in making informed decisions about test security when considering online delivery.

The Mechanics of Legal Protection

Generally speaking, legal protection of tests and assessments varies with the efforts expended to keep them secure. This rule of thumb is derived from two distinct bodies of law: copyright and trade secret. Although the least intuitive, copyright law is the most well developed in this area and will be treated first and in the greatest depth.

Copyright

It's no secret that unlike most types of creative works that derive value from publication and access, the value of nearly all tests and assessments hinge on secrecy and security.

Copyright is a legal term of art which, in the United States, has a very specific meaning derived from the Federal Constitution and statutes. The popular perception of copyright is comparable to the rights enjoyed with the ownership of a car, a PC or a home—unqualified, durable, and inviolate. The legal reality however, is somewhat different.

According to the U.S. Constitution and Supreme Court, the "primary objective" of copyright protection is "not to reward the labor of authors, but to promote the 'Progress of Science and useful Arts'" [*sic*], (Kreiss, 1996). The overarching goal of copyright, in other words, is tipped in favor of providing public access and use of works that are regarded as creative and economic catalysts, rather than of the protection of such works for the private benefit of authors, painters, or, by extension, test and assessment owners.

Realizing, however, that tests and assessments are a different species of creative work that "are particularly vulnerable to having their utility obliterated by unauthorized disclosure" (137 Cong. Rec., 1991), Congress authorized a special class of secure-test copyright. The ensuing regulations generated by the Copyright Office define a *secure test* as "a nonmarketed test administered under supervision at specified centers on specific dates, all copies of which are accounted for and either destroyed or returned to restricted locked storage following each administration" (37 C.F.R. § 202.20(b)).

For all types of copyright material, the ability to bring a legal action to enforce the right through an injunction or collection of statutorily prescribed damages[1], hinges upon the registration and deposit of the material with the Copyright Office. For the majority of copyright materials, the regulations demand the deposit of a full copy of the material in a publicly accessible archive. The specific purpose of the secure test copyright regulations is to relieve assessment owners of the deposit requirement, allowing them to deposit a portion of the test or a description of its contents in order to maintain confidentiality.[2]

In recognition of the special nature of these secure-test copyrights, however, courts called upon to enforce these rights have been deferential to the point of granting a further exemption from the legal analysis governing whether unauthorized use of assessment content constitutes an allowable or "fair use."[3] To wit, the Copyright Office's chief administrator once observed that "the courts have … been particularly solicitous in protecting these works. Indeed, so far as we are aware, the courts have never upheld a fair use claim advanced by any private entity with regard to copying of secure tests or test questions" (137 Cong. Rec., 1991).

For example, in condemning the publication of practice tests by *The Princeton Review,* which contained material copied from the Scholastic Aptitude Test (SAT) developed by the Educational Testing Service (ETS), the Third Circuit Court of Appeals in *ETS v. Katzman* (1986 p. 543) stated that "the unique nature of secure tests means that any use is destructive of ETS's rights." To emphasize its point, the court approvingly quoted *American Association of Medical Colleges v. Mikaelian* (1983): "The very purpose of copyrighting the... questions is to prevent their use as teaching aids, since such use would confer an unfair advantage on those taking a test preparation course."

Elsewhere, in actions brought to prosecute infringement and to compel disclosure of test content,[4] judicial recognition of a secure test copyright triggers the presumptions that copying or disclosure contradicts the purpose and nature of tests, and eviscerates their "potential market." In *ETS v. Katzman* (1986, p. 543), the court stated that "use of ETS' materials by Review renders the materials worthless to ETS."

The upshot of these regulatory and judicial exemptions from the normal operation of copyright law is the creation of special type of copyright which is deemed to serve the public interest by denying public access.[5] For test and assessment owners, of course, this is a welcome outcome—especially when coupled with the provisions of the Digital Millennium Copyright Act (1998), which provide all copyright owners the legal leverage to interrupt infringement on the Internet before having to institute formal legal proceedings.

Notably, secure test copyright protection is available for all types of tests and assessments, whether high or low stakes, used in education, business or elsewhere. The only prerequisite for obtaining this protection is security. Specifically, security measures reasonably contemplated to prevent assessment content from becoming generally available to the public, or to use the copyright term of art, "published."[6]

The secure-test copyright regulations reference five types of security precautions: (1) tests are not distributed in a way that might compromise their ownership or control (i.e., nonmarketed)[7]; (2) test administration is supervised or proctored; (3) tests are administered at testing centers; (4) tests are administered on an established timeline; and (5) following test administration, authorized copies of the test are closely controlled.

None of these precautions is incapable of being applied to online assessments, however this phalanx of test security measures dim those properties that make online assessments attractive to users in education and business: low-cost, remote, and asynchronous delivery. The central question, therefore, is whether these properties (which differentiate online assessments from "computerized tests") are at odds with the legal conditions of secure testing.

Unfortunately, courts have provided little direct guidance on the issue. In *GMAC v. Cuomo* (1992, pp. 139-140), for example, the court simply summarized that "by its very definition, a secure test is one over which the owner of the copyright does not relinquish ownership or control." Similarly, in *ETS v. Katzman* (1986, p. 536), the court merely observed that "ETS regards the tests as secret until they have been released ... and attempts to maintain strict secrecy with respect to these tests." It is important to note that each court regarded the tests at issue as "secure" based on the use of unspecified measures to control the dissemination of their content and otherwise maintain secrecy, as well as their registration with the Copyright Office as secure tests.

Taken together, these observations support the notion that "control" of assessment content is the threshold for judicial recognition of a test or assessment as a secure test. Because the regulations dissect the meaning of control into several concrete security precautions, the following sections will examine those requirements.

Nonmarketed and Control of Assessments Following Test Administration. The clear premise of the entire regulation is that in the absence of adequate controls, test and assessment content is less likely to be secret or of much value. The Copyright Office's circular, *Copyright Registration for Secure Tests* (U. S. Copyright Office, 2002), explains the regulation's "nonmarketed requirement to mean that "tests are not sold or distributed in a way that compromises their ownership or control."

"Ownership" is an aspect of control that establishes the legal rights to govern the distribution and use of assessment content. To establish ownership, test and assessment owners typically contract with employees and other contributors to obtain unequivocal title to the developed content. To maintain ownership, savvy owners also make it clear at the point of delivery that the test or assessment content is licensed for the exclusive purpose of measurement.

Apart from ownership, the most important aspect of control is ensuring that tests move only in very narrow and closely monitored channels to the locations where they are administered—and back. For many testing programs, this process involves employee screening and training, security detection systems, as well as insular printing facilities, bonded carriers, and storage vaults. Do online assessments (as they are defined for the purposes of this chapter: low cost, remotely, asynchronously delivered via the Internet) satisfy these criteria? Yes and no.

The many excellent online assessment management and delivery platforms that are currently available are unquestionably capable of storing and transmitting test content with little or no erosion of control. In fact, by using one or two centralized servers to deliver assessment content, rather than the distributed networks of servers used in most commercial testing centers, online assessments may be more secure in this regard. Most online assessment platforms can also frustrate attempts to "copy and paste" into other software applications and otherwise capture assessment content.

On the other hand, because the legal precautions surrounding development of test content require contracts and lawyers, they introduce cost elements at odds with the ambition of online assessment to facilitate broader use of tests and assessments by lowering their cost. Whether cost driven or otherwise, close experience indicates that the implementation of these precautions are the exception rather than the rule, even within many high-stakes testing programs.

Supervision at Specified Centers on Specific Dates. To reiterate, *online assessment* is defined, to a large degree, by use of the Internet to facilitate remote and asynchronous test delivery. As the growth of online assessment attests, these characteristics comport with the needs of many assessment users in education and business, chiefly by facilitating closer integration of learning and assessment. The question addressed here is whether this versatile method of test and assessment delivery comports with either the letter or spirit of the regulation's requirements of "supervision at specified centers on specific dates."

Unfortunately, as it is currently practiced, online assessment fails this analysis, chiefly because these "time, place, and manner" restrictions frustrate the versatility for which online assessment is valued. In the absence of effective supervision, for example, the security capabilities of online assessment platforms described in the previous section are largely powerless to prevent manual or electronic recording of assessment content for later distribution.

Supervision-related security precautions, however, are not necessarily incompatible with the goals of online assessment, chiefly because security possibilities too are expanded by the Internet. For example, because online assessments can be administered wherever Internet connectivity exists, supervision and proctoring can be provided by teachers, trainers, human resource managers, dorm supervisors, and others in classrooms, offices and similar venues—in many instances at little or no additional cost.

In sum, to reconcile the vision of making assessment as flexible and common-place as distance learning with eligibility for secure-test copyright status, test and assessment owners should explore point-of-delivery security solutions.

Trade Secret

Test and assessment owners can also look to the law governing trade secrets for the protection of their content. Unlike copyright, registration is not required to qualify for special protection, and generally any information which derives value from secrecy qualifies as a trade secret.

Until recently, trade secret protection was the exclusive province of state law, typically, through the adoption by individual states of the Uniform Trade Secrets Act (UTSA, 1985). The UTSA, defines a *trade secret* as

> *information, including a formula, pattern, compilation, program device, method, technique, or process, that: (i) derives independent economic value, actual or potential, from not being generally known to, and not being readily ascertainable by proper means by, other persons who can obtain economic value from its disclosure or use, and (ii) is the subject of efforts that are reasonable under the circumstances to maintain its secrecy.* (UTSA, National Conference Of Commissioners On Uniform State Laws, 1985, § 1(4)).

The UTSA provides protection from the misappropriation of assessment content by "improper means," such as theft, bribery, and so forth, and by "wrongful acquisition" (i.e., acquisition by a person who knows or has reason to know that the trade secret was acquired by improper means; 1985, § 1(1)-(2)).[8]

In addition to the injunctive relief and damages available under the UTSA, in 1996 Federal criminal sanctions were added to the list of possible penalties for the theft or misappropriation of trade secrets. The Economic Espionage Act of 1996 (EEA) criminalizes the theft of trade secrets for the purpose of gaining economic or commercial advantage. Individuals found guilty under the EEA can be imprisoned for up to 10 years and fined $250,000 or both, and organizations can be fined up to $5 million. Notably, in 2002 the U.S. Justice Department relied upon the EEA to obtain a guilty plea from an Internet retailer of misappropriated Microsoft certification examinations (U.S. Dept. of Justice, 2002).

The latter result notwithstanding, the few state courts that have addressed the question have shown little desire to extend trade-secret protection to test and assessment content in the specific context of requests, under public record laws, for disclosure of state educational tests. Courts in Ohio and Arizona, specifically, have held that tests and assessments are not "trade secrets" within the meaning of the UTSA, and that the routine administration of tests and assessments—even under secure testing procedures—is fatal to their secrecy (*Phoenix Newspapers, Inc., v. Keegan*, 2001; *State ex rel. Rea v. Ohio Dept. of Education*, 1998).[9]

Apart from these isolated holdings,[10] however, the question of trade secret protection for tests and assessments is an open question in most jurisdictions and is routinely asserted by major test publishers.[11] Applying the provisions of the UTSA and the EEA, the question of greatest moment for this discussion is whether the security efforts surrounding the creation and administration of tests and assessments are reasonably calculated to maintain secrecy.

In its commentary accompanying the UTSA, the National Conference of Commissioners on Uniform State Laws (Commissioners; 1985) noted that "the efforts required to maintain secrecy are those 'reasonable under the circumstances.'"

> *The courts do not require that extreme and unduly expensive procedures be taken to protect trade secrets against flagrant industrial espionage. ... It follows that reasonable use of a trade secret including controlled disclosure to employees and licensees is consistent with the requirement of relative secrecy.* (UTSA, 1985, pp. 7-8)

In what may be the clearest piece of guidance on the topic of relative secrecy, in a dissent lodged in *State ex rel. Rea v. Ohio Dept. of Education* (1998), Justice Lundberg-Stratton approvingly recited the secure test copyright regulation (i.e., nonmarketed, supervised at specific centers on specific dates) in relation to the appropriate level of security for the State's educational test. Although not binding, the analysis mirrors the opinion of the Commissioners that the intent of the UTSA is relative rather than absolute secrecy, allowing trade secret material to be used (licensed) under appropriate security without compromising its protection. The secure test Copyright Regulations, therefore, may be the best authority regarding the security measures necessary to create and maintain trade secret protection.

Because trade secret and secure-test copyright requirements arguably converge on the ownership and control requirements contained in the secure test copyright regulation, there is little need to renew the prior analysis of threshold security measures for online assessment. It may be useful to however to review how, like secure test copyright status, trade secrets can be lost through the negligence or inaction of assessment owners.

Among the "proper means" by which trade secrets can be obtained by another and lost by their owners are "discovery under a license from the owner of the trade secret," whereby the owner fails to attach confidentiality and nondisclosure provisions to the license; and "observation of the item in public use or on public display" (UTSA, 1985, p. 7). The application of these proper means of obtaining trade secret information importantly, does not require that information be generally known to the public; only that individuals or organizations who can obtain economic benefit from the information are aware.

Whether explicit or implicit, the administration of tests and assessments involves licensing to, among others, distributors, instructors and test takers to view and use their content. In the absence of agreements requiring confidentiality and nondisclosure, those same individuals can record and use the content without running afoul of trade secret law. With regard to the loss of trade secrets through public display, online assessment owners should note that the phrase contemplates an absence of screening and supervision.

The essence of the relative secrecy standard in other words, is that "except by the use of improper means, there would be difficulty in acquiring the information" (RESTATEMENT OF TORTS, 1991). To ensure that relative secrecy is maintained, online assessment owners should, according to one observer "implement physical measures and procedural practices to guard against unwitting disclosure, reduce the likelihood of theft, and raise employee awareness regarding the need to safeguard information" (Friedman & Papathomas, 2002).

Conclusion

At a time when information, including purloined test and assessment content, moves at Internet speed, it is a comfort to know that the law recognizes that secrecy and security are perquisites of valid and reliable tests, and also provides meaningful protection to those test and assessment programs which possess a modicum of each.

Online assessments owners are potential beneficiaries of trade secret and heightened secure test copyright protection. Applying a threshold security standard derived from copyright and trade secret law and its interpretation, the analyses in the prior sections have shown that in its widely practiced and current form, online assessment delivery fails to satisfy the threshold standard.

This result suggests that test and assessment owners should candidly explore in the first instance, whether additional test security would further their goals by increasing confidence in test and assessment results. Many formative assessments with very limited audiences perhaps, require little or no security. Test and assessment owners should further consider, however, whether the level of interest surrounding a test or assessment and its results, as well as investments in the creation and distribution of the instrument, warrant heightened security and legal protection against its theft, dissemination and use. Finally, test and assessment owners should consider whether the benefits of online delivery including broader access, remote and asynchronous delivery possibilities outweigh the above concerns.

Clearly, online delivery of tests and assessments is here to stay. However members of the measurement community need to understand its security limitations, and the consequences of those limitations, before deciding to put tests and assessments online, and to assist in formulating the next generation of online assessment.

References

37 C.F.R § 202.20(b)(4)

137 Cong. Rec. S13923 (daily ed. Sept. 27, 1991)

American Association of Medical Colleges v. Mikaelian, 571 F. Supp. 144, at 153 (E.D. Pa. 1983)

College Board v. Pataki, 889 F. Supp. 554 (N. Dist. N.Y.1995)

Digital Millennium Copyright Act, 17 U.S.C §512 (1998)

Economic Espionage Act, 18 U.S.C. §1832 (1996)

ETS v. Katzman, 793 F.2d 533, at 543 (1986)

Friedman, M., & Papathomas, K. (2002). *Primer on trade secrets.* Goodwin Proctor.

Graduate Management Admissions Council (GMAC) v. Cuomo, 788 F. Supp. 134 (N.D.N.Y. 1992)

Kriess, R. (1996). Copyright fair use of standardized tests. 48 *Rutgers L. Rev.* 1043

Nimmer, M. B., & Nimmer, D. (2003). *Nimmer on Copyright.* New York: Matthew Bender.

Phoenix Newspapers, Inc., v. Keegan, 35 P.3d 105 (Ariz. App. 2001)

Restatement of Torts §757 (1991)

State ex rel. Rea v. Ohio Dept. of Education, 692 N.E.2d 596 (Ohio 1998)

Uniform Trade Secrets Act, National Conference Of Commissioners On Uniform State Laws (1985)

U.S. Copyright Office. (2002, June). *Circular 64: Copyright Registration for Secure Tests*. Washington, DC: Author.

U.S. Dept. of Justice. (2002, August 23). Former Vancouver, Washington, resident pleads guilty to theft of trade secrets from Microsoft Corporation. Press release available online from *http://www.cybercrime.gov/keppelPlea.htm*

Endnotes

[1] $750 to $30,000 per infringement, up to $150,000 per infringement if "willful."

[2]   Specifically, the Copyright Office's *Circular 64: Copyright Registration for Secure Tests* (June, 2002), provides that "the examiner will make a preliminary examination of the complete test in your presence and immediately return the test to you, retaining adequate identifying portions of the deposit of the work to create an archival record." The statutory and constitutional authorization for these regulations was upheld in *National Conference of Bar Examiners v. Multistate Legal Studies, Inc., 692 F.2d 478, 482-87 (7th Cir. 1982), cert. denied, 464 U.S. 814.*

[3]   17 U.S.C. § 107 specifies four factors for the court to consider in determining whether a use is fair: (1) the purpose and character of the use, including whether such use is of a commercial nature or is for nonprofit educational purposes; (2) the nature of the copyrighted work; (3) the amount and substantiality of the portion used in relation to the copyrighted work as a whole; and (4) the effect of the use upon the potential market for or value of the copyrighted work.

[4]   State agencies, newspapers and others interested in policing standardized tests have sued to compel disclosure by testing organizations. See, for example, *College Board v. Pataki*, 1995.

[5]   Stated another way, the treatment of secure tests turns copyright law on its head by using a legal apparatus designed to promote access to creative works into a mechanism for maintaining secrecy. Kriess (1996) persuasively argued, however, that the courts have got it wrong; copyright law was not intended to reinforce secrecy but to ensure access by rewarding creators. Viewed through the lens of copyright's "primary objective"

argues Kriess, copyright law eschews secrecy in favor of access—especially where, as with test content, the private gain derived from copyright protection, outweighs the public benefit derived from access.

⁶   In *GMAC v. Cuomo* (1993, pp. 140-141), the court stated that "to concede that GMAC's tests are secure is likewise to concede that they are unpublished works within the meaning of the Act … the mere fact that GMAC administers the test, without more, is not enough to convert it into a published work for purposes of the Act."

⁷   A test is "nonmarketed" if copies are not sold and the test is distributed and used in such a manner that ownership and control of copies remain with the test sponsor or publisher.

⁸   Section 1(2) of the Uniform Trade Secrets Act provides:

> *(i) acquisition of a trade secret of another by a person who knows or has reason to know that the trade secret was acquired by improper means; or (ii) disclosure or use of a trade secret of another without express or implied consent by a person who (A) used improper means to acquire knowledge of the trade secret; or (B) at the time of disclosure or use knew or had reason to know that his knowledge of the trade secret was (I) derived from or through a person who has utilized improper means to acquire it; (II) acquired under circumstances giving rise to a duty to maintain its secrecy or limit its use; or (III) derived from or through a person who owed a duty to the person seeking relief to maintain its secrecy or limit its use; or (C) before a material change of his position, knew or had reason to know that it was a trade secret ad that knowledge of it had been acquired by accident or mistake.*

⁹   In *Phoenix Newspapers, Inc. v. Keegan* (2001), citing the amicus curiae brief of the Association of Test Publishers (ATP) the Arizona Court of Appeals held that test content is not a trade secret in the sense that test content alone does not reveal "the methodology for engineering the test."

> *A competitor would have to see the technical manual, setting forth all of the research associated with the test, before it would gain any meaningful information for competitive test-building purposes; thus, merely seeing the specific items would not disclose the trade secret, i.e., the methodology for*

engineering the test, what items are correlated to what subject matter content and for what level of difficulty. Absent full disclosure of the full research information, the kind of limited disclosure that was provided through the Arizona inspection process would not have resulted in any serious concern by the test developer.

In *State ex rel. Rea v. Ohio Dept. of Education* (1998), the Ohio Supreme court found that "the placement and use of the test within the public educational domain is sufficient to constitute public release under the statute." In a lengthy dissent, Justice Lundberg-Stratton stated that

The majority stretches to conclude that once VIML developed the [the test] for the public purpose of evaluating school students pursuant to state and federal requirements, the test was effectively disseminated into the public domain, thereby losing any trade secret status it might have had.

[10] In fact, the trial court in *Phoenix Newspapers, Inc. v. Keegan* (2001) specifically rejected the reasoning in Rea v. Ohio DOE, stating that "disclosure as part of the test taking process does not destroy their independent economic value … Disclosure of test questions as a part of the testing process has everything to do with the use of the test as a test."

[11] Harcourt Assessment, Inc., for example, unequivocally asserts on its website that "Harcourt considers its secured tests to be trade secrets" (see *http://harcourtassessment.com/haiweb/Cultures/en-US/Footer/Legal+Policies.htm*).

Chapter XI

Accessibility of Computer-Based Testing for Individuals with Disabilities and English Language Learners within a Validity Framework

Eric G. Hansen, Educational Testing Service (ETS), Princeton, USA

Robert J. Mislevy, University of Maryland, College Park, USA

Abstract

There is a great need for designers of computer-based tests and testing systems to build accessibility into their designs from the earliest stages, thereby overcoming barriers faced by individuals with disabilities and English language learners. Some important potential accessibility features include text-to-speech, font enlargement and screen magnification, online dictionaries, and extended testing time. Yet accessibility features can, under some circumstances, undermine the validity of test results. Evidence centered assessment design (ECD) is offered as a conceptual framework— providing sharable terminology, concepts, and knowledge representations— for representing and anticipating the impact of accessibility features on validity, thus helping weigh the consequences of potential design alternatives for accessible computer-based tests and testing systems.

Introduction

Computer-based tests—including Web-based tests—are likely to become much more common in the future, and it is important that they be designed in a way to be as accessible as possible to individuals with disabilities or who are English language learners.[1] Examples of features that might be considered for such systems include built-in text-to-speech (speech synthesis) with visual highlighting as text is read aloud, font enlargement, screen magnification, color and contrast modification, spelling and grammar checkers, dictionaries, extended testing time, and compatibility with external assistive technologies such as screen readers and refreshable braille displays.[2,3] However, accessibility features that may be useful in overcoming accessibility barriers can, in some instances, invalidate the results of tests. For example, a person with a spelling disability (dysorthographia) could argue that his or her use of spell-checker software would help overcome an accessibility barrier on educational tests that involve writing. Yet, if a test is intended to measure *spelling* ability, then such an accommodation will tend to invalidate the test results by providing an unfair advantage for the person who uses that feature.[4] As we will see, it is not always easy to identify the impact of an accessibility feature on the validity of test results.[5,6] There is clearly a need for a conceptual framework for determining how accessibility features impact validity, thereby clarifying decisions about: which features to provide with computer-based testing systems, whether to build or buy those features, and how much control to allow to test takers in the use of those features.

Purpose

The purpose of this chapter is to sketch out a conceptual framework—a *validity framework* that can help clarify the relationships between accessibility features and validity, thereby clarifying possible strategies for increasing accessibility without undermining validity. The first sections of this chapter lay out key concepts in the framework, and the latter sections apply the framework to considerations in computer-based testing.

In terms of the design of accessible computer-based testing systems, this chapter focuses on *laying the groundwork for establishing requirements* for computer-based testing that is more accessible.[7] Thus, the focus is *not* on development of a list of necessary accessibility features per se, but rather on a framework for evaluating possible accessibility-related features for computer-based tests and testing systems. It is hoped that such a framework can help design computer-based testing systems that will be flexible and powerful enough to be used for computer-based testing of individuals with a wide range of profiles of disability (or nondisability) or language status in many different subject areas. While many of the examples used in this chapter are relatively simple, they serve to illustrate key principles and concepts.

Following is a list of the remaining sections of this chapter:

- "An Overview of the Framework" provides an example of the need for such a framework and a brief overview of the framework, with special attention paid to a way of being more precise about what one intends to measure.

- "Basic Reasoning About Accessibility Features" outlines some key basics in reasoning about accessibility features.

- "Universal Design of Assessment" describes this concept and relates it to the concepts described in this chapter.

- "Toward a Common Understanding" provides a hypothetical discussion between team members working toward a design for an accessible computer-based testing system. This discussion underscores how assessment design inevitably requires deciding among tradeoffs and that a conceptual framework can help clarify the nature of the tradeoffs.

- "Conclusion" provides conclusions and recommendations.

This chapter draws upon material in other works (e.g., Hansen & Mislevy, 2004, in press; Hansen, Mislevy, & Steinberg, 2003; Hansen & Steinberg, 2004).

An Overview of the Framework

A National Research Council committee that examined accommodation policies for the United States' National Assessment of Educational Progress (NAEP)

and other large-scale assessments reported that determining which accommodation is right for particular circumstances is difficult.

> *The accommodation must at the same time be directly related to the
> disability or lack of fluency for which it is to compensate and be
> independent of the constructs on which the student is to be tested.*
> (National Research Council, 2004, p. 7)

An example of this difficulty concerns the use of the *readaloud* accessibility
feature, that is, having test content read aloud to the test taker by a live reader,
by synthesized speech (text-to-speech), or by prerecorded audio.[8] The NAEP
does *not* allow test content to be read aloud on their reading assessment,
considering that accommodation incompatible with the nature of the proficiency
being measured (i.e., reading).[9] (The assessment is not computerized.) NAEP
allows the readaloud accommodation by a live reader on their math assessment.[10] Yet the readaloud accommodation is *allowed* on several state assessments of reading. This discrepancy underscores the issue of identifying principles, criteria, or procedures for determining what accessibility features can be
offered without compromising the validity of the scores.

The challenge of determining the validity of the readaloud feature on assessments of reading may be indicative of current limitations in research knowledge.
A committee of the National Research Council that examined accommodation
policies for the NAEP and other large-scale assessments reported that "overall,
existing research does not provide definitive evidence about which procedures
will, in general, produce the most valid estimates of performance for students
with disabilities and English language learners" (National Research Council,
2004, p. 6).[11]

Citing as a resource the work of Hansen, Mislevy, Steinberg, and Almond (e.g.,
Hansen et al., 2003; Hansen & Steinberg, 2004; Mislevy, Steinberg, & Almond,
2003) in what is termed *evidence-centered design* (ECD), the committee then
provided "a very basic introduction to the development of an inference-based
validation argument" (National Research Council, 2004, p. 105). They concluded
by urging sponsors of large-scale assessment programs to "identify the inferences that they intend should be made from its assessment results" and to
"embark on a research agenda that is guided by the claims and counterclaims for
intended uses of results in the *validation argument* they have articulated"
(National Research Council, 2004, p. 122, emphasis added).

Evidence-Centered Assessment Design

Evidence centered assessment design (ECD) was originally formulated at Educational Testing Service (ETS) by Mislevy et al. (2003) and may be seen as part of a tradition in educational assessment that revolves around validity arguments (Cronbach & Meehl, 1955; Kane, 1992; Messick, 1989, 1994; Spearman, 1904).[12] ECD seeks to make more explicit the evidentiary argument embodied in assessment systems, thereby clarifying assessment design decisions. The key idea of ECD is to lay out the structures and supporting rationales for the evidentiary argument of an assessment. By making the evidentiary argument more explicit, the argument becomes easier to examine, share, and refine. ECD may be regarded as an "evidence-based" approach. Evidence-based approaches, as the full title indicates, have proven useful in law, science (e.g., medicine, natural resource exploration), intelligence analysis, and many other fields. Principles of logic, reasoning, and probability figure prominently in evidence-based approaches.

Recently, ECD has been extended by Hansen, Mislevy, and Steinberg to reason about how validity is affected by accessibility features provided to students with disabilities and English language learners (Hansen & Mislevy, 2004; Hansen, Mislevy, Steinberg, Lee, & Forer, 2005; Hansen, Mislevy, & Steinberg, in press). These extensions include structures for representing and reasoning about a core validity issue, specifically, the alignment between (a) what one intends to measure and (b) what one is actually measuring in operational settings, where alignment favors validity and misalignment favors invalidity. There is a great need to apply this approach in the area of computer-based testing (Hansen et al., 2004). With the continued advances in computer technology and its relentless infusion into the world of work (Bennett, 1995, 2002), computer-based testing is an increasingly important area in which to examine and apply these ECD extensions.

The ECD Framework

The essence of the ECD framework is captured in a quote from Messick (1994), who wrote,

> *[We] would begin by asking what complex of knowledge, skills, or other attributes should be assessed, presumably because they are tied to explicit or implicit objectives of instruction or otherwise valued by society. Next, what behaviors or performances should*

reveal those constructs, and what tasks or situations should elicit those behaviors. (p. 17)

Thus, the details of the ECD framework are intended to support this basic concept of having a well-defined construct (i.e., targeted proficiency) and trying to make clear and explicit the nature of the performances that should reveal those constructs and the tasks and situations that are likely to elicit these performances. As we consider issues of accessibility, we are additionally aware that if we are to obtain the same "revelation" of those constructs for individuals with disabilities or English language learners, then the tasks or situations may need to be altered in order to overcome accessibility barriers. At the same time, we need to ensure that the removal of an accessibility barrier does not simultaneously result in an unfair advantage for the individual using the accessibility feature. ECD seeks to provide a logical framework for understanding conditions that are favorable and unfavorable to test-score validity for individuals with disabilities and English language learners.

Layers of the Framework. The ECD framework may be described as consisting of four layers of design and delivery of an assessment system, specifically, three layers for *design* and one layer for *delivery*. This material draws upon separate works that focus on individuals with disabilities (Hansen et al., 2005) and English language learners (Hansen & Mislevy, 2004, in press) as well as earlier work not specifically focused on subpopulations (Mislevy et al., 2003).

Figure 1 shows the layers of assessment design and delivery, as depicted in the ECD framework. These layers are then briefly described thereafter.[13]

The four layers depicted in Figure 1 are as follows: Layer A represents Domain Analysis, in which one gathers information about the domain (e.g., the nature of proficiency, situations in which the proficiency is used in the real-world). Layer B represents Domain Modeling, in which one organizes information about the domain in terms of assessment arguments. In Layer C, one creates the Conceptual Assessment Framework (CAF), which is a blueprint for the "machinery" of operational tests. Layer D represents the processes of an operational assessment. This includes, of course, the events that occur *during* test administration, labeled D2 in the figure. Equally important to the assessment argument are events that take place *before* test administration, labeled D1, (e.g., determination of eligibility for accommodations, provision of familiarization materials such as those found in test bulletins), and those that take place *afterwards,* labeled D3 (e.g., score reporting).

Layers A and D are probably the most familiar to people who are not experts in testing. Layer A seeks to define the nature of the proficiency, drawing upon

Figure 1. Some relationships between layers of assessment design and assessment operation. The three double-headed hashed arrows are intended to signify (a) that domain modeling informs the development of the operational assessment delivery system and (b) that issues that one expects to arise in the various phases of the delivery system inform the domain modeling process. The three single-headed solid black arrows signify that the specifications for assessment construction detailed in the CAF address all three phases of the delivery system. (From Hansen and Mislevy, in press)

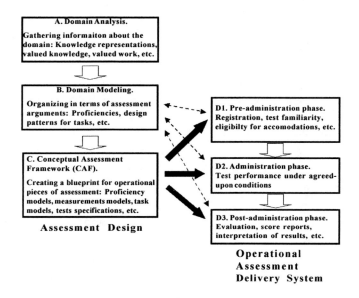

experience and intuition about the nature of proficiency in a domain, especially "how you know it when you see it." Layer D is the operational process of testing, wherein are the processes of delivering items, scoring performances, reporting scores, and so on. The challenge is getting from a desire to carry out assessment somehow grounded in Layer A and implemented in the processes of Layer D, in a way that is fair, efficient, and valid.

The Importance of Layer B: Domain Modeling

Layer B is the key. This is where the assessment argument is structured. An assessment argument can be summarized as comprising (a) a claim about a person possessing at a given level a certain targeted proficiency; (b) the data (e.g., scores) that would likely result if the person possessed, at a certain level, the targeted proficiency; (c) the warrant (or rationale, based on theory and experience) that tells why the person's level in the targeted proficiency would lead to occurrence of the data; and (d) "alternative explanations" for the person's high or low scores (i.e., explanations other than the person's level in the targeted proficiency). The existence of *alternative explanations* that are both significant and credible might indicate that *validity* has been *compromised* (Messick, 1989). Much of the analysis that is the focus of this chapter has to do with these alternative explanations, factors that can hinder an assessment from yielding valid inferences. When such alternative explanations are recognized at the earliest stages of test design, then rework and retrofitting can be avoided.

Basic Reasoning About Accessibility Features

This section deals with the basics of reasoning about accessibility features, addressing several issues that are associated with Layer B (Domain Modeling) in the ECD framework. Efforts in Level B involve determining the set of features that will be offered, including the levels of control over those features that will be offered to test administrators and to examinees. It should be noted that the steps outlined below may be carried out iteratively and perhaps in a different order.

Consider the Target Population and Their Needs and Preferences

One may begin by considering the diversity that exists within the target population and the range of their accessibility needs and preferences. Consider, for example, the seven student profiles (e.g., disability and language status groups) shown in column 1 of Table 1.[14] Are there any of these profiles that would *not* be found within the target population of a specific test? If one were developing a professional exam for airline pilots, there may be *no* individuals who are completely blind (the first and second student profiles). However, for most educational tests, few of the following profiles would be *not* found among a large sample from the target population.[15]

Notice that each student profile (row) may be described by the range of values for various knowledge, skills, and other attributes (KSAs) (columns) that may be found within that profile. To keep the illustration simple, individuals with complex

Table 1. KSA values for seven student profiles

Student Profile	KSA and Its Possible Values						
	Reason (poor, okay, good, very good)	Know math vocabulary in English (none, poor, okay, good, very good)	Know nonmath vocabulary in English (none, poor, okay, good, very good)	Decode (poor, okay, good)	See (no, partial, yes)	Hear (no, yes)	Know braille codes (no, yes)
1. Blind and able to read braille	Okay or good or very good	Poor or okay or good or very good	Okay or good or very good	Okay or good	No[a]	Yes	Yes[a]
2. Blind, nonreader of braille	Okay or good or very good	Poor or okay or good or very good	Okay or good or very good	Okay or good	No[a]	Yes	No[a]
3. Low vision	Okay or good or very good	Poor or okay or good or very good	Okay or good or very good	Okay or good	Partial[a]	Yes	No
4. Deaf	Okay or good or very good	Poor or okay or good or very good	Okay or good or very good	Okay or good	Yes	No[a]	No
5. Dyslexia	Okay or good or very good	Okay or good or very good	Okay or good or very good	Poor[a]	Yes	Yes	No
6. Nondisabled English language learner	Okay or good or very good	Poor or okay or good or very good	Poor[a]	Okay or good	Yes	Yes	No
7. Nondisabled native speaker of English	Okay or good or very good	Poor or okay or good or very good	Okay or good or very good	Okay or good	Yes	Yes	No

profiles (e.g., multiple disabilities or disabilities coupled with English language learner [ELL] status) are not considered in this table.

Note that Table 1 characterizes English language learners as having poor knowledge of nonmath vocabulary in English but poor-or-better knowledge of math vocabulary in English. While this characterization is a simplification, it does capture the idea that English language learners (in English speaking countries) often receive academic instruction in English and therefore may have greater knowledge of English academic-content vocabulary (e.g., math vocabulary) than knowledge of English noncontent vocabulary.

Now consider, as detailed in Table 2, the range of accessibility features that might be helpful or necessary in overcoming accessibility barriers faced by individuals having the various student profiles.[16] Here we are concerned with information access in general rather than access to tests per se. Text-to-speech is important to individuals who are blind or have dyslexia. Braille is important for individuals who are blind and read braille. Visual alteration (enlargement and

Table 2. Information access features likely to be desired or needed by individuals with different student profiles

Student Profile	Type of Accessibility Feature				
	1. Visually Displayed Text	2. Visual Alteration (Enlargement, Contrast, Color) of Text or Graphics	3. Braille and Tactile Graphics	4. Readaloud/ Audio (Text-to-Speech, Prerecorded Audio, Live Reader)	5. Linguistic Helps (Dictionary, Spell-Checker, Linguistic Simplification and Alteration)
1. Blind and able to read braille			High	High	
2. Blind, nonreader of braille				High	
3. Low vision	High	High		Medium	
4. Deaf	High				High
5. Dyslexia	High	Medium		High	High
6. Nondisabled ELL	High			Medium	High
7. Nondisabled native speaker of English	High				

Note. "High" indicates that the feature (or class of features; columns numbered 1 through 5) is likely to be desired or needed for providing access to general computer-based information (not tests per se) for many individuals with the specified given student profiles (rows numbered 1 through 7). "Medium" indicates a lesser likelihood. The table is not intended to indicate which features should be allowed for any given computer-based test.

alteration of contrast and color) is important for individuals with low vision. Linguistic helps, including dictionary, spell checker, linguistic simplification, or presentation of content via manual communication (e.g., American Sign Language) may be helpful for individuals with hearing disabilities, dyslexia, or who are English language learners. As suggested previously, the use of such accessibility features within *test settings* may not necessarily be *conducive to validity*.

Define the Targeted Proficiency

Perhaps the most important activity during Layer B (Domain Modeling) is that of defining what one intends to measure—essentially the "claim" that one wishes to make about what a person knows or can do. The outcome that we typically want when we administer an educational test is an accurate estimate of some targeted proficiency. To design an assessment to measure the targeted proficiency, we need to be as precise as we can in its *definition*.

We begin by defining targeted proficiency as consisting of one or more *focal KSAs*. A *focal* KSA is a KSA that is an *essential part* of what one wishes to measure. Thus, a person *must possess all the focal KSAs* in order to be considered as possessing a "good" (as opposed to a "poor") level in the targeted proficiency. On the other hand, any KSA that is *not* a focal KSA is an *ancillary* KSA.[17] For example, a very simple definition of reading comprehension might specify two focal KSAs—*comprehend* and *decode*. The ability to "see," as well as many other abilities, would be ancillary KSAs.[18] Designating certain KSAs as ancillary adds precision to the definition, beyond that provided by merely identifying focal KSAs.

It is important to note that the targeted proficiency is *defined* (or specified) rather than *estimated, calculated,* or *discovered*. The process of defining a targeted proficiency should generally be informed by considerations such as the KSAs actually needed to perform well in criterion situations.[19] In general, the decision or choice about the definition of the targeted proficiency should be based on multiple considerations, including the assessment's purpose, validity, fairness, and feasibility.

Although the binary *focal/ancillary* distinction is sufficient for some purposes, further refinements to the definition are very helpful. We do this by further defining the *level of the focal KSAs*.[20] Let us consider Table 3, which portrays the case of Omar, an English language learner who receives a math test in default (standard) conditions. Column 1 of Table 3 lists several KSAs and their possible levels (e.g., "Reason" has four levels: poor, okay, good, very good). Column B-1 shows that each of these KSAs has been designated as either a focal KSA or an ancillary KSA. Now note that column B-2 that for each *focal* KSA, a *specific level* has been designated. For example, for someone to be considered as having a "good" level of the *targeted proficiency*, he or she must possess at least "good" reasoning, "okay" knowledge of math vocabulary in English, and "poor" knowledge of nonmath vocabulary in English.[21] (Table 3 will be explained in greater detail later in this chapter.)

Identify Test-Taker KSA Levels

Next we identify the test-taker's levels in the various KSAs. In this form of analysis, we do this in "what-if" fashion, as though we knew these values. Of course, in an operational use of a test, not all this information would be known. For example, the test-taker's levels in the targeted proficiency (which is composed of focal KSAs) would generally not be known, because inferring that

Table 3. Derivation of the false-negative outcome for a math test administered to "Omar" (an English language learner) under default conditions.[a]

A. KSAs and Possible Levels [Input]	B. Definition of the Targeted Proficiency [Input]		C. Test-taker KSA Levels. What are the KSA levels of the test taker? (In "what-if" analyses, we treat these as if known) [Input]	D. Possession of Focal KSAs. Does the test taker possess the focal KSA (see column C) at the levels specified in the definition of the targeted proficiency (column B-2)? ("Yes", "No"; else "n/a" where KSA is ancillary) [Intermediate output]	E. Requirements of the Testing Situation. What level of the KSA is required of the test taker in order to have "good" effective proficiency in this operational testing situation? (This is "n/a" if the KSA is "not required," i.e., if the lowest level of the KSA is sufficient) [Input]	F. Satisfaction of Requirements. Does the test-taker's ability (column C) satisfy the requirement (column E)? ("Yes", "No"; else "n/a" where column E is "n/a") [Intermediate output]
	1. "Focal" or "Ancillary" KSA	2. Level, If Focal KSA (Level of focal KSA; else "n/a" where KSA is ancillary)				
1. Reason (poor, okay, good, very good)	Focal	Good	Good	Yes	Good	Yes
2. Know math vocabulary in English (none, poor, okay, good, very good)	Focal	Okay	Okay	Yes	Okay	Yes
3. Know nonmath vocabulary in English (none, poor, okay, good, very good)	Focal	Poor	Poor	Yes	Okay[b]	No
4. Decode (poor, okay, good)	Ancillary	n/a	Okay	n/a	Okay	Yes
5. See (no, partial, yes)	Ancillary	n/a	Yes	n/a	Partial[c]	Yes
6. Hear (no, yes)	Ancillary	n/a	Yes	n/a	Yes[d]	Yes
7. Know braille codes (no, yes)	Ancillary	n/a	No	n/a	n/a	n/a

G. Level in the Targeted Proficiency ("Good" if column D has zero No's and at least one Yes; otherwise "poor") [Intermediate output] →	H. Effective Proficiency ("Good" if column F has zero No's and at least one Yes; otherwise "poor") [Intermediate output] →
Good	Poor

I. Outcome. Result of comparing the test-taker's level in the targeted proficiency with the effective proficiency [Output] →
False-negative

[a] Default conditions, in this case, include presentation via visually-displayed test (instead of audio or braille) and without any linguistic helps (dictionaries, etc.).

[b] This requirement for knowledge of nonmath vocabulary in English ("good") is an example of an excessive focal requirement.

[c] A person must have at least partial vision in order to see the test.

[d] In this case, the requirement for hearing derives from spoken test directions from the proctor.

proficiency is the job of the test. Other information might be known, having been established, for example, by documentation about the test-taker's disability or language status.

As mentioned previously, we are examining the case of Omar, an English language learner who receives a math test in default (standard) conditions. Default conditions include presentation via visually displayed test (instead of audio or braille) and without any linguistic helps (e.g., dictionaries). As shown in column C of Table 3, Omar has "good" reasoning, "okay" knowledge of math vocabulary in English, and (per our definition of "English language learner") a "poor" knowledge of nonmath vocabulary in English; as one can see from column B, these are focal KSAs. On the other hand, with regard to the ancillary KSAs, Omar has "okay" decoding; has the senses of sight ("see = yes") and hearing ("hear = yes") but does not know braille codes ("know braille codes = no").

Determine if the Test Taker Possesses the Targeted Proficiency

If the test taker possesses the focal KSAs at levels specified in the definition of the targeted proficiency, then he or she possesses a "good" level in the targeted proficiency. In Table 3, this can be ascertained by examining all the rows in column D, which displays, for each focal KSA, whether the test-taker's level (column C) meets-or-exceeds the level specified in the definition of the targeted proficiency (column B-2). Specifically, if in column D there are zero No's and at least one Yes, then test-taker's level in the targeted proficiency is "good" and otherwise the level is "poor." In this case, we have zero No's and three Yes's, yielding a "good" level in the targeted proficiency, as shown immediately to the right of the cell labeled "G. Level in the Targeted Proficiency."[22] As noted earlier, in an operational testing situation (as opposed to a "what-if" analysis), this value would be inferred rather than supplied as a given.

Identify the Requirements Imposed by Operational Testing Situations

We now need to examine the KSAs that are *required* (or demanded) for good performance in operational test settings. A KSA that is a *requirement* is one that the test taker must *satisfy* in order to *perform well* in a specific *operational* test setting. Analyzing features of tasks for their impact on requirements for various KSAs is an important enterprise that may include empirical investigation (Sheehan & Ginther, 2001); however, for the purposes of this chapter, it is sufficient to focus on the end product of that effort, an understanding of the

KSAs required to perform well. It is important to note that the investigation of requirements is distinct from (but related to) the effort to define the construct (which distinguishes between focal and ancillary skills).

Requirements for focal KSAs are called "focal requirements" and requirements for ancillary KSAs are called "ancillary requirements."[23] Column E of Table 3 shows requirements for a math test administered in default conditions (e.g., visually displayed text as opposed to audio or braille).[24]

Focal Requirements

We naturally expect a well-constructed operational test to impose *focal requirements* upon the test taker—after all, that is its job. In other words, if we have declared that *math vocabulary* is a *focal* KSA for a math test, then we naturally expect the test items (questions) to impose a *focal requirement* for knowledge of math vocabulary. Of course, a convenient way to impose this focal requirement is *by using math vocabulary* in the test items.[25]

Ancillary Requirements

The targeted proficiency (composed of focal skills) is what we are really interested in measuring, but is rarely possible in educational testing to obtain evidence about the targeted proficiency without requiring some ancillary skills by which the test taker will apprehend tasks or provide responses. For example, tests in which seeing, hearing, and typing are defined as ancillary skills, then operational testing conditions may well impose requirements (i.e., *ancillary requirements*) for these ancillary skills due to the necessity of apprehending the tasks (e.g., see, hear) and of recording one's answers via the computer keyboard.[26]

Ancillary requirements come in great variety and, in general, the object of accessibility efforts is to ensure that each test taker can satisfy the ancillary requirements with which he or she is faced. It must be emphasized that defining a specific skill that a test requires (e.g., seeing, hearing, typing) as ancillary or focal depends strictly on the definition of the targeted proficiency.

Determine if the Requirements are Satisfied

Next we determine if the focal and ancillary requirements imposed by operational testing conditions are *satisfied* by the test-taker's abilities. As shown in column F of Table 3, we can ascertain this by determining whether the test-taker's level meets or exceeds the level required in column E. Column F of Table 3 shows that "Omar" satisfies the focal requirements for reasoning, and knowledge of *math* vocabulary but not for the focal requirement for knowledge of *nonmath* vocabulary. Omar satisfies the ancillary requirements for decoding, sight, and hearing.

Determine Effective Proficiency

Once we have reasoned about each individual KSA as discussed in previous steps, we then determine what we term "effective proficiency," which is the *capability* of the test taker to perform well when faced by the combination of requirements (focal or ancillary) imposed by a specific testing conditions.[27] Thus, while the person's targeted proficiency is what one *intends to measure*, the person's effective proficiency is basically what one *actually measures* under a specific set of operational testing conditions.[28] A person is said to have a good (as opposed to a poor) level of effective proficiency if their own levels of KSAs satisfy both the focal requirements and the ancillary requirements. Otherwise, he or she possesses a poor level of effective proficiency.[29]

What level of effective proficiency results from the situation portrayed in Table 3? Omar (the ELL individual) satisfies all ancillary requirements but does not satisfy all the focal requirements.[30] Specifically, Omar's knowledge of nonmath vocabulary ("poor", column C) does not satisfy the test's requirement for that KSA ("okay"; column E) and, hence, effective proficiency is "poor."

Reason About Validity by Comparing Targeted Proficiency to Effective Proficiency

Having defined both the targeted proficiency and effective proficiency, we are now ready to consider the concept of validity.[31] We would argue that an important aspect of validity has to do with the alignment between (a) what one intends to measure and (b) what one is actually measuring, wherein alignment favors validity and misalignment favors invalidity.

Insofar as the notion of targeted proficiency captures what one intends to measure, and the notion of effective proficiency captures what one is actually measuring, then a comparison (or match) between the targeted proficiency and effective proficiency captures an important aspect of validity (or alignment).

Thus, at the heart of this ECD approach to Layer B (Domain Modeling) are structures for representing and reasoning about a core validity issue, specifically, whether a test-taker's effective proficiency matches their level in the targeted proficiency.[32] The analyses are typically run in a "what-if" fashion, as though the test-taker's state in the focal and ancillary KSAs were known. See Hansen, Mislevy, and Steinberg (in press) for more detail regarding the strengths and weaknesses of this analytical approach and of this particular index of validity (i.e., match/mismatch between the test-taker's targeted proficiency and their effective proficiency for a specific set of testing conditions).

Consider in a bit more detail the possible results of comparing a test-taker's level in the targeted proficiency to their effective proficiency. If the outcome of comparing the targeted proficiency with a test-taker's effective proficiency is a match, then a score will tend to be valid, at least with respect to this index of validity. Such a match could occur where the outcome is either true-positive (good targeted proficiency and good effective proficiency) or true-negative (poor targeted proficiency and poor effective proficiency). On the other hand, mismatches suggest invalid outcomes and can occur if the outcome is either (a) false-positive (poor targeted proficiency and good effective proficiency), representing unfair advantage for the test taker or (b) false-negative outcome (good targeted proficiency and poor effective proficiency) representing unfair disadvantage for the test taker.

As shown in Table 3, in the cell to the right of the cell labeled "G. Level in the Targeted Proficiency," Omar actually possesses a good level in the *targeted proficiency,* despite having poor effective proficiency in the situation portrayed in Table 3. Omar has a good level in the targeted proficiency as indicated because he possesses each of the focal KSAs (see column D) at the levels specified in the definition of the targeted proficiency (column B-2). As noted earlier, the mismatch between Omar's good level in the targeted proficiency and his poor effective proficiency indicates a false-negative outcome and suggests that Omar is at an unfair disadvantage in this testing situation.

Key causes for mismatch include (a) unsatisfied ancillary requirements and (b) focal requirements that are either excessive or insufficient relative to the KSAs that are specified in the definition of the targeted proficiency. Unsatisfied ancillary requirements and excessive focal requirements tend to yield false-negative outcomes (unfair disadvantage) and insufficient focal requirements tend to yield false-positive outcomes (unfair advantage).

Determine How to Address Invalidity

The next step is to determine how to address areas of invalidity. An examination of Table 3 suggests that the invalidity (which was indicated by the false-negative outcome) has a fairly obvious source—an excessive focal requirement for knowledge of nonmath vocabulary in English. Specifically, despite the fact that the definition of the construct specifies that merely a "poor" knowledge of nonmath vocabulary in English is adequate for a good level in the targeted proficiency, the testing conditions impose a higher—or "okay"—requirement for that KSA, thus constituting an excessive focal requirement. The sensible solution would be to lower that requirement, such as by using simpler nonmath vocabulary (Abedi, Courtney, & Leon, 2003; Abedi & Dietel, 2004; Abedi, Hoffstetter, & Lord, 2004).[33] Another approach would be to provide a dictionary of nonmath vocabulary.[34] If by one of these means we could reduce the focal requirement for knowledge of nonmath vocabulary to the proper level (i.e., the level specified in the definition of the targeted proficiency; "poor"), then Omar's ability would have satisfied this requirement and his effective proficiency would have been good, yielding a true-positive outcome (i.e., a valid outcome relative to this validity index).

It is beyond the scope of this chapter to discuss in detail all the possible strategies for dealing with threats to validity. Following are a few for consideration, based in part on the foregoing discussion:

1. **Manage ancillary and focal requirements:** Take steps to manage ancillary and focal requirements, an overarching theme of this chapter that may be broken down into two sections:

    a. Ensure that each test taker can satisfy the ancillary requirements of tests. Where necessary and feasible, provide accessibility features (e.g., accommodations or universal design features) to eliminate excessive ancillary requirements.

    b. Keep focal requirements at the levels specified in the definition of the targeted proficiency.

2. **Expand the range of accessibility features:** Consider expanding the range of accessibility features. A carefully thought-out set of accessibility features—including accommodations and universal design features—may increase the diversity and number of individuals who can meaningfully participate in the test. Expanding the range of accessibility features could involve modifying or switching methods and test delivery. Consistent with the preceding point, note which KSAs each feature requires and each feature circumvents.

3. **Refine the definition of the targeted proficiency:** Consider refining or otherwise modifying the definition of the targeted proficiency, if feasible, to make it more amenable to the valid use of accessibility features. For example, by defining *reading* (or *reading comprehension*) as *excluding* decoding, one makes the test more amenable to the readaloud accommodation.[35]

4. **Increase test-taker's ancillary abilities:** Increase the test-taker's capacity to satisfy ancillary requirements. For example, if test-takers are having difficulty handling a high ancillary requirement for knowledge of how to handle the test format, then consider taking steps to improve the quality and effectiveness of practice and familiarization materials.

5. **Consider using additional evidence to allow expanded use of accessibility features:** Consider obtaining and admitting additional evidence about targeted proficiencies as a way to increase the use of accessibility features. Normally, in operational use of a test, one relies on the test-taker's score(s) as the major direct evidence of his or her level in the targeted proficiency. However, by allowing the use of additional evidence, one may be able to expand the use of accessibility features. Consider, for example, a case in which reading comprehension proficiency consists of two focal KSAs—*comprehend* and *decode*. Normally, one would *not* allow the readaloud accommodation in this case, since it would open the possibility that a person with little or no decoding ability would perform well, yielding an unfair advantage that person. However, if one could use an independent and prior measure of decoding as evidence, then one might allow the readaloud accommodation only for individuals who possessed adequate decoding ability and therefore would not be unfairly advantaged by the use of the accommodation. Such an approach obviously requires careful consideration of the purpose the test and its intended uses.

6. **Consider modifying eligibility rules for taking the test:** Consider modifying the eligibility rules to allow only individuals who can satisfy the ancillary requirements of a test. Of course, in those rare cases in which the target population for the assessment is composed only of such individuals, eligibility rules would be unnecessary. In considering this option, one must realize that this may actually have the effect of decreasing inclusion and opportunity for English language learners and individuals with disabilities.

7. **Consider modifying reporting practices:** Consider modifying reporting practices to help test score users interpret results of using accessibility features. For example, one might provide additional information that would allow test score users to know, for example, about the specific accessibility features or category of features used. For example, based on work by Thurlow, House, Boys, Scott, and Ysseldyke (2000), Sheinker, Barton, and Lewis (2004) have placed testing accommodations into three categories, based on the likelihood that such accommodations would alter the interpretation of scores (e.g., *are likely to alter* [category 3], *may alter* [category 2], or *are not expected to alter* test-score interpretations [category 1]). Sheinker et al. (2004) noted, however, the tentativeness of such categorizations and emphasize the need for protecting test-taker's confidentiality.

Summary

Based on the foregoing discussion, we can identify three major kinds of threats to validity that relate to accessibility of tests and assessments. While the first of these threats has a clear and obvious relationship to accessibility, the other two are somewhat subtler.

Unsatisfied Ancillary Requirements: "Classic" Accessibility Problem

Unsatisfied ancillary requirements are "classic" accessibility problems, which can yield false-negative outcomes and therefore unfair disadvantage for the test taker. The best illustration of this is that a person possesses a good level in the targeted proficiency, but his or her performance on the test is poor because they cannot satisfy an ancillary requirement that is excessive relative to the test-taker's abilities. For example, suppose that for a particular test as delivered under its default (standard) conditions, the sense of sight is an ancillary requirement, that is, the use of a computer monitor imposes the requirement for *seeing* the test content. The test taker who is blind cannot satisfy that ancillary requirement and therefore faces such an accessibility barrier. If this individual actually possesses a good level in the targeted proficiency (math, let us say), then the outcome is false-negative.[36]

Arguably the main purpose for accessibility features (including testing accommodations) is to reduce or eliminate ancillary requirements for abilities that were adversely impacted by disabilities or language status. Generally, for a person who is blind, the accessibility feature typically eliminates the ancillary requirement for sight. For person who is deaf, the accessibility feature typically eliminates the ancillary requirement for hearing; for example, test directions ordinarily spoken aloud by a proctor may be signed via some form of manual communication (e.g., ASL) to a test taker who is deaf. It should be emphasized that ancillary requirements are not a problem unless the test taker cannot satisfy them. That ancillary requirements are numerous and high will not, in principle, hinder good measurement, unless an individual cannot satisfy those ancillary requirements.[37]

Excessive Focal Requirements

Another way to have a false-negative outcome and unfair disadvantage is through excessive focal requirements. These are focal requirements that are excessive relative to those experienced by individuals in known cases of valid measurement (i.e., focal requirements that are excessive relative to those specified in the definition of the targeted proficiency). As noted in the example of Omar, an excessive focal requirement for knowledge of nonmath vocabulary depressed his score, resulting in an unfair disadvantage (false-negative outcome).

Excessive focal requirements may constitute an accessibility problem, especially if the excessive requirements have a much more negative impact on subgroups than with nondisabled native speakers of English. Nevertheless, the problem of excessively difficult vocabulary may affect both nondisabled individuals and subgroups, which tends to suggest a general test development issue rather than simply an accessibility issue. The practice of removing unnecessary linguistic complexity from test items can help address both the issue of excessive requirements for focal skills (e.g., esoteric knowledge of math vocabulary on a math test where that skill has been defined as a focal KSA) and the issue of excessive ancillary requirements (e.g., difficult-to-decode words where decoding is an ancillary KSA).

Insufficient Focal Requirements

In contrast to the previous two threats that tend to result in unfair disadvantage to the test taker, the third threat, that of insufficient focal requirements, tends to result in unfair advantage for the test taker. Continuing our math example, this third threat to validity would result from using math vocabulary that imposed a requirement for only the lowest level of knowledge of math vocabulary (poor) in order to perform well (when the next higher level, okay, should have been required). Insufficient focal requirements tend to result in unfair advantage for the test taker, manifesting itself in unduly good effective proficiency, and hence, unduly good performance.

Consider an example of insufficient focal requirements that would occur as a result of an accessibility feature. Suppose that reading proficiency is defined as involving focal KSAs of comprehension (comprehend) and decoding (decode). The readaloud accessibility feature would tend to eliminate the focal requirement for decoding, since content is generally read aloud a word at a time rather than requiring the test taker to decode (i.e., decipher words from individual characters).[38] Thus, in this case, the readaloud would tend to cause an unfair advantage, notably for individuals with poor decoding ability and good comprehension who would perform well despite having poor reading proficiency, per the definition of reading comprehension. Thus, while excessive ancillary requirements and excessive focal requirements might be thought of as causes of accessibility barriers, insufficient focal requirements can arise as a side-effect of attempting to address accessibility barriers.

A couple points are worth noting. First, note that among the three threats to validity, there is no threat called "insufficient ancillary requirements." This is because ancillary requirements can arguably never be too low. Second, note that although these three threats seem most relevant to accessibility issues, they also appear to have more general application; attention to these threats might also be helpful in refining general test development practices.

Universal Design of Assessments

An important and relatively new concept that deserves careful consideration in the development of accessible computer-based tests in the concept of universal design and its application to the area of testing and assessment as universal design of assessments.

The concept of universal design was defined by Ron Mace (an architect who was a wheelchair user) as "the design of products and environments to be usable by all people, to the greatest extent possible, without the need for adaptation or specialized design." Mace promoted the idea that rather than promoting special purpose designs that tend to segregate and stigmatize people, we should design environments that work for diverse populations. For example, a door handle could be designed to work not only for people who could easily use a doorknob, but also for children or individuals with disabilities who would be unable to use (i.e., turn) a regular doorknob.

With funding from the U.S. Department of Education, a cross-disciplinary group convened by the Center for Universal Design at the University of North Carolina set forth the following seven principles for universal design (i.e., the "design of products and environments to be usable by all people, to the greatest extent possible, without the need for adaptation or specialized design"; Center for Universal Design, 1997):

1.   Equitable use ("The design is useful and marketable to people with diverse abilities")
2.   Flexibility in use
3.   Simple and intuitive use
4.   Perceptible information

5.   Tolerance for error

6.   Low physical effort

7.   Size and space for approach and use

Universal Design in Instruction

The concept of universal design is beginning to be applied in the area of instruction:

> In terms of learning, universal design means the design of instructional materials and activities that makes the learning goals achievable by individuals with wide differences in their abilities to see, hear, speak, move, read, write, understand English, attend, organize, engage, and remember. Universal design for learning is achieved by means of flexible curricular materials and activities that provide alternatives for students with differing abilities. These alternatives are built into the instructional design and operating systems of educational materials—they are not added on after-the-fact. (ERIC/OSEP Special Project, 1999)

Wiggleworks, an early literacy program from Scholastic, is cited as an example of how universal design features can be implemented in an instructional application (ERIC/OSEP, 1999):

> Students read stories and respond to activities in the reading program. All of the text can be enlarged, changed in color or highlighted, or read aloud by the computer. Children can navigate the software's learning activities via mouse or keyboard. A single switch can turn on a built-in scanning feature. Wiggleworks activities also offer a variety of options for expression, such as writing, drawing, and recording. When in "Write," students can begin a composition by typing text, by recording themselves speaking or drawing, or by placing words from a word list into their text. (ERIC/ OSEP Special Project, 1999)

As may be discerned from the discussion thus far, the concept of universal design is more "at home" with the concept of direct usability (through built-in features) than it is with indirect accessibility (through external assistive technology).

Universal Design of Assessments (UDA)

The idea of universal design has begun to be applied to the area of assessments. As the National Center on Educational Outcomes (NCEO) says, "The term 'universally designed assessments' refers to assessments that are designed and developed from the beginning to be *accessible and valid* for the widest range of students, including students with disabilities and students with limited English proficiency" (NCEO, 2004, emphasis added). This definition seems to appropriately emphasize the fundamental issues of validity and accessibility rather than what is arguably a less-important issue—whether an accessibility feature is built-in or added on.

Thompson, Johnstone, and Thurlow (2002) of the NCEO proposed seven elements of UDA that were derived from a review of the literature on universal design, assessment and instructional design, and research on topics such as assessment accommodations (see Table 4).

As Thompson, Thurlow, and Malouf (2004) noted, UDA is not a panacea and does not address the important issue of ensuring opportunity to learn. Furthermore, "the specific criteria for putting all the universal design considerations

Table 4. Seven elements of UDA

Number	Element	Explanation
1	Inclusive Assessment Population	Tests designed for state, district, or school accountability must include every student except those in the alternate assessment, and this is reflected in assessment design and field testing procedures.
2	Precisely Defined Constructs	The specific constructs tested must be clearly defined so that all construct irrelevant cognitive, sensory, emotional, and physical barriers can be removed.
3	Accessible, Nonbiased Items	Accessibility is built into items from the beginning, and bias review procedures ensure that quality is retained in all items.
4	Amenable to Accommodations	The test design facilitates the use of needed accommodations (e.g., all items can be brailled).
5	Simple, Clear, and Intuitive Instructions and Procedures	All instructions and procedures are simple, clear, and presented in understandable language.
6	Maximum Readability and Comprehensibility	A variety of readability and plain language guidelines are followed (e.g., sentence length and number of difficult words are kept to a minimum) to produce readable and comprehensible text.
7	Maximum Legibility	Characteristics that ensure easy decipherability are applied to text, to tables, figures, and illustrations, and to response formats.

Note. Adapted from Thompson, Thurlow and Malouf (2004) and used with permission of the National Center on Educational Outcomes (NCEO) and the Association of Test Publishers (ATP).

together have not yet been figured out—we do not know when the right balance has been achieved." In other words, we need to learn how the elements of UDA interact with each other and with other facets of test design to produce an accessible and valid operational assessment. Thompson, Thurlow, and Malouf further noted that

> *it is difficult to anticipate what accessibility issues will arise when a test is delivered on a variety of different systems with a variety of assistive technologies (e.g., screen readers). Trying to anticipate these issues is important, however, and trying to design computer-based testing systems in a more universally accessible manner is an endeavor worth pursuing.* (p. 10)

ECD (as described in this chapter) may be seen as an attempt to provide a framework for identifying when the "right balance has been achieved." Such a framework is expected to be useful when designing accessible systems for computer-based testing, as well as specific tests to be delivered by such systems.

Defining Universal Design Features

For the purposes of this chapter, we refer to *universal design features* as accessibility features that are available to all or to virtually anyone who needs or desires them. In contrast to the idea of an "accommodation," which involves a "departure" from established protocol (National Research Council, 2004, p. 1), an arguably defining characteristic of a universal design feature might be that it is *not* a departure from established protocol per se but rather part of a new or refined protocol that seeks to be more inclusive and attentive to individuals' accessibility needs and preferences. By contrast, processes for determining appropriate *accommodations* tend of focus more on the examinee's accessibility needs rather than on his or her personal preferences.

Herein lies one of the great challenges in achieving the ideals of UDA (i.e., assessment that is both accessible and valid)—that of establishing a new kind of protocol, a more flexible protocol that lowers accessibility barriers, meets more personal preferences, and minimizes the number of occasions in which an accommodation is required, but without undermining validity. As suggested previously, ECD appears to have potential as a framework for establishing such new protocols.[39]

Toward a Common Understanding

This part of the chapter illustrates how people who are exploring ways to make computer-based assessment more accessible might benefit from a common understanding of ECD concepts to aid them in their communication. It also illustrates that test design always involves trade-offs. Following is an excerpt of part of a hypothetical conversation between members of a team that is preparing to gather requirements for a computer-based testing system that they anticipate developing. This conversation presupposes that all participants share some common understanding of ECD accessibility-related concepts.

- **Moderator:** From a test developer's point of view, what are some of the things that seem helpful in making tests more accessible or "universally designed"?

- **Test Developer:** Basically, we try to think about accessibility early in the test development process and then take steps to make sure that our thinking gets reflected in the operational test. This means recognizing the diversity that exists within our target population and that some individuals will require accessibility features in order to obtain valid scores. As test developers, we have long tried to be explicit in defining the construct to be measured, but taking a hard look at the diversity within our target population and the accessibility features that some individuals have come to expect challenges us to think more precisely about how we can maximize accessibility without reducing the rigor of our tests. Part of that precision comes in trying to be even more precise in our definition of the construct. For example, defining a construct clearly involves specifying the focal skills—skills that are essential parts of the construct. But it also seems valuable to designate some skills as ancillary, that is, as *not* being essential parts of the construct. It seems especially important to make these focal or ancillary designations with regard to skills that are affected by disabilities, such as sight, hearing, and so on. Furthermore, where certain linguistic abilities, such as vocabulary knowledge, are part of the construct, as they often are, we try to be more explicit about the actual level of that knowledge that is part of the construct. Putting in that extra effort up front as we define the construct helps us write test specifications that avoid implying that certain sensory, cognitive, physical, linguistic, or other abilities are part of the construct of the test, when in fact they are not. With a cleaner test specification, we are less likely to get items that have accessibility problems. Another thing that we are trying to do is plan better for alternate test formats, such as audio and braille, because, for our tests, no one approach or system seems able to handle the needs of every test taker.

- **Moderator:** What features should be considered in our planned accessible computer-based testing system?

- **Accessibility Person:** Some of the most important features are providing text equivalents for nontext elements such as images, audio, video, animations, and audio descriptions for video clips (Chisholm, Vanderheiden, & Jacobs, 1999).

- **Measurement Person:** The system should give test administrators a high degree of control over the kind of accessibility features that they grant to students. Otherwise, we might inadvertently compromise validity by allow-

---

Standards for Educational Testing

The Standards for Educational and Psychological Testing (Standards), which were developed by the AERA, the APA, and the NCME (1999), serves as a key reference point for educational testing, clarifying the role of validity as the central consideration and unifying concept. The Standards define validity as the "degree to which accumulated evidence and theory support specific interpretations of test scores entailed by proposed uses of a test" (AERA, APA, & NCME, 1999, p. 184). Chapter 10 of these standards pertains to test takers with disabilities, and Chapter 9 pertains to individuals with diverse linguistic backgrounds.

---

ing use of an accessibility feature that reduces requirements for focal KSAs (AERA, APA, NCME, 1999; see the text box on Standards for Educational Testing).

- **Accessibility Person:** Wouldn't a test or assessment be fairer or more equitable to all test takers if accessibility features are available to all who *need or desire* them rather than keeping a tight reign on access to accessibility features? Wouldn't that make assessments more "universally designed"? The IMS Global Learning Consortium's Accessibility for Learner Information Package (ACCLIP) specification defines a way of structuring information about individuals' *needs* and *preferences* for a wide range of accessibility features (IMS Global Learning Consortium, 2003). [See text box on IMS Global Learning Consortium]

IMS Global Learning Consortium

The IMS Global Learning Consortium focuses on developing specifications that will result in online learning systems—including online tests, as well as online instruction, computer-based instructional management systems, and so on—that are interoperable in the sense of being able to run on diverse systems and platforms. For example, by packaging instructional content or test content according to IMS specifications, the content might be deliverable using learning management systems from several different vendors.

An IMS accessibility group has developed two major specifications. First is the Accessibility for Learner Information Package version 1.0 (ACCLIP; IMS Global Learning Consortium, 2003) specification that allows the accessibility needs and preferences of individuals to be stored. The set of ACCLIP need/preference settings for a given student constitutes a student profile. The second specification is the AccessForAll Meta-data 1.0 (ACCMD) specification, which specifies mark-up that will allow the accessibility-related significance of the content to be recognized (IMS Global Learning Consortium, 2004). With the student profile provided by ACCLIP and the resource profile provided by ACCMD, a system has some basis for matching accessible resources to people who need or desire them, thus helping automate the delivery of accessible online content. A white paper on accessible online learning systems (IMS Global Learning Consortium, 2002) developed by the accessibility group includes a section on testing and assessment (Heath & Hansen, 2002).

- **Measurement Person:** Well, I find it *harder* to think about validity in the context of accessibility *preferences* than in the context of well-documented *needs*. I am accustomed to thinking about accessibility needs that that been reviewed by experts to ensure that accommodations that are intended to address those needs do so without undermining validity. It is harder for me to think about learner-selected *preferences*. What happens when a test taker *prefers* a feature that would arguably give them an unfair advantage, such as having test content read aloud when decoding is a focal KSA? Or what happens when the test taker chooses an accessibility feature *unwisely,* and it wastes their time and energy in a high-stakes testing situation? One way to think about this is that allowing a test taker to choose between an array of alternate accessibility features imposes new requirements—presumably ancillary requirements—for knowing how, when, and if to use those accessibility features. It seems to me that a computer-based testing system needs to have some way to control access to many if not all accessibility features, not only to avoid giving unfair advantage to the test taker, but also to protect individuals from excessive requirements for knowledge about how, when, and if to use a potentially confusing array of accessibility features. Does ACCLIP provide methods for controlling access to accessibility features?

- **Accessibility Person:** To a certain extent. ACCLIP 1.0 does provide an eligibility section that provides the beginnings of such a capability (IMS Global Learning Consortium, 2003).[40]

- **Measurement Person:** I agree that we need to think about inclusion and accessibility from the earliest design stages, but we have to make sure that accessibility features do not undermine validity.

- **Accessibility Person:** I think that we are in agreement. Validity is an important part of the universal design of assessments, a concept that is intended to make test more accessible. The NCEO makes clear that universally designed assessments are designed and developed from the beginning to be both *accessible and valid* for the widest range of students (NCEO, 2004, emphasis added). So *validity* is recognized as a key issue in universally designed assessments.

- **System Designer:** The concept of universal design of assessments is wonderful and important, but in designing an accessible computer-based testing system, we need to move toward this ideal in a practical way, one step at a time. I know that there is great diversity among test takers with disabilities and English language learners. Isn't there some way that we can design the first version of our accessible computer-based testing system to have some limited set of features that is capable of meeting the accessibility needs of a limited set of disability and language status categories? If we can do this efficiently and effectively, then we can develop a more capable and accessible system in the next version. Can we identify a set of high priority features—ones that everyone agrees are both feasible and very useful?

- **Accessibility Person:** Well, there are guidelines and standards that may be helpful. For example, the World Wide Web Consortium's Web Content Accessibility Guidelines are among the most cited accessibility guidelines of which I am aware (Chisholm et al., 1999). They were also very influential on the development of the U.S. government's Section 508 standards (Architectural and Transportation Barriers Compliance Board, 2000). The portion of the standards that deals with Web-based applications are closely patterned after the lowest (least rigorous) level of accessibility (level A) of the W3C Content Accessibility Guidelines (Chisholm et al., 1999).[41] For guidelines specifically related to computer-based testing, there are guidelines for accessible computer-based testing from NCEO (Thompson, Thurlow, Quenemoen, & Lehr, 2002) and the American Printing House for the Blind (Allen, Bulla, & Goodman, 2003).[42]

- **System Designer:** Okay. I will look at these documents. For now, could we focus on a single feature? How about the readaloud feature? I would

---

Accessibility Guidelines of the World Wide Web Consortium

The World Wide Web Consortium (W3C) provides several accessibility guidelines, the best known being the W3C Web Content Accessibility Guidelines (WCAG), version 1.0 (Chisholm et al., 1999), which guide Web content developers in developing accessible content. The guideline's first "checkpoint"—which is required for any of the conformance levels—requires that a content developer provide "text equivalents" for nontext content (e.g., images, audio, video/animations; Chisholm et al., 1999, checkpoint 1.1). A text equivalent consists of text that is written to convey essentially the same meaning as the nontext content that it describes. Text equivalents can be output using several different presentations modes (e.g., visually displayed text [large font, modified color and contrast, as necessary] for individuals who are deaf or nondisabled or read aloud via synthesized speech, prerecorded audio, or live reader[43]). The User Agent Accessibility Guidelines (UAAG) version 1.0 (Jacobs, Gunderson, & Hansen, 2002) is directed to developers of Web browsers and media players, although much of the material seems relevant to anyone seeking to develop accessible computer-based systems. The UAAG emphasizes among other things, keyboard operability, user control over a wide array of accessibility features, and compatibility with assistive technologies.

Section 508 Standards

The U.S. government's Standards for Electronic and Information Technology—typically called the Section 508 standards—provide technical criteria specific to various types of technologies such as software applications and operating systems; Web-based information or applications; telecommunications products; video or multimedia products; self contained, closed products such as information kiosks and transaction machines, and computers. Of particular interest are the standards for Web-based information or applications, which, for the most part, are quite consistent with the WCAG 1.0 conformance level A (i.e., the priority 1 checkpoints; see Thatcher, 2004, for a side-by-side comparison). Although the Section 508 standards pertain to U.S. government procurement, its influence is being felt more widely. For example, many states and other countries have adopted Section-508-like regulations or laws.

---

think that we would want the system to be able to speak the test content to a person who is blind. Does it matter if the voice is synthesized instead of prerecorded?

- **Researcher:** In tests of listening comprehension, I think that the stimulus should be prerecorded human speech. For other tests, I don't think we know enough about the advantages and disadvantages of synthesized versus prerecorded human speech. Fortunately, the quality of synthesized speech is gradually improving.

- **System Designer:** If the person is blind, how do they control the computer?

- **Accessibility Person:** Many individuals who are blind interact very comfortably with the computer keyboard. So keyboard operability is very important. Actually, it might be better to say that our system ought to have a full keyboard interface that allows one to control all functionality of the computer either via keyboard or an alternate method that generates text as if a keyboard had been used. At least the functionality that is available to any nondisabled test-taker needs to be accessible via the keyboard (Jacobs et al., 2002). Of course, in order to use the keyboard, the person who is blind generally needs to be able to navigate through and hear the test content, usually via text-to-speech, that is, synthesized speech.

- **System Designer:** What are the best approaches for implementing text-to-speech?

- **Accessibility Person:** There are basically two major ways of implementing synthesized speech: via screen reader or a self-voicing application. Screen reader programs are commercial software packages that allow a person who is blind or has certain other specific disabilities to interact with a variety of other software—such as word processors, spreadsheets, and Web browsers—using synthesized speech output. Thus, the screen reader is an assistive technology that is *external* to the other software applications. On the other hand, the self-voicing approach involves providing built-in speech output capabilities for a given application. The user of a self-voicing application does not need to start another software application (the screen reader) to navigate through the content. In principle, a self-voicing application can have a simplified, easier-to-learn interface because the speech and navigation capabilities are designed with the application itself. The user of a self-voicing test delivery application need not have more commands than those necessary for test delivery. A self-voicing application also has the advantage over the screen reader approach in that the developer of the self-voicing application, such as a testing organization, has a very high degree of control over the experience of the test taker. This high level of control allows the testing organization to better ensure consistency of experience across test takers. A possible limitation of the self-voicing approach is that the navigation scheme for a self-voicing test may be unlike that scheme for any other application and would have to be learned. While some approaches may seem to straddle the distinction between the screen reader approach and the self-voicing approach, the distinction seems to remain quite useful.

- **Researcher:** Screen reader packages are used by many individuals who are blind, but they tend to be quite challenging to learn to use well. Many screen reader users who are blind tend to have difficulty with Web pages that use "frames" or other complex structures (Hansen et al., 2004). I am concerned about trying to develop high-stakes computer-based tests that rely on ordinary commercial screen readers. If one is not concerned about having computer-based testing's capabilities for automated scoring and reporting, then there are several possible solutions, including simply placing the test content in a Microsoft Word document and then having the test taker who is blind interact with the document via a screen reader, typing their answers into the same document. But if we want those special advantages of computer-based testing, then we need to look seriously at building or buying a self-voicing testing application (gh, 2003; Hansen et al., 2004). If we decide that we need to develop a self-voicing application, we ought to investigate a range of potential solutions. For example, the Digital Talking Book standard is intended to allow someone to navigate easily through audio content (DAISY Consortium, 2004). Although DTB is just a presentation medium, it may be extended to include data capture capabilities. (See the following text box on Digital Talking Book.)

- **System Developer:** I would be interested to see what levels of configurability and control the user has with the Digital Talking Book. If we

---

Digital Talking Book

The Digital Talking Book (DTB) standard (e.g., ANSI/NISO Z39.86-2002) specifies a way of structuring prerecorded audio content and synchronizing it with text, enabling quicker and more convenient navigation than is possible using audiocassette (DAISY Consortium, 2004). A DTB player can play each segment of prerecorded audio, visually displaying the text for each sentence as the prerecorded audio is playing. The visual appearance of the text in a DTB player can be modified, such as by enlarging the font. Using a somewhat more advanced DTB player, a user who prefers synthesized speech instead of prerecorded audio can have a commercial screen reader read the text aloud and the prerecorded audio will be temporarily disabled. The screen reader can also be used to output the text to a refreshable braille display. A limitation of the current ANSI standard for DTB is that it is essentially a display method rather than being capable of recording user responses (e.g., answers to test questions). Nevertheless, in the future, modules might be added to provide such capabilities. In July of 2004, the U.S. Department of Education endorsed the use of this ANSI/NISO Z39.86-2002 standard as a voluntary national file format for creating multiple student-ready transformations such as braille and Digital Talking Book formats (Center for Applied Special Technology, 2004a, 2004b).

---

design our own self-voicing test delivery application, issues of control and configurability are important. Among the features that the examinee would be allowed to control regarding the readaloud feature—such as the rate and gender of the synthesized speech or the font size, color, or contrast—we need to decide which features they can set (or reset) at any time versus that can be set only before starting the test. By the way, I suppose that since we don't plan on testing people's vision, the sense of sight will virtually always be an ancillary KSA. On the other hand, because decoding ability will sometimes be a focal KSA, we are likely to need the system to allow us to limit examinee's access to the readaloud capability.

- **Researcher:** That seems right. The better we can define the target audiences, the range of targeted proficiencies that we will want to assess, and the amount of resources we have for research, development, and deployment of the system, the better decisions we can make in the design of an accessible test delivery system.

- **System Developer:** Eventually we need to know enough about these different parameters that we can make specific decisions about the level of control and configurability that is needed by test takers, administrators, and so on. We need to determine the granularity of the control and configuration options, that is, whether its effect is global or if pertains to a very specific resource (Jacobs et al., 2002). What other information resources are available to help us?

- **Researcher:** Other useful resources include the IMS guidelines for accessible online learning (IMS Global Learning Consortium, 2002) and guidelines on making educational software and Web sites accessible from National Center for Accessible Media (NCAM) at WGHB/Boston; Freed, Rothberg, & Wlodkowski, 2003). It is also worthwhile for us to learn about other efforts to implement accessible prototypes (gh, 2003; Hansen et el., 2004; Landau, 2004; Shute, Graf, & Hansen, in press) and to become acquainted with current practice and empirical research related to the validity of testing accommodations (Koretz & Barton, 2003; National Research Council, 2004 ; Pitoniak & Royer, 2001; Thurlow, House, Boys et al., 2000[44]; Thurlow, House, Scott, & Ysseldyke, 2000; Willingham et al., 1988). Also, we cannot ignore the legal issues (Phillips, 2002). Somehow our approach needs to be comprehensive enough to encompass not only computer-based tests, which are the main focus of our efforts, but also the alternative approaches that will still be needed to handle special cases. We need to develop a body of knowledge—essentially a body of cases regarding what has worked and what has not worked—that will allow us to improve best practices in accessible computer-based testing; we also need to develop useful ways of representing and communicating these cases.[45] I am hopeful that through conceptual tools like ECD we can find ways to organize all this knowledge in ways that inform the design of the tests that are more accessible, feasible, and protective of the integrity of the test results.

Conclusion

This chapter has described an ECD approach for understanding accessibility issues involving computer-based testing. The approach described in this chapter seeks to integrate thinking about accessibility features (i.e., accommodations, universal design features), task design, and validity—all in a framework of sharable terminology, concepts, and knowledge representations. Such a framework, we believe, can allow designers and developers of computer-based tests and testing systems to more accurately anticipate the impact on validity of changes in assessment designs, particularly those changes that are intended to improve accessibility for subpopulations such as English language learners and individuals with disabilities. By helping reconcile the sometimes-competing demands of accessibility and validity, the framework may be helpful in increasing the diversity of individuals whose proficiencies can be validly measured by computer-based tests.

This chapter has emphasized several keys to ensuring the accessibility without undermining validity of test results. Among these are (a) consideration of the diversity of the target population and a concerted effort to increase inclusion through accessibility features, (b) precise definition of the targeted proficiency, (c) avoiding ancillary requirements that are excessive relative the abilities of the test taker, and (d) avoiding focal requirements that are either excessive or insufficient relative to those experienced by test takers in known cases of valid measurement (i.e., the levels specified in the definition of the targeted proficiency).

Among the key goals for future computer-based test delivery systems would be to provide more built-in accessibility features and better communication with assistive technologies, such as screen readers and refreshable braille displays. Such computer-based testing systems should be built with the understanding that an accessibility feature that is appropriate for one test might be inappropriate for another test. For example, the capability of reading aloud the test to the student may be valid for a test where decoding has been defined as an ancillary KSA and invalid for a test where decoding is a focal KSA.

Another key goal would be to find ways to deal with the issue of excessive linguistic demands important for both English language learners and some individuals with disabilities. Regardless of whether the excessive linguistic demands pertain to focal or ancillary KSAs, their effect will be to depress student performance and to undermine validity. Computer-based testing systems have tremendous potential for providing easy access to various kinds of linguistic help, but such features need to be implemented in ways that do not reduce the essential rigor or validity of the tests (Abedi et al., 2003; Abedi & Dietel, 2004; Abedi et al., 2004; Hansen & Mislevy, 2004).

ECD can inform accessibility-related decisions regarding computer-based tests and testing systems. Following are some key questions and some basic suggestions for individuals who are considering the development of accessible computer-based tests and testing systems.

- **Question:** What is the most fundamental step of developing an ECD-based design?

- **Suggestion:** Although opinions may differ, it is difficult to identify a step more important than that of defining what one intends to measure, that is, the targeted proficiency. Ideally, the constructions should not only make explicit the KSAs that are essential parts of what one wants to measure (focal KSAs), but also selected KSAs that are not essential (ancillary KSAs). Basic KSAs regarding sensory abilities (see, hear, etc.) and language abilities (decode, know terminology of the field) are worthy of serious consideration as part of the definition of a construct. It may be important to revisit the definition of the targeted proficiency at various points in the process.

- **Question:** What are possible basic criteria for determining what accessibility features to provide for a given test?

- **Suggestion:** Provide features that (a) maintain the focal requirements and (b) minimize ancillary requirements or suit them to the students' abilities. This basic recipe will necessarily be somewhat shaped by practical and legal considerations as well as by considerations of accessibility and validity.

- **Question:** How can one identify the most important accessibility features for computer-based testing systems?

- **Suggestion:** Try to identify reasonable boundaries for the challenges that your accessible computer-based system is seeking to address, at least for the initial version of the system. Boundaries might pertain to kinds of subject areas for the tests, the characteristics of the students (age, disability status, language status), number of targeted assistive technologies, diversity of delivery platforms, and so forth. Be certain that accessibility considerations are considered as well as issues of validity and feasibility. Use a framework such as ECD to organize the knowledge within an argument structure to clarify how prospective design features affect requirements for focal and ancillary KSAs. Draw upon existing standards and guidelines where feasible. Improve the system on succeeding versions.

To summarize, there is a great need for designers of computer-based tests and testing systems to build accessibility into their designs from the earliest stages, yet to do so in a way that avoids undermining validity. Achieving these goals necessitates a conceptual framework for determining how accessibility features impact validity. ECD is offered as a conceptual framework—providing sharable terminology, concepts, and knowledge representations—for representing and anticipating the impact of accessibility features on validity, thus helping weigh the consequences of potential design alternatives for accessible computer-based tests and testing systems.

Acknowledgments

The authors acknowledge Russell Almond, Malcolm Bauer, Daniel Eignor, and Lois Frankel for their reviews of this chapter. Any errors are the responsibility of the authors.

References

Abedi, J., Courtney, M., & Leon, S. (2003). *Effectiveness and validity of accommodations for English language learners in large-scale assessments* (Center for the Study of Evaluation [CSE] Report 608). Los Angeles: University of California, National Center for Research on Evaluation, Standards, and Student Testing.

Abedi, J., & Dietel, R. (2004). *Challenges in the No Child Left Behind Act for English language learners* (CRESST Policy Brief No. 7). Los Angeles: University of California, National Center for Research on Evaluation, Standards, and Student Testing.

Abedi, J., Hoffstetter, C. H., & Lord, C. (2004). Assessment accommodations for English language learners: Implications for policy-based empirical research. *Review of Educational Research, 74*(1), 1-28.

Allen, J. M., Bulla, N., & Goodman, S. A. (2003). Test access: Guidelines for computer-administered testing. Louisville, KY: American Printing House for the Blind. Retrieved December 1, 2004, from *http://www.aph.org/tc/access/*

Allman, C. (2004). Making tests accessible for students with visual impairments: A guide to test publishers, test developers, and state assessment personnel (2nd ed.). Louisville, KY: American Printing House for the Blind. Retrieved December 1, 2004, from *http://www.aph.org/tc/access2/*

American Educational Research Association, American Psychological Association, & National Council on Measurement in Education. (1999). *Standards for educational and psychological testing.* Washington, DC: American Educational Research Association.

Architectural and Transportation Barriers Compliance Board. (2000). *Electronic and information technology accessibility standards (Section 508).* Retrieved October 4, 2002, from *http://www.access-board.gov/sec508/508standards.htm*

Bennett, R. E. (1995). *Computer-based testing for examinees with disabilities: On the road to generalized accommodation* (ETS RM-95-1). Princeton, NJ: Educational Testing Service.

Bennett, R. E. (2002). Inexorable and inevitable: The continuing story of technology and assessment. *Journal of Technology, Learning, and Assessment, 1*(1), 2002. Retrieved October 27, 2004, from *http://www.bc.edu/research/intasc/jtla/journal/v1n1.shtml*

Center for Applied Special Technology. (2004a). National file format initiative at NCAC. Retrieved August 25, 2004, from *http://www.cast.org/ncac/NationalFileFormat3138.cfm*

Center for Applied Special Technology. (2004b). National instructional materials accessibility standard report—Version 1.0: NIMAS report executive summary. Retrieved August 25, 2004, from *http://www.cast.org/ncac/nimas/executive_summary.htm*

Center for Universal Design (1997). The principles of universal design: Version 2.0-4/1/97. Retrieved July 28, 2004, from *http://www.design.ncsu.edu/cud/univ_design/principles/udprinciples.htm*

Chisholm, W., Vanderheiden, G., & Jacobs, I. (Eds.). (1999*). Web content accessibility guidelines.* Retrieved May 5, 1999, from the World Wide Web Consortium site *http://www.w3.org/TR/WAI-WEBCONTENT*

Cronbach, L. J., & Meehl, P. E. (1955). Construct validity in psychological tests. *Psychological Bulletin, 52,* 281-302.

DAISY Consortium. (2004, August 25). Frequently asked questions. Retrieved August 20, 2004, from *http://www.daisy.org/about_us/g_faq.asp*

ERIC/OSEP Special Project. (1999, Fall). Universal Design: Ensuring access to the general education curriculum. *Research Connections, 5.* Retrieved July 28, 2004, from *http://ericec.org/osep/recon5/rc5sec1.html*

Freed, G., Rothberg, M., & Wlodkowski, T. (2003). Making educational software and web sites accessible: Design guidelines including math and science solutions. Retrieved March 24, 2004, from the WGBH Educational Foundation Web site: *http://ncam.wgbh.org/cdrom/guideline/*

gh. (2003). gh, LLC and Educational Testing Service working to provide historic computer-voiced standardized test for people with visual disabilities. Retrieved August 6, 2004, from *http://www.ghbraille.com/microsoftnewsrelease.html*

Hansen, E. G., Forer, D. C., & Lee, M. J. (2004). *Toward accessible computer-based tests: Prototypes for visual and other disabilities* (RR-78). Princeton, NJ: TOEFL Research Reports.

Hansen, E. G., & Mislevy, R. J. (2004, April 13). Toward a unified validity framework for ensuring access to assessments by individuals with disabilities and English language learners. Paper presented at the annual meeting of the *National Council on Measurement in Education (NCME)*, San Diego, CA. Retrieved October 27, 2004, from the ETS Web site *http://www.ets.org/research/conferences/aera2004.html#frame work*

Hansen, E. G., & Mislevy, R. J. (in press). *Toward a unified validity framework for ensuring access to assessments by English language learners.* Princeton, NJ: ETS Research Report Series.

Hansen, E. G., Mislevy, R. J., & Steinberg, L. S. (2003). Evidence-centered assessment design and individuals with disabilities. E. G. Hansen (Organizer), Assessment design and diverse learners: Evidentiary issues in disability, language, and non-uniform testing conditions. Symposium presented at the annual meeting of the *National Council on Measurement in Education*, Chicago, IL. Retrieved October 27, 2004, from the ETS Web site*www.ets.org/research/dload/ncme03-hansen.pdf*

Hansen, E. G., Mislevy, R. J., & Steinberg, L. S. (in press). *Evidence centered assessment design for reasoning about testing accommodations in NAEP reading and math.* Princeton, NJ: ETS Research Report Series.

Hansen, E. G., Mislevy, R. J., Steinberg, L. S., Lee, M. J., & Forer, D. C. (2005). Accessibility of tests for individuals with disabilities within a validity framework. *System: An International Journal of Educational Technology and Applied Linguistics, 33*(1), 107-133.

Hansen, E. G., & Steinberg, L. S. (2004). *Evidence-centered assessment design for reasoning about testing accommodations in NAEP reading and math.* Paper commissioned by the Committee on Participation of English Language Learners and Students with Disabilities in NAEP and Other Large-Scale Assessments of the Board on Testing and Assessment (BOTA) of the National Academy of Sciences/National Research Council.

Heath, A., & Hansen, E. (2002). Guidelines for testing and assessment. In IMS Global Learning Consortium (Ed.), *IMS guidelines for developing acces-*

sible learning applications. Retrieved August 6, 2004, from *http://ncam.wgbh.org/salt/guidelines/sec9.html*

IMS Global Learning Consortium. (Ed.). (2002). IMS guidelines for developing accessible learning applications. Retrieved August 6, 2004, from *http://ncam.wgbh.org/salt/guidelines/*

IMS Global Learning Consortium. (Ed.). (2003). IMS learner information package accessibility for LIP information model. Version 1.0 final specification. Retrieved August 24, 2004, from *http://www.imsglobal.org/accessibility/acclipv1p0/imsacclip_infov1p0.html*

IMS Global Learning Consortium. (Ed.). (2004). IMS AccessForAll Meta-data — Information Model. Version 1.0 final specification. Retrieved August 25, 2004, from *http://www.imsglobal.org/accessibility/accmdv1p0pd/imsaccmd_infov1p0pd.html*

Jacobs, I., Gunderson, J., & Hansen, E. (Eds.). (2002). User agent accessibility guidelines 1.0. W3C Recommendation 17 December 2002. Retrieved October 27, 2004, from the World Wide Web Consortium Web site: *http://www.w3.org/TR/UAAG10/*

Jensen, F. V. (1996). *An introduction to Bayesian networks*. New York: Springer-Verlag.

Kane, M. (1992). An argument-based approach to validation. *Psychological Bulletin, 112,* 527-535.

National Research Council. (2004). Keeping score for all: The effects of inclusion and accommodation policies on large-scale educational assessment. In J. A. Koenig & L. F. Bachman (Eds.), *Committee on Participation of English Language Learners and Students with Disabilities in NAEP and Other Large-Scale Assessments*. Washington, DC: National Academy of Sciences.

Landau, S. (2004, February). *Development of an audio-tactile accommodation for delivery of standardized tests to students who are blind or visually impaired—Interim Report* (U.S. Department of Education, SBIR Grant ED-02-CO-0054). New York: Touch Graphics.

Linn, R. L. (2002). Validation of the uses and interpretations of results of state assessment and accountability systems. In G. Tindal, & T. M. Haladyna (Eds.), *Large-scale assessment programs for all students* (pp. 109-148). Mahwah, NJ: Erlbaum.

McConnell, S. (1998). *Software project survival guide*. Redmond, WA: Microsoft Press.

Messick, S. (1989). Validity. In R.L. Linn (Ed.), *Educational measurement* (3rd ed., pp. 13-103). New York: Macmillan.

Messick, S. (1994). The interplay of evidence and consequences in the validation of performance assessments. *Education Researcher, 23*(2), 13-23.

Mislevy, R. J. (1994). Evidence and inference in educational assessment. *Psychometrika, 59,* 439-483.

Mislevy, R. J., Steinberg, L. S., & Almond, R. G. (2003). On the structure of educational assessments. *Measurement: Interdisciplinary Research and Perspectives, 1,* 3-67.

National Assessment Governing Board. (2002). *Reading framework for the 2003 National Assessment of Educational Progress.* Washington DC: Author.

National Center on Educational Outcomes. (2004). Special topic area: Universally designed assessments. Retrieved August 24, 2004, from *http://education.umn.edu/NCEO/TopicAreas/UnivDesign/UnivDesign_topic.htm*

Phillips, S. E. (2002). Legal issues affecting special populations in large-scale testing programs. In G. Tindal, & T. M. Haladyna (Eds), *Large-scale assessment programs for all students* (pp. 109-148). Mahwah, NJ: Erlbaum.

Pitoniak, M. J., & Royer, J. M. (2001). Testing accommodations for examinees with disabilities: A review of psychometric, legal and social policy issues. *Review of Educational Research, 71,* 53-104.

Sheehan, K. M, & Ginther, A. (2001, April). What do multiple choice verbal reasoning items really measure? An analysis of the cognitive skills underlying performance on TOEFL reading comprehension items. Paper presented at the annual meeting of the *National Council on Measurement in Education (NCME)*, Seattle, WA.

Sheinker, A., Barton, K. E., & Lewis, D. M. (2004). *Guidelines for inclusive test administration 2005.* Monterey, CA: CTB/McGraw-Hill.

Shute, V. J., Graf, E. A., & Hansen, E. G. (in press). Designing adaptive, diagnostic math assessments for sighted and visually-disabled students. In L. PytlikZillig, R. Bruning, & M. Bodvarsson (Eds.). *Technology-based education: Bringing researchers and practitioners together.* Greenwich, CT: Information Age.

Spearman, C. (1904). "General intelligence" objectively determined and measured. *American Journal of Psychology, 15,* 201-293.

Thatcher, J. W. (2004). Side-by-side: WCAG v. 508. Retrieved August 25, 2004, from *http://www.jimthatcher.com/sidebyside.htm*

Thompson, S. J., Johnstone, C. J., & Thurlow, M. L. (2002). Universal design applied to large scale assessments (NCEO Synthesis Report 44). Minneapolis, MN: University of Minnesota, National Center on Educational Outcomes. Retrieved October 27, 2003, from *http://education.umn.edu/NCEO/OnlinePubs/Synthesis44.html*

Thompson, S. J., Thurlow, M. L., & Malouf, D. B. (2004). Creating better tests for everyone through universally designed assessments. *ATP Journal*. Retrieved July 28, 2004, from the Association of Test Publishers Web site: *http://www.testpublishers.org/Creating_Better_Tests%20Final%20Revision%205.15.04.pdf*

Thompson, S. J., Thurlow, M. L., Quenemoen, R. F., & Lehr, C. A. (2002). *Access to computer-based testing for students with disabilities* (NCEO Synthesis Report 45). Retrieved February 6, 2004, from the National Center on Educational Outcomes Web site: *http://education.umn.edu/NCEO/OnlinePubs/Synthesis45.html*

Thurlow, M., House, A., Boys, C., Scott, D., & Ysseldyke, J. (2000). *State participation and accommodation policies for students with disabilities: 1999 update* (Synthesis Report No. 33). Minneapolis, MN: University of Minnesota, National Center on Educational Outcomes. Retrieved January 1, 2005, from *http://education.umn.edu/NCEO/OnlinePubs/Synthesis33.html*

Thurlow, M. L., House, A. L., Scott, D. L., & Ysseldyke, J. E. (2000, Fall). Students with disabilities in large-scale assessments: State participation and accommodation policies. *Journal of Special Education*. Retrieved August 27, 2004, from *http://www.findarticles.com/p/articles/mi_m0HDF/is_3_34/ai_76157521/pg_1*

Wainer, H., & Thissen, D. (1994). On examinee choice in educational testing. *Review of Educational Research, 64*, 159-195.

Willingham, W. W., Ragosta, M., Bennett, R. E., Braun, H., Rock, D. A., & Powers, D. E. (Eds.). (1988). *Testing handicapped people*. Boston, MA: Allyn &Bacon.

Endnotes

[1]   Although this chapter refers to native speakers of "English," virtually all of what is discussed might be similarly applied in the context of the dominant language of virtually any country or region.

[2]   Screen reader programs are commercial text-to-speech software packages that allow a person who is blind (or has certain other specific disabilities) to interact with a variety of other software—such as word

processors, spreadsheets, and Web browsers. The most popular screen reader programs are very feature rich, but they often require considerable practice to learn how to use them well.

3   A refreshable braille display has hardware that raises and lowers braille dot patterns on command from a computer. Users can read the braille dot patterns by moving their fingertips across the Braille panel, much as they would read the braille dot patterns on paper. Refreshable braille displays are the primary means of access to computers for individuals who are deaf-blind.

4   In the context of educational testing, the term *accommodation* has been defined as "any action taken in response to a determination that an individual's disability or level of English language development requires a departure from established protocol" (National Research Council, 2004, p. 1; definition adapted from AERA, APA, & NCME, 1999, p. 101). Accommodations are typically categorized into five categories: setting (e.g., separate testing location, individual administration); timing and scheduling (e.g., extended testing time, frequent breaks); presentation (e.g., reading aloud by a live reader, prerecorded audio, or synthesized speech; font enlargement); response (e.g., student dictates answer to scribe, student types instead of writing by hand); and "other" (e.g., nbq use of bilingual word lists, dictionaries).

5   As Linn (2002) wrote: "The purpose of an accommodation is to remove disadvantages due to disabilities that are irrelevant to the construct the test is intended to measure without giving unfair advantage to those being accommodated."

6   Validity refers to the "degree to which accumulated evidence and theory support specific interpretations of test scores entailed by proposed uses of a test" (American Educational Research Association, American Psychological Association, & National Council on Measurement in Education [AERA, APA, & NCME], 1999, p. 184).

7   In other words, this chapter might be helpful in the "requirements development" stage of developing computer-based testing systems. Regarding that stage of software development, McConnell (1998) wrote,

> *I depart somewhat from the common terminology in referring to these activities as requirements development. These activities are often referred to as "specification," "analysis," or "gathering." I use the word development to emphasize that requirements work is usually not as simple as writing down what key users of the software want the*

software to do. Requirements are not "out there" in the users' minds waiting to be gathered in the same way that iron ore is in the ground waiting to be mined. Users' minds are fertile sources of requirements, but the project team must plant the seeds and cultivate them before the requirements can be harvested. (p. 114)

The importance of both validity and accessibility and the idea of a validity framework within which to understand accessibility might be regarded as "seeds" that may be planted, cultivated, and yield a crop of requirements.

[8] The readaloud feature, especially when implemented with prerecorded audio, is often referred to as the "audio" feature or presentation modality. The term *readaloud* seems to be most common at the kindergarten through Grade 12 level.

[9] The Reading Framework for the 2003 NAEP indicated that "because NAEP is a reading comprehension assessment, test administrators are not allowed to read the passages and questions aloud to students" (National Assessment Governing Board, 2002, p. 3). However, the document does not provide a detailed rationale.

[10] The NAEP math assessment provides the readaloud accommodation with a live reader. The readaloud capability is important for individuals with visual impairments and certain learning disabilities.

[11] It should be noted that the investigation that led to this conclusion included a review of research on what is termed the "interaction hypothesis," the idea that a valid accommodation "will improve scores for students who need the accommodation but not for the students who do not need the accommodation" (National Research Council, 2004, pp. 87, 96). Studies based on the interaction hypothesis were deemed as not providing evidence sufficient for determining the validity of an accommodation. See Koretz and Barton (2003) for a brief but useful discussion of research on accommodations, including comments on the research based on the interaction hypothesis.

[12] The terms *Evidence Centered Design* and *Evidence Centered Assessment Design* both use the acronym ECD.

[13] This description is adapted from Hansen et al. (2005). For the sake of simplicity, this figure does not portray important test construction activities (e.g., item writing, field testing, calibration) that occur after the assessment design and before the full operational assessment delivery system.

[14] These profiles are described with respect to seven KSAs (i.e., reason, know math vocabulary in English, know nonmath vocabulary in English, decode, see, hear, know braille codes) that may be important to consider

regarding a test of *math*. This set of seven KSAs includes both (a) skills that may be important components skills of the targeted proficiency that the test attempts to measure (reason, know math vocabulary, etc.) and (b) other skills that, although not part of the targeted proficiency, may be essential for test takers to possess to perform well on the test in one or more conditions under which it may be administered (some of which involve accessibility features). Solidifying the distinction between "a" skills and "b" skills will occur in a later step.

[15]   This statement may not be true in some societies or settings in which there is not an affirmative effort to remove barriers to educational opportunity for subpopulations.

[16]   Table 2 is indicative of tendencies and does not represent the full diversity that can exist within a test-taker profile.

[17]   Given that the set of *ancillary* KSAs consists of *all those that are not focal* KSAs, the number of ancillary KSAs could be extremely large. In analyses of such KSAs, one should select ancillary KSAs that deserve special attention. For example, for tests of typical academic subjects, the abilities of knitting with yarn ("knit") and running ("run") would be designated as ancillary KSAs but are probably not worth special attention because they are not only *not* part of the targeted proficiency, but are also quite unlikely to be required or demanded of the test taker under any set of operational testing conditions for typical academic subjects.

[18]   These definitions are not intended as indications of what math and reading comprehension constructs *should* consist of—whether decoding ought to be a focal or ancillary in a given assessment application depends on the purpose and target population of the assessment—but rather to illustrate ways of making more explicit specifically what one intends to measure.

[19]   In considering KSAs needed to perform well in criterion situations, it seems sensible not only to consider KSAs needed by nondisabled native speakers of English but *also* KSAs needed by individuals with disabilities and English language learners *who are receiving reasonable accommodations* in the criterion situations.

[20]   The convention in this presentation is to construct KSA scales into discrete levels such that designating the lowest level of the scale as part of the targeted proficiency is functionally equivalent to saying that the KSA is nonessential and is therefore an ancillary KSA. Scales constructed in this fashion capture all essential lower end variation within the test taking population for that test.

[21]   In another analysis (Hansen & Mislevy, in press), knowledge of nonmath vocabulary was considered ancillary. Again, the point of this chapter is not

to specify the definition of targeted proficiencies but rather to help one reason about accessibility in the presence of specific definitions of targeted proficiencies.

22   An assumption undergirding this analysis is that the model is sufficiently complete and accurate in terms of capturing and representing the related KSAs. Let us suppose that this assumption is met.

23   Note that although the *definition of a targeted proficiency* is fundamentally a *choice* or *policy*, the determination of *requirements* of the operational testing situation is arguably more amenable to resolution through *empirical research*. That empirical research will generally be organized around some model or theory of the psychological or physiological processes underlying performance. One may be able to strengthen the coherence of the argument of an assessment by making that model or theory explicit and explaining evidence for its accuracy and adequacy.

24   Although not specifically emphasized in the analyses of this chapter, one needs to be alert to requirements for knowledge about *how* and *when* to use accessibility features (e.g., "know how and when to use readaloud feature"). There may be difficult learning curves associated with the use of certain assistive technologies (e.g., braille, screen reader, tactile graphics). To avoid having these requirements inappropriately depress scores, sometimes testing organizations may specifically stipulate that test takers must have used the assistive technology in instructional settings in order to be granted an accommodation in a testing situation.

25   Thus, we see the connection between (a) the definition of the construct, (b) the KSAs actually required to perform well, and (c) the task performance features (use of math vocabulary) in order to elicit or drive those requirements.

26   The term *ancillary requirements* has a meaning similar to the term *access skills,* when considered in relation to a particular assessment and manner of administration.

27   Obviously, only a very poorly designed test would impose no focal requirements.

28   It may be useful to some readers to have a somewhat more technical description of "effective proficiency." Effective proficiency is a latent variable representing the capability for performance on a test or item under a specific set of testing conditions. This makes effective proficiency tantamount to what is termed the *true score* in standard psychometrics (i.e., an expected score for performance on a test under the conditions it is administered). As implied in its similarity to the true score of standard psychometrics, effective proficiency is modeled as excluding measurement

error. In this chapter, we sometimes refer to effective proficiency as related to performance under operational conditions; this is a somewhat imprecise convention, because we would normally think of performance as visible rather than hidden (latent). "Expected performance" would be more accurate, but it adds a layer of complexity to the presentation. Despite the imprecision, this convention underscores the idea that although a person's level in the *targeted* proficiency is independent of operational conditions, one's *effective* proficiency is closely related to the performance in operational conditions because effective proficiency is a reflection not only of the test-takers' levels in focal and ancillary KSAs, but also the focal and ancillary requirements imposed by the operational performance conditions.

[29]   We ignore measurement error (i.e., random error) in this simple assertion. In addition, this does not account for shortcomings of the model (e.g., sets of focal or ancillary requirements that are either incomplete or inaccurately modeled). Despite these simplifications of reality, this paradigm serves to illustrate many of the interactions between different parts of an assessment argument. Developing useful models of effective proficiency involves consideration not only the definition of the targeted proficiency, but also an understanding of how the range of task features (including accessibility features) interact with or impinge upon the cognitive, sensory, and physical abilities of test takers.

[30]   Note the excessive focal requirement for knowledge of nonmath vocabulary in English would not, arguably, undermine the validity of a test administered to a nondisabled native speaker of English, since that person has "okay" or better level in that KSA (per the definition of nondisabled native speaker of English) and is therefore able to satisfy the excessive focal requirement. Thus, excessive focal requirements may foster invalidity, yet the seriousness of the actual impact depends (among other things) on the characteristics of the specific test taker.

[31]   See Hansen and Mislevy (2004, in press), Hansen, Mislevy, Steinberg, Lee, et al. (in press), and Hansen et al. (2005) for a fuller version of this paradigm.

[32]   Just as any analytical approach has strengths and weaknesses, so does any single index of validity (Hansen, Mislevy, & Steinberg, in press). As a rule, one cannot guarantee or prove that a score will be valid; however one can identify situations that are virtually certain to yield invalid results.

[33]   Regarding tests of content areas such as math and science (as opposed to language arts), Abedi and Dietel (2004) wrote the following: "Modifying, sometimes simplifying the language of test items…has consistently resulted in ELL performance improvement without reducing the rigor of the tests." Without using the terms *focal* or *ancillary*, they seem to be asserting that

for tests of math and science, language ability is an ancillary ability, that the language requirements (driven by factors such as linguistic complexity and vocabulary rarity) cannot be satisfied by ELL students, and that these ancillary requirements can be reduced, thereby reducing or eliminating an accessibility barrier, without reducing the rigor (e.g., the focal requirements) of the assessments. Another variant interpretation is that linguistic ability is focal but that in the original test the requirement for that ability was excessive relative to the level defined in the definition of the targeted proficiency and that reducing the requirement to the proper level eliminated a barrier that had an adverse effect only for the English language learners.

[34]   See Hansen and Mislevy (2004) for a fuller treatment of this issue; this work analyzed issues associated with the use of different kinds of dictionaries as well as re-writing items to reduce unnecessary linguistic complexity.

[35]   Generally, it appears that a test with the label *reading* is more likely to have decoding as part of the construct than a test with the label *reading comprehension.* Also decoding is more likely to be considered part of the reading-related test at the lower grade levels (e.g., elementary) than a reading-related test at the higher grade levels (high school). Nevertheless, the focus of this chapter is not on prescribing what these definitions should consist of, but rather on the importance of and methods for making those definitions explicit and useful parts of one's reasoning about validity and accessibility.

[36]   The fact that there are virtually always ancillary requirements is especially obvious in most educational testing, which is typically a form of "mental" testing. In educational tests, mental or cognitive abilities are typically focal, whereas the sensory and physical abilities required to receive test content or to record one's answers are typically ancillary. Since valid testing cannot occur without receiving test content or recording one's answers, educational testing virtually always involves some ancillary requirements.

[37]   A significant issue in the use of accessibility features concerns how much effort one can afford to expend in identifying the ancillary abilities of each test taker and in matching the ancillary requirements to suit those abilities. The difficulty surrounding this issue is compounded by uncertainties in the diagnosis or identification of disabilities or other special status (Koretz & Barton, 2003; Pitoniak & Royer, 2001). Such issues bear upon policies regarding eligibility to take the test at all or to receive accommodations.

[38]   This is one of several possible definitions of the term *decoding.* For some analyses, it may be important to change or refine this definition.

[39]   In planning for universal design features, it is important to realize these features may impose new requirements for knowing how and when to use the features. Care must be taken to ensure that these requirements are not

excessive. As has been discussed, if the requirements are ancillary (as would often be the case), then test implementers should take steps to ensure that the test taker can satisfy these requirements, such as by having adequate practice and familiarization activities. It should noted, however, that some universal design features, such as elimination of unnecessary linguistic complexity, may induce no important new requirements (Abedi & Dietel, 2004).

[40]   "In addition to the <accessForAll> element, an extension to the LIP <eligibility> section is included here. The <accommodation> element allows a description of the accommodations made for interactions with a particular learning object (or set of them). Also included is a means to represent who authorized this accommodation, when it was authorized, and when it expires. These extensions represent the start of a more systematic approach to describe eligibility and accommodations" (IMS Global Learning Consortium, 2003).

[41]   See the text boxes on the Accessibility Guidelines of the World Wide Web Consortium and on the Section 508 Standards.

[42]   See also Allman (2004) for guidance from the American Printing House for the Blind on accessible testing for individuals with visual disabilities, not specifically focused on computer-based delivery.

[43]   Even when a human reader is used in a computer-based testing situation, it can be very important to provide a text equivalent—especially where the reader has not had an opportunity to preview the nontext content (e.g., graphics for a math test). (The lack of a preview opportunity would be common where the computer-based test is adaptive or otherwise constructed by the computer from a large pool so that the selection and ordering of items could not be known in advance.)

[44]   Sheinker et al. (2004) provided a useful adaptation from Thurlow, House, Boys, et al. (2000).

[45]   The researchers have found Bayes net editing systems to be useful tools in representing complex argument structures produced in ECD domain modeling. A Bayes net consists of a set of variables, a graphical structure connecting the variables, and a set of conditional distributions. One adds "evidence" to a Bayes net by setting variables to particular values. Adding evidence can take the form of either (a) observing the values of certain variables and then studying the implications for other variables in the network or (b) hypothetically treating certain variables as if their values were known in order to carry out "what-if" analyses that illuminate implications for other variables in the network. Our work has focused on this latter form. Changes made to these values propagate according to

Bayes theorem, yielding updates (posterior values) for each of the other variables. Once such models have been constructed in Bayes net software, it is possible to quickly work through the validity implications of various combinations of test-taker profiles, task characteristics, definitions of the targeted proficiency, and so forth. See other works (Mislevy, 1994; Jensen, 1996; Hansen & Mislevy, in press; Hansen et al., 2003; Hansen et al., 2005) for information on the use of Bayes nets.

Section IV

Security, Authentication, and Support

Chapter XII

Delivering Computerized Assessments Safely and Securely

Eric Shepherd, Questionmark Corporation, USA

John Kleeman, Questionmark Corporation, USA

Joan Phaup, Questionmark Corporation, USA

Abstract

The use of computers to assess knowledge, skills, and attitudes is now universal. Today, distinguishing between the various delivery and security requirements for each style of assessment is becoming increasingly important. It is essential to differentiate between the different styles of computerized assessments in order to deploy assessments safely, securely, and cost effectively. This chapter provides a methodology for assessing the security requirements for delivering computer-based assessments and discusses appropriate security measures based on the purpose and nature of those assessments. It is designed to help readers understand the issues that need to be addressed in order to balance the need for security with the need for cost effectiveness. The authors hope to give readers a working knowledge of the technological innovations that are making it easier to ensure the safety and security of a wide range of computerized assessments including online tests, quizzes, and surveys.

Introduction

Security is a pressing concern for organizations using computer-based assessments, but so is cost effectiveness. It is essential to maintain the integrity of item pools, keep test scores secure, and minimize the possibility of cheating—but not at an exorbitant price. The thoughtful use of technology, combined with careful consideration of an assessment's purpose and consequences, makes it possible to deliver assessments safely, securely, and economically.

Overengineering low-stakes assessments can result in needless expense and wasted time. Underengineering high-stakes assessments can erode confidence, jeopardize organizational processes, and call into question the face validity of an assessment.

This chapter provides a methodology for assessing the security requirements for delivering computer-based assessments and discusses appropriate security measures based on the purpose and nature of those assessments. It explores the means and technologies that can ensure the safety and security of a wide range of computerized assessments. The following topics will be covered:

- Assessment definitions
- Considerations in deploying assessments
- Delivery environments
- Technologies for deploying assessments safely and securely
- Future trends

Background

The use of computers to assess knowledge, skills, and attitudes is now universal. Today, distinguishing between the various delivery and security requirements for each style of assessment is becoming increasingly important. It is essential to differentiate between the different styles of delivering computerized assess-

Table 1. Types of assessment

Term	Definition within the Computerized Assessment Context	Measure or Learn	Consequences or Stakes
Assessment	Any systematic method of obtaining evidence by posing questions to draw inferences about the knowledge, skills, attitudes, and other characteristics of people for a specific purpose.		
Exam	A summative assessment used to measure a student's knowledge or skills for the purpose of documenting his or her current level of knowledge or skill.	Measure	High
Quiz	A formative assessment used to measure a student's knowledge or skills for the purpose of providing feedback to inform the student of his or her current level of knowledge or skill.	Promote learning	Low
Survey	A diagnostic assessment to measure the knowledge, skills, and attitudes of a group for the purpose of determining needs that must be met in order to fulfill a defined purpose.	Measure	Low
Test	A diagnostic assessment to measure a student's knowledge or skills for the purpose of informing the student or his or her tutor on his or her current level of knowledge or skill.	Measure to promote learning	Medium

ments in order to deploy assessments safely, securely and cost effectively. One important distinction that governs the delivery of an assessment is whether the assessment is used to measure knowledge, skills, attitudes, and personality traits, or if it is used to promote learning and minimize forgetting.

Assessments may also be distinguished by their consequences. As a result of an assessment, people might be hired, fired, promoted, demoted, graduated, not graduated, released from custody, and authorized or certified to perform a particular job. The potential outcomes of an assessment could tempt a participant to cheat and consequently affect the degree of security that needs to be applied.

Specific assessments take the form of quizzes, tests, exams, or surveys, depending on whether the purpose of the assessment is to measure or promote learning. Table 1 defines the types of assessments that will be discussed in this chapter.

In addition to this terminology, Table 2 classifies assessments as follows: formative, summative, diagnostic, and reaction.

Table 2. Classifying assessments

Name of assessment	Uses
Formative	These assessments are designed to help students turn information into knowledge by asking questions that tend to increase remembering and reduce forgetting.
Summative	These assessments are designed to measure knowledge, skills and attitudes by asking questions and measuring the responses.
Diagnostic	These assessments are designed to provide a diagnosis and prescription to help people reach their objectives.
Reaction	These assessments are designed to assess the reactions and opinions of students about their learning experience.

Table 3. Consequences and motivation

Type of Assessment	Consequences	Motivation to Complete Assessment
Exam	Consequences for passing or failing an exam might extend to being hired, fired, promoted, demoted, released from custody, authorized or certified, and graduating or not graduating.	High
Quiz	Passing or failing a quiz has minimal consequences. Quizzes are normally formative assessments used to promote learning and reduce forgetting rather than form a summative judgment.	Medium
Survey	There should never be direct consequences to a respondent (i.e. the person answering a survey). However, as results are tabulated for the group, training courses, job aids, and other interventions may be planned. Sometimes the opinion might have consequences for others, such as an instructor in the context of a course evaluation, or managers in the context of an employee attitude survey.	Low The challenge is to get people to respond.
Test	The consequences for passing or failing an individual test are not high. However, consistent failures might extend to not being hired, being fired, being demoted, or not graduating.	Medium

The method of delivering a computerized assessment depends not only upon the use and consequences of the results, but also on the motivation of the participant to complete the assessment (see Table 3).

This chapter does not describe all of the consequences, legal liabilities, validity, reliability, and planning issues that are involved with writing a suitable assessment. Many other publications (see the Recommended Reading list) detail the processes and procedures used to produce accurate and reliable assessments. However, as Figure 1 shows, all of these issues are related to the stakes of an assessment.

Figure 1. Issues related to stakes of assessment

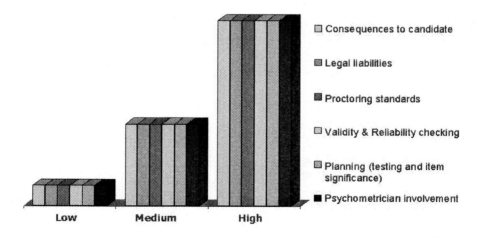

Table 4. Delivery issues

Type of Assessment	Delivery Issues to Consider
Exams	o Face validity (trustworthiness of the results) o Motivation to cheat is high o Overexposure of content o Content is expensive or time consuming to produce o Content protection o Authentication of candidate o Correct authorization (prerequisites completed) o Time window to limit access o Preventing repeated access to exam o Consistency of delivery o Secure player required o Answers saved regularly in case of a technical failure
Quiz	o Content protection o Environment similar to performance situation o Motivation to cheat is medium
Survey	o Ballot rigging o Anonymity of respondent
Test	o Face validity (trustworthiness of the results) o Motivation to cheat is high o Over exposure of content o Content protection o Environment similar to performance situation o Time window to limit access o Limiting the number of times a person can have access to test o Consistency of delivery o Secure players are recommended o Answers saved regularly in case of a technical failure

Assessments can be delivered safely and securely without over- or underengineering their delivery and deployment. Table 4 details some of the issues to be considered.

Environments for Delivering Computerized Assessments

The delivery environment for an assessment is driven by the stakes of the assessment. The higher the value of the assessment, the more care must be taken to control the environment and to supervise the assessment. Because assessments are attached to a range of consequences, there is also a variety of ways and places to deliver them (see Figure 2).

Figure 2. Ways to deliver assessments

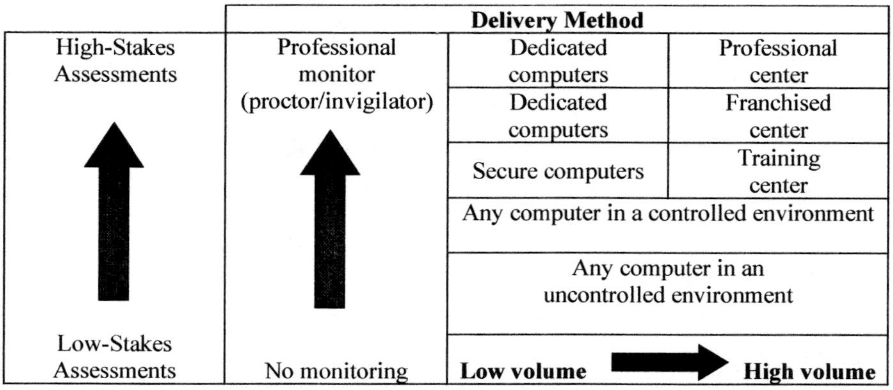

		Delivery Method	
High-Stakes Assessments ↑	Professional monitor (proctor/invigilator) ↑	Dedicated computers	Professional center
		Dedicated computers	Franchised center
		Secure computers	Training center
		Any computer in a controlled environment	
		Any computer in an uncontrolled environment	
Low-Stakes Assessments	No monitoring	**Low volume** ➡ **High volume**	

Assessment Centers

Assessment centers are controlled environments designed to provide groups of users with a venue that meets the conditions stipulated by the organization that is administering the assessment. This means that all candidates take the assessment under identical conditions, even if they go to different assessment centers.

High-stakes assessments must offer an environment that is free from distractions, together with some form of monitoring. Low-stakes assessments do not require such rigor. The greater the rigor that is expected, the more costly an assessment will be to deliver. The physical environment, technology, monitoring standards, and authentication methods for delivering an assessment are stipulated by the organization that owns the assessment. Following are the essential ingredients of a formal assessment center:

* **Physical Environment:** Assessment centers should be enclosed rooms without through traffic. They should be quiet and free of distractions, well lit, and at a comfortable temperature throughout the year. Computers should be separated from one another by acoustic partitions, and the equipment should offer similar operation, accessibility options and performance. Candidates should not be able to print or capture the assessment content and should be prevented from accessing resources, such as Web pages, that might unfairly assist them during the assessment process.

- **Technology Environment:** Assessments should be delivered on consistent and comparable technology. Screen sizes, processor speeds, and network connections do not have to be the latest and greatest, but they do need to be comparable.

- **Monitoring:** Assessment centers can be monitored (or proctored and invigilated) in several ways. Candidates should be in an environment that permits constant visual monitoring. This can be achieved directly by a person in the room or via video surveillance equipment.

- **Authentication:** It is important to confirm that the person taking the test or exam is truly the person authorized to do so. This can be done by confirming a candidate's identity via a photographic ID provided by the government, academic institution, or employer (MCAT and LSAT require a thumbprint). In some cases a candidate's identity is confirmed directly by the employer or by the candidate's signature.

To aid the following discussion, we have presented the issues to consider and established the following grades to help readers quickly understand the suitability of each delivery method for a particular style of assessment.

Quick Reference Guide for Suitability	
Suitability for style of assessment	Grade
Very suitable	A
Suitable	B
Suitability will vary by circumstance	C
Unsuitable	D
Definitely not suitable	F

Types of Dedicated Assessment Centers

Computerized assessment centers come in various forms that include the following:

- **Professionally Controlled Centers:** Professionally controlled centers provide a very consistent and highly controlled environment in which a candidate could expect to receive exactly the same treatment and have the same experience in every assessment center. This consistency would start with the reception and continue through the assessment process to the time that he or she leaves the premises. These centers are used for very high-stakes tests such as nursing and medical exams and are expensive to use and maintain. Such centers provide the consistency and rigor required for a very high-stakes, government-regulated certification exams. These centers are rarely used to deliver nonregulatory exams due to high running costs.

Professionally Controlled Centers	
Form of assessment	Grade
Very high-stakes exams	A+
High-stakes exams	A
Test	D
Quiz	F
Survey	F

- **Franchised Centers:** Franchised centers are similar to professionally controlled centers, but they cannot enforce the rigor required for very high-stakes assessments. However, these centers are less expensive and are great places to offer high-stakes tests in a reasonably consistent environment. These centers are far more commonplace than professionally controlled centers and deliver both government-regulated and non-government-regulated exams. There are approximately 5,000 franchised centers around the world.

Franchised Centers	
Form of assessment	Grade
Very high-stakes exams	B
High-stakes exams	A
Test	C
Quiz	F
Survey	F

- **Higher Education Centers:** Many colleges and universities make their testing facilities available to local companies and are sometimes part of a franchised assessment network. These centers resemble professionally controlled centers, and they enforce the rigor required for their institution. Such centers are less expensive and are suitable for offering high-stakes tests in a reasonably consistent environment. These centers are ideal, not only for the institution's use but also for local companies that need to administer high-stakes assessments in partnership with a local institution.

Higher Education Centers	
Form of assessment	Grade
Very high-stakes exams	B
High-stakes exams	A
Test	C
Quiz	D
Survey	F

- **Corporate Centers:** Increasingly, large organizations are establishing their own assessment centers to provide a consistent, confidential, and cost effective method for delivering high-stakes assessments. These centers might be testing potential employees, partners, or current employees on issues pertaining to safety or government regulation. These centers are equipped similarly to professionally controlled or franchised centers.

Corporate Centers	
Form of assessment	Grade
Very high-exams	A
High-stakes exams	A
Test	B
Quiz	C
Survey	C

Nondedicated Assessment Centers

Most assessments—computerized or otherwise—are administered in nondedicated assessment centers. Moreover, as the features, quality, and sophistication of assessment software grows and methods of authentication improve, then more and more higher-stakes assessments are being administered within this type of environment. Nondedicated assessments centers can be any of the following:

- **Training rooms doubling as assessment centers:** Training rooms may not offer such a tightly controlled environment as might be found at a dedicated assessment center. For example, a training room is unlikely to offer screening between computers, but these limitations can be easily overcome with a little thought and by deploying professional assessment software. Packages such as Questionmark™ Perception™ and Questionmark Secure can provide the technology to convert training rooms into assessment centers. Professional software packages can prevent printing, capturing assessment content, exiting inappropriately, and task switching. They also make it possible to randomize the distracters in multiple choice and response questions or even create an assessment from randomly drawn questions, thereby making it difficult to cheat.

Training Rooms	
Form of assessment	Grade
Very high-stakes exams	B-
High-stakes exams	A-
Test	A
Quiz	B
Survey	A

- **Closely supervised in workplace:** Of all the issues that affect the delivery of computerized assessments, monitoring remains critical for delivering high-stakes certification exams. Widely dispersed candidates can be monitored by their supervisor or manager, making supervised assessments in the workplace a practical, valid, and cost effective way to provide high-stakes assessments. Beyond that, the advances in assessment software are now proving that even very valuable assessments can be scheduled at an assessment taker's own desk with limited supervision. Assessments can open at a specified time, making sure that the candidate cannot get a "sneak preview."

Closely supervised in workplace	
Form of assessment	Grade
Very high-stakes exams	C
High-stakes exams	B+
Test	A
Quiz	C
Survey	C

- **Remotely supervised in workplace:** With the advances of technology it is now possible to monitor candidates using video cameras. Although not suitable for very high-stakes assessments, this can provide a valid way to administer exams remotely.

Remotely Supervised in Workplace	
Form of assessment	Grade
Very high-stakes exams	C
High-stakes exams	C+
Test	B
Quiz	C
Survey	C

- **Unsupervised in workplace:** Low- and medium-stakes, unsupervised assessments at a participant's desk or other work venue are growing in importance as organizations strive to evaluate the effectiveness of their training and the need for training. Such assessment also provides new ways of promoting the learning process. Although this form of assessment does not require the same monitoring, it does require assessment software that prevents participants from printing or capturing confidential information.

Unsupervised in Workplace	
Form of assessment	Grade
Very high-stakes exams	F
High-stakes exams	D
Test	A
Quiz	A
Survey	A

- **Supervised at home:** Some distance-learning institutions and small-scale certification authorities are now allowing students to take higher stakes assessments at home. The candidate must provide the organization with a choice of professionals (e.g., a doctor, lawyer, or accountant) to supervise the assessment. Once again, the success of this kind of delivery lies in the assessment software and the selection methods for the monitor.

Supervised at Home	
Form of assessment	Grade
Very high-stakes exams	C
High-stakes exams	A
Test	A
Quiz	C
Survey	C

- **Unsupervised at home:** Most commonly used for delivering low-stakes self-assessments aimed at promoting learning and reducing forgetting.

Unsupervised at Home	
Form of assessment	Grade
Very high-stakes exams	F
High-stakes exams	F
Test	B
Quiz	A
Survey	A

- **Unsupervised in a public place (library):** Most commonly used for delivering low-stakes self-assessments aimed at promoting learning and reducing forgetting.

Unsupervised in Public	
Form of assessment	Grade
Very high-stakes exams	F
High-stakes exams	F
Test	B
Quiz	A
Survey	A

Creating the Right Environment

Delivering assessments securely is becoming easier as the quality of networks and assessment software improves. Secure assessment players and monitoring solutions are pushing back the barriers of what types of assessments may be safely and securely delivered either remotely, at the desk, or in a training room.

When close monitoring and a consistent environment are essential for geographically dispersed candidates, there are many excellent assessment centers available.

Assessment software now permits many organizations to create and run high-stakes assessments within their own environments. As the quality of software improves, and monitoring and network solutions grow, this trend will certainly continue.

Case Study
Not all organizations believe that it is essential to have exams proctored.

A U.S. software company decided that it was more important to ensure that certification was available to everybody who wanted it than ɔ insist that all certification exams be proctored. Setting proctored assessments in specific locations at specific times inevitably limits the opportunities for taking the assessments.

Instead, following the successful completion of the software training courses, candidates can use a credit card for the exam to become a certified professional in the use of its products. They then take the appropriate assessments at home or at their convenience. The assessments are authored, delivered and reported on using Question*mark* Perception software.

The company claims that people who wish to dispute the validity of a certified professional can arrange to serve as proctor and request that the person retake the test. As a result, the company has found it much easier to run its certification program and is confident that the system is not being abused.

Creating the Right Physical Environment

The ideal assessment environment would incorporate the following characteristics:

- Comfortable, clean, and free of smoke
- Adequate lighting, ventilation, comfortable seating and work surfaces
- No external distractions
- No opportunity to use printers, fax machines, copiers, or telephones during an assessment
- A viewing window or video surveillance system
- If a monitor is present, it should have an unobstructed view of each candidate in the assessment room
- IT equipment should be consistent, robust, and in compliance with the local accessibility rules and regulations

Deploying Software to Deliver Secure Assessments

Although many security issues can be addressed within a physical environment, new software features can ensure that assessments are delivered in a secure manner. The following features are essential in deploying assessments safely and securely:

- **Authoring secure assessments:** Many safeguards can be taken during the assessment authoring process to protect the security of the assessment. One of the most popular is to shuffle the order of the choices. The questions themselves can also be delivered in a random order. Both of these features help prevent cribbing when users are sitting in nonscreened assessment centers.

- **Encrypted communications:** The majority of computerized assessments are now delivered via the Internet or an intranet. Communications between the browser and server must be encrypted. This means that someone who "sniffs" the network will not be able to see the information sent from server to browser and vice versa. Secure assessment management systems use Secure Sockets Layer (SSL), a protocol that allows the browser and Web server to encrypt their communication. This means that anyone who intercepts the communication will not be able to read it.

- **Scheduling assessments:** Scheduling enables users to set start times of assessments, the length of assessments, and the number of times that assessments may be taken. It also enables user names to be specified and password-protects the assessments.

- **Monitoring assessments:** A candidate cannot start a monitored assessment until after a proctor has logged on. The monitor can also be limited to a range of IP addresses to ensure that a certain physical location is used to administer the assessment.

- **Secure browsers:** One of the key recommendations is to "lock down" computers that are being used in assessments to keep users from accessing other applications and Web sites while completing a medium- or high-stakes assessment. This lockdown is usually performed via a secure browser and prevents candidates from printing, capturing screens, accidentally exiting the assessment viewing source, task switching, and so forth. The secure browser must be authenticated by the server to ensure that the server isn't tricked into sending secure content into an insecure environment.

- **Analytics:** One key aspect of delivering assessments safely and securely is to analyze the results. Analyzing the results make it possible to determine if assessments are useful, effective, and secure.

Having considered the alternatives for delivering assessments, it is worth examining the assessment software, tools, and environments that organizations might need to create, deploy, and report on their own secure assessments.

Technologies for Deploying Assessments Safely and Securely

The table on the following page details technological features and environments for deploying assessments safely and securely.

Future Trends

Computers have been used to conduct assessments for more than two decades, but assessment technologies are still in their infancy. Computerized assessments were once delivered using green screen terminals, then PC DOS, Macs, Windows, and now, browsers. As different device profiles evolve, we will see new and exciting delivery systems. Phones, PDAs, tablets, and wireless and wearable devices will soon be used to perform assessments. These devices make assessment delivery more accessible and are likely to be used within the environment in which an individual is required to perform at work, thus making the measurement more authentic. Risks associated with security, content protection, reliability, and validity will come along with these benefits.

Authentication was originally handled by each computer application; this was streamlined by central repositories of user names and passwords that applications could reference. Now Single-Sign-On Portals (SSOPs) allow people to sign on in one place and access the applications they need by reentering their user names and passwords. These SSOPs are currently organization specific, but they will become interorganizational so that the portal a worker uses for their office e-mail might also authenticate them for night school classes.

Imagine everyone having a DNA sample taken at birth, which would be used to authenticate them on any and every system that they ever needed to use. While this idea is problematic, it would offer the simplicity of having a single system to authenticate people for an endless range of activities.

The methodologies discussed in this chapter develop new dimensions as assessments are authenticated by SSOPs and delivered to these new devices. Consider

Type of Assessment	Features to Use

Exams	o Secure browsers
	   o Keep the exam content confidential
	   o Prevent printing, capturing content, task switching, and so forth
	o Secure sockets
	   o Prevent network pirates intercepting content
	o Proctors–Invigilators–Monitors
	   o Authenticate that the right candidate takes the exam
	   o Monitor the candidate during the exam
	o Monitoring the exam from a specific range of IP addresses
	   o Ensures the correct room and computer are used
	o Exam only available during a limited time window
	   o Reduces opportunities for people to relay information about the test
	o Shuffling choices
	   o Prevents candidates seeing which choice another candidate selected
	o Randomizing questions
	   o Candidates get different questions, which reduces the possibility that they will see the answers from another candidate
	o Jumps can stop tests for poor candidates
	   o Prevents exposing content to candidates who obviously will not pass

Test	o Secure browser
	   o Keeps assessment content confidential
	   o Prevents printing, capturing content, task switching, and so forth
	o Secure Sockets
	   o Prevent network pirates intercepting content
	o Monitors
	   o Authenticate candidate
	   o Monitor the candidate
	o Monitor from a specific range of IP addresses
	   o Ensures the correct room/computer is used
	o Assessment only available during a limited time window
	   o Reduces opportunities for people to relay information about the test
	o Shuffling choices
	   o Prevents candidates seeing which choice another candidate selected
	o Randomizing questions
	   o Candidates get different questions, which reduces the possibility that they will see the answers from another candidate
	o Jumps can direct students to helpful content
	   o Prevents premature exposure of content

Quiz	o Secure browsers
	   o Keep the assessment content confidential
	   o Prevent printing, capturing content, task switching, and so forth
	o Secure Sockets
	   o Prevent network pirates from intercepting content
	o Assessment only available during a limited time window
	   o Reduces opportunities for people to relay information about the test and short-circuit the cognitive process
	o Randomizing questions
	   o Candidates get different questions
	   o Provides repeated search and retrieval practice

Survey	o Respondents' details (name, IP address) can be erased to
	   o Maintain anonymity
	   o Eliminate ballot rigging

this: even the machines we use every day could assess our knowledge, skills, abilities, and attitudes. Why not have phones use voice recognition technologies to determine the level of customer service skills? Could a person's car monitor and evaluate their driving, and relay that information to their insurance company? Unskilled and inexperienced drivers would oppose the idea, and it would be difficult to correlate knowledge, skills, abilities, and attitudes with safe driving, but such a development is conceivable.

As the future unfolds, we will discover that capturing data becomes less of a challenge as technological innovations and new tools enhance our ability to analyze it. Such tools will allow us to understand the data points that do correlate, thus ensuring the validity and reliability of tests and exams. More varied accessible delivery devices will make enable us to analyze data to tease out the knowledge, skills, and behaviors that drive performance. We will also be able to detect potential fraud. It is possible, using statistical analysis, to determine if someone is cheating. For instance, it would be suspicious if someone spent the same amount of time on easy questions as on hard ones. Using these and other measurements, fraud can be detected and the culprit identified. This reduces but does not eliminate the requirement for proctoring high-stakes assessments.

In addition to these trends, we will see the following new developments:

- Work flow environments will facilitate the secure creation and approval of assessments and their contents. At the moment, items have to be disseminated through networks and computers that are not secure. Workflow systems with centralized authentication and authorization will only expose test items that the test developers are allowed to see.

- The use of sound, video, and virtual reality will make the assessment environment simulate the performance environment, thus producing a more authentic measurement.

- Integration with commonly used technologies will provide frictionless access to the right assessment at the proper time and place, with the appropriate security. For instance, a text message might remind a student who is away for the weekend to take a quiz. The student could simply respond by using the text messaging feature of the phone.

- Analytics will help organizations effectively monitor deployments of higher stakes assessments, enabling rapid response to any security breaches.

Conclusion

Computerized assessments provide a powerful means of delivering exams, tests, quizzes, and surveys. The underlying technologies for delivering different types of assessments are similar, but each assessment must be evaluated in terms of its particular outcome to determine which delivery method will best serve the purpose of the assessment.

Organizations have a wide variety of assessment environments, authoring practices, and technologies that make it possible for them to deliver all types of assessments safely, securely, and economically. It is essential to consider the nature and purpose of assessments and use appropriate delivery environments in order to balance security requirements with the need for cost effectiveness.

Recommended Reading

Godwin, J., & Shepherd, E. (2004). White paper: Assessments through the learning process. Available online from *http://www.questionmark.com/communities/getresource.asp?file=Assessments%20Through%20the%20Learning%20Process.pdf&group_id=5*

Hale, J. (1999). *Performance-based certification: How to design a valid, defensible, and cost-effective program.* San Francisco: Jossey-Bass.

Hall, P., Kleeman, J., Miller, G. V., Squires, P., & Shepherd, E. (2002). White paper: Creating and deploying computerized level 1 assessments: Trainee reaction surveys that count. Available online from *http://www.questionmark.com/communities/getresource.asp?file=Level%201%20assessments.pdf&group_id=5*

Kirkpatrick, D. (1998). *Evaluating training programs: The four levels.* San Francisco: Berrett-Koehler.

Shrock, S. A., & Coscarelli, W. C. (2000). *Criterion referenced test development: Technical and legal guidelines for corporate training and certification.* Silver Spring, MD: International Society for Performance Improvement.

Thalheimer, W. White paper: The learning benefits of asking questions. Available online from *http://www.questionmark.com/communities/getresource.asp?file=LearningBenefitsOfQuestions.pdf&group_id=5*

Westgaard, O. (1999). *Tests that work.* San Francisco: Pfeiffer.

Chapter XIII

Securing and Proctoring Online Assessments[1]

Jamie R. Mulkey, Caveon Test Security, USA

John Fremer, Caveon Test Security, USA

Abstract

This chapter examines five main aspects of delivering secure tests and examinations via online methods. First, the importance of understanding the problem of test cheating is discussed. Existing guidelines that help shape computerized test delivery are reviewed for how they manage test administration processes. A taxonomy of different types of cheating is presented and solutions to important security issues in online assessment are considered. Last, future trends in deterring and detecting cheating are proposed. Furthermore, the authors hope that understanding the threats to online test security and considering the options to solve these problems, will provide a forum for reducing test theft and increasing integrity for test takers and test sponsors alike.

Introduction

This chapter examines five main aspects of the issue of delivering secure tests via online methods:

- Growth in importance of the problem of cheating
- Guidelines for delivering computer-based tests
- A taxonomy of types of cheaters
- Solutions to important security problems in online assessment
- Future trends in deterring and detecting cheating in the online assessment setting

We first point out that cheating appears to be on the upswing in all phases of life, and our experience in online testing merely reflecting broader social developments. Measurement professionals have been sensitive to the growing problem of cheating and have been developing and applying professional guidelines designed to encourage uniform and effective testing practices.

We offer a 10-part categorization of types of test cheaters, doing our best to come up with memorable names that might cause test proctors and others to be more vigilant about observing the kinds of misbehavior that all too frequently occurs in the online testing setting.

The next section provides specific examples of solutions to common security problems. We close by speculating about the future directions that will strengthen the security of online assessments.

The Growing Importance of the Problem of Cheating

Rising costs for developing online assessments have made their security and protection even more crucial in today's marketplace. A test that has been developed to meet professional psychometric standards can cost a test sponsor many hundreds of thousands of dollars to develop. Once a test is developed for online assessment, all precautions must be taken to maintain its security.

Cheating on tests is on the rise. Over the last 30 years, the amount of cheating has gone from 23% to 49% in both high school and college settings (Cizek, 1999). Research conducted by McCabe, Trevino, and Butterfield (1999) suggested that as many as 75% of college students admit to some form of cheating (McCabe et al., 1999). Recent research suggests that college students who are associated with a university's "academic ethic," that is, having an academic locus of control, high class attendance, high GPA, and a low level of going to parties or drinking, were less likely to engage in academic dishonesty (Pino & Smith, 2003).

Not only has cheating on tests become more prevalent, but cheaters are also getting more sophisticated in their methods of cheating. For example, fraudulent test takers now use a variety of technology-based tools to cheat on tests; cell phones are used to text message answers to friends, walkie-talkies are used to transmit and receive answers, and microscopic or cell-phone-based cameras are used to transmit pictures of questions to others outside of the testing environment (Fox, 2004).

Part of what the administrator of a testing program faces as he or she strives to deliver tests that yield fair and valid results is that cheating in every facet of our lives is on the rise. According to David Callahan (personal communication, May 2004), there are a whole host of reasons why individuals cheat. However, there seem to be some recurring themes:

- *"**Whatever it takes, I must win.**"* Rewards for performance have grown exponentially over the past 20 years, especially for those individuals at the top of the pyramid, be it in sports, school, or business. The result is that an individual will do whatever it takes to be a winner.

- **"I'm really worried about money."** In recent years, there has been an increased concern regarding the security of one's future. Individuals who should feel secure about their futures do not. As a result, individuals will cheat to save money, cheat on tests to get good grades, or falsify their own job performance to gain a competitive advantage.

- **"Let sleeping dogs lie."** The idea that one will probably be able to get away with cheating has become more widespread. Watchdog agencies such as the Internal Revenue Service (IRS), the Securities and Exchange Commission (SEC), and state regulatory boards have been shown to be ineffective in the enforcement of monitoring and sanctions for cheaters. In addition to high-profile examples of failures of oversight, key agencies have endured budget cuts at a time when the need for systematic monitoring has gone up.

- **"Go on, indulge yourself."** Last, Callahan suggested that there is more cheating in today's society because "our culture indulges it. We live in a more dog-eat-dog society, where greed and cutthroat competition are often encouraged by role models and television shows (e.g., The Apprentice)" (D. Callahan, personal communication, May 2004)

It is this pervasive climate of cheating that makes it essential that those who develop and manage testing programs have proven processes, policies, and procedures in place that prevent and detect cheating so that appropriate sanctions can be applied.

Guidelines for Computerized Testing

Fortunately, testing professionals have devoted a good deal of attention to creating guidelines for the development and delivery of computerized tests. These guidelines provide the backdrop for processes and procedures that need to be followed in the development and delivery of online assessments in order to produce results that can be counted on. Two very useful resources are the *Guidelines for Computer-Based Testing* (Association of Test Publishers [ATP], 2002) and the *International Guidelines on Computer-Based and Internet Delivered Testing* (International Test Commission [ITC], in press)

According to the *Guidelines for Computer-Based Testing* (ATP, 2002), managing a secure testing program requires attention to topics such as

- **Secure Testing Environments:** Ensuring that test-taker workspaces are clear from extraneous materials, testing cubicles are private so that computer screens can be seen only by the individual taking the test, and sufficient proctoring is provided depending on the exam stakes level. Finally, the ability to encrypt and decode test material and test-taker scores and reports must be made available so that candidate results are stored and maintained in a secure environment.

- **Trained Test Administrators:** Ensuring that proctors understand both the significance of security as well as processes that must be followed during test administration.

- **Appropriate Testing-Environment Infrastructure and Data-Transmission Arrangements:** Confirming that consistent platforms are used for test delivery and that secure test data transmission exists for test content and candidate data.

- **Investigation of Test Delivery Anomalies:** Ensuring processes are available for recognizing and reporting any anomalies such as system failures, irregular test-taker behavior, or errors in transmitting results. The reports for such anomalies need to be preserved for further use as required.

- **Incidence Reporting:** Ensuring that individuals whose job it is to proctor online assessments are trained to recognize abnormal test-taking behaviors and to report them as necessary.

- **Copyright of Testing Materials:** Providing a means to protect the copyright of the testing material during both the development and delivery of online tests. Agreements between the test sponsor and the test taker must be made to bind the test taker to the confidentiality of the test's content.

The ITC is in the process of creating additional guidelines for Internet and computer-based testing (ITC, in press). Although they are similar in content to the ATP guidelines, the ITC guidelines are focused on internationally developed and recognized standards for computer-based and Internet delivery of all types of assessments. The ITC guidelines look at test administration from three perspectives: those of the test developer, the test publisher, and the test user. Specific sections on test security include the following:

- **Supervision of Testing:** The level of supervision required depending on the level of the test delivered.
- **Control of Item Exposure:** Limiting pilot testing and making sure there are large banks of items to support multiple parallel forms.
- **Authentication and Cheating:** Determine the level of authentication required, based on the stakes level of the test. Test takers must have valid forms of picture identification, a test administrator should be present, and surveillance equipment such as video and audio monitoring should be made available to specified tests. It is important to identify the threats to validity that exist if test control is not maintained. Honor policies for unsupervised testing candidates should be considered as a means to avert cheating.
- **Test Taker Commitments:** Test takers should be required to acknowledge in writing they will adhere to the specified conditions for taking the test. If any conditions are breached by the test taker, consequences must result.

Although we know that cheating is inevitable, we also realize there are proactive and reactive activities that can be done to minimize cheating, detect its occurrence, and deter fraudulent test-taking behaviors.

Background

Research in the area of computerized test administration has not generally focused on issues related to test administration, such as security (Bartram, in press). Many studies, on the other hand, have addressed the comparability of paper and pencil and computer versions of tests. Results from these studies suggest that both delivery formats can meet validity and reliability standards (Wainer & Dorans, 2000). However, there is little detail on the administration of computerized tests, either for ability, clinical, or selection processes. (Bartram, in progress).

There are many forms of cheating on tests. Typically, the higher the exam stakes, the greater the likelihood of cheating. Research in this area suggests that college students are more likely to cheat if higher stakes are involved (Cizek, 1999). Elementary school teachers have been found to cheat on behalf of their students for standardized exams where there are negative consequences for the teachers whose students perform poorly (Jacobs & Levitt, 2002) Similarly, it was found that teachers who had irregularities in test administration, including inaccurate timing, altering answer sheets, coaching, teaching the test, scoring errors, and student cheating obtained less accurate test results (Gay, 1990).

Bartram (2001) suggested that there are three areas that test developers should be concerned about ensuring the security of exam results:

1. **The test itself** (i.e., item content, scoring rules, norms, report-generation algorithms, report content)
2. **The test-taker's identity** (i.e., authenticating the person's identity and preserving their confidentiality)
3. **The test results** (i.e., only those eligible to access the test scores are able to do so)

In the administration of computerized high-stakes ability testing, cheating falls into two main categories: exam piracy and exam fraud. Exam piracy deals with the theft or receiving of test-item content, either directly or through a third party. Exam fraud, also known as proxy testing, is the practice of having an individual take an exam on behalf of another individual (Foster, 2003).

Cizek (1999) provided a robust taxonomy of different types of cheating that occur in the testing environment. Most of these descriptions relate to cheating that takes place within paper-and-pencil testing environments. For example, there are sections on "giving, taking, and receiving" (p. 40) of information, use of "forbidden materials" (p. 43), and "taking advantage of the testing process" (p. 48) (Cizek, 1999). These same ideas can be applied to computerized assessments: smuggling calculators into the testing room, memorizing items to bring test content back to others, or knowing test questions and answers ahead of time to achieve a higher test score.

Delivery of assessments for purposes of self-assessment, recruitment, post-hire selection, and psychological evaluation raises some of the same issues related to potential security breaches. There are individuals who will steal items for the purpose of providing them to others as well as individuals who do so in an effort to improve their own future results.

Computerized testing can be planned and managed to provide a more secure environment than paper-and-pencil delivery of high-stakes tests. Tests can be securely transmitted to testing sites using encrypted files. Biometrics such as digitized thumbprints or facial scans can be used to verify a candidate's identity. Test items within an exam can be scrambled so that each individual receives a unique version of the test.

In other cases, test sponsors of paper-and-pencil tests are concerned about moving their tests to a computerized testing environment because they feel their exams are more prone to test security problems. For example, candidates may have the ability to spend an abnormal amount of time on particular items, review an entire test versus just the section they are completing, and be able to retake a failed attempt multiple times within a short period of time.

Computerized test administration also means convenience for the candidate. Computerized test administration is typically continuous meaning a test candidate can make an appointment to take a test any day of the week, whereas paper-and-pencil tests are typically scheduled events.

A Taxonomy for Types of Cheaters

How, then, are those responsible for maintaining and enhancing testing programs to thwart cheaters? To contend with this vulnerable situation, we must shine the spotlight on cheaters in many different settings, using a variety of communication styles. In this section, we borrow a practice long followed by law enforcement agencies and give you our personal "10 most wanted cheaters" list. Each of these test-taking offenders has been given a name as a shorthand way of highlighting some prevalent methods of cheating.

The Top 10

1. **The Impersonator:** When a person who registered for a test under the name "Jane R. Smith" shows up to take the test and presents her identification materials, how can we be sure who is really taking the test? Impersonators, also known as proxy test takers, try to mislead proctors for virtually all major high-stakes examinations. The larger the exam, and the broader its global scope, the more serious the problem tends to be. For example, individuals were caught proxy testing for the Test of English as

a Foreign Language (TOEFL) exam. This exam for individuals who have learned English as a second language is given to more than 500,000 examinees per year and is used as a college entrance requirement. When proctors personally know all test takers, the task of checking and verifying identification is very different from a situation in which the proctor has never seen any of the test takers before the actual administration of the test.

Impersonators can range from the friend or family member who is "helping out" on a one-time basis, to dedicated professionals, some of whom are brazen enough to advertise their services on the Internet and elsewhere.

2. **The Smuggler:** The practice of carrying forbidden materials across borders in violation of laws and regulations has been observed across countries and cultures over the millennia. In the testing world, the smuggler brings into the test setting materials or devices intended to provide an advantage over honest examinees. A packet of facial tissues may contain formulas, diagrams, or prepared answers to free-response questions. A pencil, a pair of eyeglasses, or a wrist watch that may appear innocent enough can serve as a storage device for enormous amounts of information that can be used by the cheater. A cell phone may be employed to send (or receive) answers to or from someone else sitting for the exam. Any test program sponsor can recall an anecdote or two about a smuggler's ingenious method for carrying some version of a crib sheet into a testing room.

3. **The Storyteller:** The label of the "storyteller" is used for the individual who memorizes test items only to retell them later to others. The storyteller is not likely to memorize an entire exam in one sitting; there simply is not enough time. Instead, storytellers take an exam multiple times and memorize a few items in each testing session. The sharing of items from the storyteller may range from telling just a few colleagues, to selling items over the Internet – often boasting to would-be buyers about having authentic items from the actual test

4. **The Chain Gang:** Whereas storytellers work alone, another group of evildoers has been dubbed "the chain gang," borrowing from the concept of prisoners linked together to work on roads or other projects. In the case of a testing chain gang, the group memorizes and sells items, typically through the Internet. As is the case with the storytellers, they collect and memorize items. The difference is that the chain-gang members work in concert, often employing carefully developed and coordinated strategies for stealing particular parts of the test or item pool.

5. **The Time Traveler:** In the movie *The Sting*, a key element of the movie is the creation of a situation in which it appears that it is possible to transmit the results of horse races to a bettor before the race is announced to the

betting house. If one could bet on a sure thing, there would be many unscrupulous individuals who would relish the ability to know key information ahead of everyone else.

In our global testing environment, where many exams are given during the same day, across multiple time zones, the illegitimate sharing of test questions, keys, and similar test content information is aimed at producing a "big score," or at least a passing or qualifying one for those in earlier time zones. In some instances, examinees have been apprehended with complete answer keys at the start of an exam that proved to be derived from tests given within the last few hours elsewhere in the world. Of course, just as in *The Sting*, not all evil schemes turn out as planned. For example, the answer key supplied to an individual may be appropriate for a version of the test different than the one they are taking. This is the case wherein different forms of the test are used in each time zone, so that the key for the later time zone is not the same as the key for the earlier time zone.

6. **The Collaborators:** A great deal of important work is accomplished by teams of people. Much as most individuals applaud productive cooperation, though, they scorn those who try to pass off the contributions of others as their own work. In the academic environment as well as other settings, the label *plagiarist* is one that is very damaging to an individual's reputation.

    In the testing arena, when we set out to test an individual's knowledge of a domain, we want the results to accurately reflect his or her own knowledge. Some cheaters devote extraordinary effort to employ collaborators within a testing site so that test administrators are unable to determine the actual level of proficiency and knowledge of each examinee.

    One family of collaboration strategies can be likened to the device that school children use to learn about Paul Revere's midnight ride. How did he know what route the British forces would follow? He simply observed how many lanterns were lit—"One if by land and two if by sea." Many cheaters use hand signals or their equivalent to communicate within group-testing settings as they work to undermine the efforts to maintain a level playing field for all.

    The not-too-distant cousins of the collaborators are the photographer and the radioman. These individuals bring to the test setting electronic devices that can be used to steal the items. Unlike collaborators or storytellers, these individuals do not depend on memory but use modern technology to steal the items.

7. **Robin Hood:** Some cheating occurs after testing is completed, *without* the active participation of the test taker. A teacher or other grader reviews an answer sheet for a paper-and-pencil-based test and "fixes" responses to clarify what the student really intended. The opportunity is great to give an

advantage to one or more students over others. This cheater may view himself or herself as a modern-day Robin Hood—giving answers to the poorer students to inflate their test standings, often at the expense of all the other students who sat for the exam.

A similar problem can occur when a test is being administered orally to an individual or group. The reader's facial expressions, tone of voice, or body language can easily lead test takers toward particular answers. This problem can also occur when the test administrator or proctor is also the trainer, whose pay depends on maintaining a minimum pass rate for a test. Such individuals may walk around the test administration site and "assist" examinees to insure that they meet or exceed their passing quota.

8. **The Hacker:** In the movie *Sleepless in Seattle*, a precocious, computer-savvy daughter of a travel agent helps a young boy fly across the United States alone by inserting misleading information into his travel profile. When the boy protests that he just doesn't look old enough, the young lady says "Don't worry, they will believe anything if it is in the computer."

    Hackers have devoted enormous amounts of effort to infiltrating computer systems, including ones very carefully designed and tested to be resistant to such manipulation. Managers of testing programs face a high-tech version of the classic practice of dishonest students borrowing the teacher's grade book or school file in order to alter test scores or grades. Students once turned F's into B's by essentially adding the number three to the original (F plus 3 in the same space on a form equals B, try it and you will see). Nowadays, hackers who obtain unauthorized access to records can make virtually any change within systems that do not have the proper safeguards or internal consistency checks.

9. **The Ticket Scalper:** The ticket scalper sits for a computerized exam's beta test for the sole purpose of obtaining a free voucher. The voucher, once obtained, is then resold for less than the legitimate price of the exam. It is not uncommon to find exam vouchers for sale on eBay and other electronic auction houses. The ticket scalper is, in some ways, a less serious offender than others on our list. However, the results of their actions are quite substantial to testing programs in terms of test development and delivery costs. This is principally because many ticket scalpers do not make an honest attempt at answering the beta test questions correctly, thus providing misleading information about the quality of the test or test items.

10. **The Insider and the Fence:** To close this list, there are a pair of cheaters who represent some of the most common and most damaging classes of attackers of test security: the insider, who steals test content, and the fence, who sells the results. Sometimes, of course, it is the same person or group playing both roles.

It is unfortunate, but test content can be compromised by people participating in the development or administration of the exams. When controls and restrictions are not carefully developed and maintained, the repercussions of this problem can be staggering. When groups of items are compromised, the overall test or bank can gradually become virtually useless as a fair gauge of knowledge and skills.

Solutions to Security Problems

Threats to computer-based test security have been discussed, and it is now time to speak about their remedies. The best way to stop cheating is to prevent it before it starts. This means a robust security planning process to protect a testing program's intellectual property, create deterrents for would-be cheaters, and to set policies for when cheating does occur. Additionally, there are two, broad categories of security used to diminish the impact of test theft and fraud on item exposure, test validity, and reliability: physical security and Virtual Security™.

Preventing cheating before it starts means having a strong security plan in place. A security plan is a document providing a set of instructions for managing the security of a testing program. Just as a test manual is designed to be a kind of road map for creating and delivering a test, a security plan is a kind of road map for designing and implementing security within a testing program.

Elements of a Test Security Plan

A security plan contains elements such as the following:

- **Responsibility for security:** Elements of a security plan should describe who in the organization is responsible for test security. Specifically, who leads and who contributes to the testing program's security. For example, does everyone in the organization understand the role they play in terms of protecting test content? Is everyone trained on test security measures? And who is responsible for budgeting for test security costs?

- **Budgeting:** Budgeting will also be a consideration for security, and questions needs to be answered, such as, Does the program have adequate financial resources for security? What happens if test items are compromised? Is there enough money in the budget for item redevelopment, investigation of potentially fraudulent scores, or detection of item piracy?

- **Legal precautions and agreements:** These are something else one should consider covering in the security plan. Is there a nondisclosure agreement in place for candidates to sign prior to testing? How about agreements between the test sponsor and the test delivery provider, do they exist? And what about item writers and test developers, are they required to sign transfer of ownership and confidentiality agreements prior to participating in a test development effort?

- **Test Development:** Are test developers trained on security precedures for test development? What type of a repository is being used for item creation? Is access secure? What happens to all those items that do not get used for the test—are they properly disposed of or used for practice tests?

- **Security Breaches:** These include the different types of breaches that can occur and the various remedies a test program uses to reduce the risk of a breach. For example, if someone is discovered cheating, will his or her score be canceled? Will a test taker be prevented from testing again? If an organization is sharing nearly identical items on a Web site, how will this behavior be stopped? What legal action will be required?

Physical and Virtual Security

Physical security manages those elements of the test administration process that deal with physically detecting cheating or test fraud. Tools such as on-site proctoring, video monitoring, and biometrics certainly serve to guard, deter, and dissuade against test theft and fraud.

Virtual security manages those elements of the test administration process that rely on the analysis of test data to determine patterns of cheating or fraud within an individual's test performance. Software that can detect aberrant response patterns, that is, response patterns to answering test questions that are so far out of the range of normal, the likelihood of them happening by chance alone is very rare. Examples of aberrant response patterns include candidates who complete items too quickly or too slowly, and thus these response patterns can be indicators of abnormal testing behaviors that are indicative of cheating and fraud (Maynes, 2004).

Additionally, source-detection software—software programs that search the Internet for existing, published content—provide great tools for detecting instances of plagiarism from students and test takers submitting written papers as part of required assignment.

Physical Security

The role of the proctor in a computer-based testing environment is certainly a necessary one when high-stakes exams are being delivered. The proctor is responsible for admitting the candidate, verifying identification, and verifying that the test candidate proceeds into the testing environment with the right materials. Although proctors in a paper-based testing environment play similar roles, they typically conduct proctoring for a testing event—testing that happens four times a year, for example. Rather the role of a computer-administered test proctor is someone who administers tests on an ongoing, daily basis.

According to Killorin, the role of the computer-based test proctor is changing. In times past, a testing proctor would admit the individual, verify identification with a valid photo identification card, and seat the test taker in the appropriate cubicle to begin testing. Today, test proctors of a computer-based testing environment wear multiple hats. They are test administrator, test police, and data forensic analyst all rolled into one (Killorin, personal communication, December 2003). This individual admits the test taker, checking against two forms of valid picture identification. Depending on how high the stakes are on the exam being administered, proctors conduct biometric scans of fingerprints, handprints, or facial prints. They must decipher the results of analysis and make a decision about whether to admit the testing candidate. They must have the individual sign a test-center regulations form, which specifies all processes and regulations of the testing process and sanctions for violation of the agreement. Proctors discuss the conditions of the agreement with the examinee, have the individual sign the agreement, and then escort the test taker to the assigned seat.

At the same time, proctors monitor other individuals through video or audio surveillance. Most dedicated computerized testing centers use media equipment to provide additional eyes and ears for proctors as they monitor test-taking activities. Proctors are also being trained to look for abnormal examinee behaviors. For example, is an examinee getting up to go to the lavatory too frequently? Is the candidate nervously looking around the testing room? Does the candidate appear to be talking to him- or herself? These behaviors, as well as a host of others, may be indicative of test-candidate misconduct.

Proctors are in a precarious situation when it comes to handling examinee wrongdoing. If a proctor believes there is any test misconduct going on, the proctor may stop a candidate before the completion of a testing event. Needless to say, proctors may find themselves in some difficult situations should an examinee aggressively protest having the exam delivery halted. Any incidents spotted by the proctor are reported in writing and the appropriate individuals within the test delivery organization are notified. Records are kept of these incidents for a specified amount of time.

Remote, Internet-based testing is a continual challenge to test sponsors. The question of test-taker identity and exam stakes must be considered. Should a high-stakes test be delivered remotely without a proctor present? Are there ways of successfully and securely delivering high-stakes computerized tests? Are the consequences of individuals passing or failing the exam limited, so the require-ment of a proctor is not an issue?

The technology of remote proctoring is feasible for Internet-based testing scenarios. That is, based on the development of high speed networking technologies, there may be situations in which test takers will be allowed to take a high-stakes test in the convenience of their home or office. A variety of solutions has been proposed. For example, providing continuous communications links between the examinee's computer and the test's host server, periodic retinal scans of the test taker, and video kiosks that monitor test takers remotely. Non-technology-based ideas of sending proctors to homes and businesses have been conceived. As the discussion continues, designers are hard at work trying to define appropriate technology-based solutions.

Biometrics

The emerging technology of biometrics is providing strong possibilities for managing physical security. Biometrics is technology that uses biological information sources to verify an individual's identity (Middlemiss, 2004). Specifically, it scans a portion of the anatomy such as a finger, hand, eye, or face and measures predefined points to confirm identity. In verifying identity for test registration, common forms of biometrics include fingerprint and facial print recognition. When identity verification is critical to the stakes of an exam, biometrics is a robust solution.

Data Forensics™

According to ITC (in press) guidelines, another way in which computerized administration can enhance psychological assessment is through the use of online diagnostics. These diagnostics evaluate the test taker's responses and can flag whether an administration is in some way invalid. It may be that the test taker has not understood the instructions for an ability test, or that answers to a personality test suggest an abnormal pattern of responding.

Whatever the reason, these statistical processes, referred to as Data Forensics™ (Maynes, 2004), use an individual's response time on each item and correlate that with the mean time of other testers to determine test taker aberrance. "Aberrance is a term used to describe something unusual about the way an examinee has responded to the test questions. Aberrant test responses can arise in a variety of ways and are usually a result of the test taker's behavior" (p. 1). Aberrance takes into consideration factors such as candidate ability level, test-taker performance, and item difficulty in determining irregular test response patterns. Additionally, the results of these analyses can be depicted in a number of different ways: through testing site behavior, individual test candidate behavior, and test performance over time.

Web Patrolling™

Cheating sometimes take the form of stealing items with the intent of providing them to others, either for free or for fee. Web sites commonly known as "braindump sites" are hosted by individuals or organizations that have set up Internet sites directly for the purpose of providing or selling stolen test items to others. Braindump sites are most commonly associated with information technology (IT) certification programs, although testing programs in a variety of genre are not immune to this form of item exposure. Research conducted by Smith (2003) suggested that items posted to braindump sites accurately reflect actual item content and can readily be found approximately 8 months from the release date of a test.

Some braindump sites reside on forums, wherein individuals share information about a related topic. A variety of individuals will list questions they can remember from taking a particular exam. However, the greater risk to a test program are braindump sites that are run as a business; these sites boast of having certification exams from a myriad of testing programs. Each test is sold for approximately $50 to $100. The candidate pays for the stolen exam questions and uses them as preparation to take the actual exam.

According to Maynes, (personal communication, January 2004) Web Patrolling™ technology uses an automated, systematic search process to identify braindump sites and catalog them according to their level of proposed test security risk. Sites are ranked from 1 to 5, with 5 being the most serious infraction. For example, if a braindump site is ranked a 5, it is highly likely that the stolen exam content being advertised can be found on the Web site. By continuously monitoring the Internet, sites can be identified and shut down so that fewer stolen test items are made available for purchase. Web Patrolling™ can also be performed in multiple languages to attack braindump sites that are in languages other than English.

Source-Detection Software

Plagiarism of written, copyrighted materials has been a recognized problem for quite some time. Recent developments in Internet search capabilities have accelerated the ability to detect copying of published works. Source-detection software searches the Internet for word strings that match already-published works (Heldref Publications, 2003). Typically, the source-detection software is looking for comparative word strings that are a minimum of eight words in succession. When the software finds these matches, it highlights the text within the reviewed document and provides a report back to the teacher or reviewer.

There are several software products on the market today that provide these capabilities. Typically, electronic versions of essays and papers are sent to software analysis companies that provide the service of source detection. Results are generally turned around within a 24-hour period.

It is important to note that the strings found in a given essay or paper are considered indicators of cheating and should be considered among other factors before categorizing a writer's paper as invalid or plagiarized. This type of software is used in many school settings, from elementary grade classrooms to college level courseware. Source-detection software has been a great deterrent for would-be cheaters. Knowing that a school will use this software to monitor students' abilities to create their own work motivates them to work harder to create original thoughts and also helps to teach students what is and is not considered copying when reviewing the published work of others.

Future Trends

What do future trends for timely, accurate, and secure test administration look like? For starters, we can count on there being more testing done via the Internet and increases in technology-based detection tools and software to solve the multifaceted problem of cheating.

Internet Testing

As the need for testing continues to grow in all facets of assessment—clinical, occupational, certification, and educational—so will the demand for technology-based testing. Decisions will need to be made as to whether these tests will be delivered in proctored and unproctored environments. That is, what level of stakes is affiliated with a particular test? How critical is it that a test taker be monitored and his or her identify validated? If protection of the public is of high concern, proctoring will still be a necessary requisite for test delivery.

Technology-Based Detection

What would happen if we could catch test cheaters in the act? That is, be able to stop a computer-based exam during the examination based on response patterns of the test taker. Real-time data forensics is closer than we think. At the time of this writing, capabilities for this technology are being developed and tested. The premise for real-time data forensics assumes that test-taker response patterns provide indicators as to whether individuals are cheating as they are actually taking a test. Imagine being able to stop a test taker if they were completing test items at an interval of 1 second per item. This means 1 second per item, whereas the average response time is 30 seconds. In addition, this swift individual may be getting all of the items correct. Robust statistical algorithms in combination with test delivery software suggest that tracking test takers in this manner is possible (Maynes, 2004).

Another means for catching test thieves will be through real-time Web patrolling. The ability to use real-time search tools that continuously monitor the Internet for the sale of test items and alert test owners when a breach has occurred will be critical in timely detection of stolen items.

Frequency jamming of electronic devices may start to find their way into computerized testing environments. Reflection of electromagnetic energy may be used to jam cell phone, audio equipment, or camera usage in a testing environment. This could prevent the use of text-messaging capabilities from candidates to outside sources and prevent the use microphones and speakers to communicate test content information.

Test Development Practices

Test development tools may also provide preventative measures for item exposure issues. Software, which can create test item clones in an automated fashion, may provide test developers with "throw away" test items and tests. The result is a test that is only delivered once. This means that candidates who intend to memorize items or get stolen items from someone who has memorized them will not be able to do either, because each test item is never delivered more than one time. In work conducted by Baron, Miles, and Bartram (2001), a numeric ability test was generated using a Web-based system, which randomly generated test items within three different levels of difficulty. The result was the ability to generate over several million different tests. The outcome of these tests was used in conjunction with other data to make hiring decisions about candidates applying for a financial-services-based job.

Performance-Based Testing

Performance-based testing may provide another venue to deliver secure tests. A performance-based test is one in which an individual must perform an actual task (Mulkey & O'Neil, 1999) In a computer-based testing environment, this is typically done through either simulation exercises or live-application testing. Taking the example of a computer network administrator, the test taker may be asked to add a user to a system, create a new group or directory, or move a user from one group to another. Because these test items actually require a test-taker's performance, no amount of memorization about factual information will help; the test taker must perform the skill.

It is clear that proctors and proctoring will continue to be needed. If an exam is considered to be high stakes, identity verification and test monitoring will be necessary to maintain a test's integrity in a secure testing environment. Advances in technology may help in providing remote proctoring capabilities. For example, remote cameras and IP address polling used in combination may be an acceptable solution for remote proctoring.

Conclusion

In this chapter, we have discussed the preponderance of cheating in today's society and its influence in the field of assessment. It appears that cheating is on the increase, which causes testing professionals to take more proactive measures to protect test items and deter would-be cheaters. Guidelines do exist to help us formulate specific ideas of where we should consider secure computerized test administration environments. These guidelines are helpful in bridging the gap between what we know about paper-and-pencil testing and what needs to be considered for online assessments.

It has also been helpful to understand the different approaches and forms cheating takes. Categorizing these wrongdoings can help shape solutions for detecting and preventing test-item theft. Solutions to these important security problems use technology to assist in the systematic and accurate collection of inferential cheating data. Both physical and Virtual Security™ tools such as biometrics, frequency jamming, Data Forensics™, Web-Patrolling™, and Source-detection software offer effective and efficient solutions to detect cheating and exam piracy.

Finally, the vision for test theft detection in the future holds great promise. Technologies that provide real-time detection and exponential item generation will help to prevent and deter the widespread occurrence of cheating.

References

Association of Test Publishers. (2002). The guidelines for computer-based testing. Retrieved July 15, 2004, from *http://www.testpublishers.org/ MembersOnly/CBT%20Guidelines.pdf*

Baron, H., Miles, A., & Bartram, D. (2001, April). Using online testing to reduce time-to-hire. Paper presented at the annual meeting of the *Society of Industrial and Organizational Psychology.*

Bartram, D. (2001, June). The impact of the Internet on testing for recruitment, selection, and development. Paper presented at the meeting of the *4th Industrial/Organizational Psychology Conference*, Sydney, Australia.

Bartram, D. (in press). Computer-based testing and the Internet. In A. Evers, O. Smit-Voskuyl, & N. Anderson (Eds.), *The Handbook of Selection.* Oxford, UK: Blackwell.

Callahan, D. (2003). *The cheating culture: Why more Americans are doing wrong to get ahead.* Orlando, FL: Harcourt.

Cizek, G. J. (1999). Cheating on tests: How to do it, detect it, and prevent it. Mahwah, NJ: Erlbaum.

Foster, D. (2003) Test piracy: The darker side of certification. Retrieved July 15, 2004, from *http://www.certmag.com/articles/templates/cmag _feature.asp?articleid=2&zoneid=8*

Fox, L. (2004, August 24). Putting deceit on hold. Retrieved August 28, 2004, from *http://www.fortwayne.com/mld/newssentinel/9481130.htm*

Gay, G. H. (1990) Standardized tests: Irregularities in administering of tests which affect test results. *Journal of Instructional Psychology*, (17), 93-104.

Heldref Publications. (2003). Gotcha now, cheaters [Electronic version]. *Science activities,* (40), 40.

International Test Commission. (in press). International guidelines on computer-based and Internet delivered Testing. Retrieved July 15, 2004, from *http:/ /www.intestcom.org/itc_projects.htm#ITC%20Guidelines%20 on%20Computer-Based%20and%20Internet%20 Delivered%20Testing*

Jacob, B.A. & Levitt, S.D. (2003). Rotten apples: An investigation of the prevalence and predictors of teacher cheating. Retrieved December 11, 2003, from *http://www.nber.org/papers/w9413*

Maynes, D. (2004) The TAO of aberrance. Retrieved June 28, 2004, from *http:/ /www.caveon.com/tao.htm*

McCabe, D. L., Trevino, L. K., & Butterfield, K. D. (1999). Academic integrity in honor code and non-honor code environments: A qualitative investigation. *Journal of Higher Education, 70,* 211-234.

Middlemiss, J. (2004, July). Biometrics add security in insecure times. Retrieved July 30, 2004, from *http://www.wallstreetandtech.com/ showArticle.jhtml?articleID=18402883*

Mulkey, J. R., & O'Neil, H. F., Jr. (1999). The effects of test item format on self-efficacy and worry during a high-stakes computer-based certification examination. *Computers in Human Behavior,* (15), 495-509.

Pino. N. W., & Smith, W. L. (2003). College students and academic dishonesty. *College Student Journal*, 37, 490-501.

Smith, R. W. (2004, April). The impact of braindump sites on item exposure and item parameter drift. Paper presented at the annual meeting of the *American Education Research Association*, San Diego, California.

Wainer, H., & Dorans, N. J. (Eds.). (2000). *Computerized adaptive testing: A primer* (2nd ed.). Erlbaum.

Endnote

[1]   Virtual Security™, Data Forensics™, and Web Patrolling™ are all trademarks of Caveon Test Security.

Chapter XIV

Securing and Proctoring Online Tests

Bernadette Howlett, Idaho State University, USA

Beverly Hewett, Idaho State University, USA

Abstract

Online course delivery has introduced a new spectrum of opportunities not only for innovative pedagogical approaches, but also for cheating. This chapter provides instructors with methods to deter students from cheating in online assessments either by limiting the opportunity to cheat or by reducing their motivation for doing so. Through an extensive discussion of research literature, we provide an exploration of cheating that includes: definitions, cost and effects, ethical considerations, motivations for cheating, role of organizational policy, history and recent examples. In this exploration, both technological solutions and instructional design solutions to reduce cheating are examined. This chapter looks closely at the capabilities and limitations of online testing and the tools technology provides to reduce cheating. We emphasize the role of instructional design in securing online tests. We conclude with a discussion on future trends.

Securing and Proctoring Online Tests

Online course delivery has introduced a new spectrum of opportunities not only for innovative pedagogical approaches, but also for cheating. As such, online assessment poses a new challenge to today's instructor. However, it is critical for all instructors to recognize that paper-based testing and oral examinations are also highly susceptible to cheating tactics. An interesting, and perhaps disconcerting, read on the subject is provided by Bob Corbett in his book, *The Cheater's Handbook* (1999), which offers 142 pages of methods to cheat on paper-based tests and writing assignments. Another author, Curry (1997), provided a particularly salient example of cheating in an environment most instructors would consider highly secure, the oral exam. In the example, the exam was administered before a dean and two other professors. According to Curry, one student cheated in the following manner:

> *Wearing his long hair loose, he hooked a bud earphone up to his cellular phone and hid a small microphone in his tie. A friend well-versed in the subject called him right before the exam, ready to read him the answers he needed. But his cell phone ran out of batteries halfway through the test and started beeping.* (p. 7)

After reading about or even encountering the varied methods of cheating, there are questions online instructors need to consider when planning an online assessment activity. A few of these questions include the following:

- How can a teacher know if the person taking the exam is the person receiving credit?

- What can be done to prevent students from looking up answers in the textbook or online during an exam?

- Is it possible to prevent students from calling one another during an online test or from using other electronic gadgets to store cheat sheets?

- How does a professor prevent students in one section of a course from printing a test and giving it to students in another section of the course?

- And, most important, what motivates students to cheat, and how can we reduce cheating by decreasing that motivation?

Though there is no guaranteed way to prevent cheating, there are techniques instructors can use in online assessment to make cheating difficult and time consuming. This chapter will provide instructors with methods to deter students from cheating either by limiting the opportunity to cheat or by reducing their motivation for doing so. This will be accomplished by introducing significant issues related to securing online tests and by discussing the various types of solutions available to reduce cheating in online assessments. After discussing the significant issues, the chapter will focus on solutions, which are broken into two major categories, technological and instructional design. Technological solutions are those that utilize equipment of some type, such as a fingerprint scanner or software to limit the student's use of the computer. Instructional design solutions are methods for designing and delivering instruction and assessments that will reduce the ability or motivation to cheat, such as using a large bank of randomly drawn questions or changing the assessment method to multiple types of student performance of course outcomes.

Background

Significant Issues

There is a variety of significant issues which may, at first, seem unrelated, but are closely connected to online assessment. The foremost of these issues include student versus faculty definition of cheating, the relationship between the cost of a cheating intervention and the results it produces, ethical questions about using technology to track student activities, the motivation for students to cheat, and the role of organizational policy in cheating. These issues establish an environment unique to each teaching situation that both affect the pedagogical approaches available to instructors and suggest possible solutions to cheating in online assessment. It is important to recognize, however, that what works in one classroom to reduce cheating may not be an option in another one.

Definition of Cheating and Why We Should Be Concerned

An instructor's perspective on cheating is critical when considering the design of an assessment and selection of assessment methods. For this reason we will explore the definition of the term *cheating* and its meaning in our society. The definition of cheating has been a subject of debate among experts and a generally accepted definition has yet to emerge. However, for purposes of this chapter, the widely cited Pavela definition offered by Whitley and Keith-Spiegel (2002) will be used:

> Cheating is "intentionally using or attempting to use unauthorized materials, information or study aids in any academic exercise. The term academic exercise includes all forms of work submitted for credit or hours (p. 78). Thus, cheating includes such behaviors as using crib notes or copying during tests and unauthorized collaboration on out-of-class assignments. (p. 17)

The definitions of cheating encountered in the literature do not seem to take into account the harm produced by cheating. This is important because there is a difference between students' perception of harm and the perception of faculty members. This perception difference is one of the factors that influence students' decisions to cheat. Students who cheat often do so because they perceive a benefit will accrue to them and there is little chance of being caught (Finn & Frone, 2002).

Faculty members often hold the position that cheating is harmful to students as well as to others. Students who cheat potentially deprive themselves of knowledge and of the satisfaction of doing well, and they harm others by degrading the value of the education programs in which they participate. Whitley and Keith-Spiegel (2002) offered additional consequences of the harm caused by cheating. For example, when one student cheats and another knows about it, if the cheating is not discovered then the witnessing student may decide that cheating is acceptable. Cheating also harms instructors in at least two ways. First, many take personal offense, feeling "violated and mistreated by their students" (p. 5), and second, instructors describe dealing with cheating as, "one of the most stressful aspects of their jobs" (p. 5).

One example that shows the far-reaching effects of cheating and the political complexity of the topic comes from Piper High School, Piper, Kansas, in the fall of 2001. Approximately one fourth of the students in a biology course plagiarized sections of a project worth half of the final biology grade (Trotter, 2002). The teacher failed the students, who themselves and their parents had signed a contract confirming their understanding of the rule against plagiarism. Nonetheless, the upset parents of the failed students petitioned the school board to overturn the teacher's grading structure and they won, allowing their children to pass the course despite the plagiarism. "The grading change allowed 27 of the 28 students to escape an F, but it also pulled down the grades for about 20 students whom had not plagiarized" (Trotter, 2002, p. 5).

The teacher at the center of the controversy quit her job, the school principal also resigned, and "the district attorney filed civil charges against the district's seven-member school board, accusing the members of violating the Kansas open-meetings law" (Trotter, 2002, p. 5). The students of Piper also paid a price for their actions. According to Trotter,

> *at an interscholastic sporting event involving Piper, signs appeared among the spectators that read "Plagiarists." Students reported that their academic awards, such as scholarships, [were] derided by others. And one girl, wearing a Piper High sweatshirt while taking a college-entrance exam, was told pointedly by the proctor, "There will be no cheating." (p. 5)*

Considering the scope of harm caused by cheating, many instructors view cheating as a simple right-versus-wrong issue. We cannot allow any room for interpretation if we are to protect students, ourselves, and our institutions. With a disparity between student and faculty perspectives, a helpful first step is for instructors to consider the student perspective. We (the authors) suggest that in our society, cheating is very often more like a continuum. At one end of the continuum are forms of cheating in which many engage and often justify, at the other end are forms of cheating that violate the law and cause material harm to others.

An example of the "low cheating" end of the spectrum comes from the United States Internal Revenue Service (IRS). The IRS reported, "of every dollar that's legally due, just 83 cents is handed over 'voluntarily,'" (Henry & Barr, 1998, p. 122) for a total of about $100 billion dollars in unpaid taxes. Some common types of income that are not reported come from small-stakes gambling, such as through office pools and bingo games, prizes from contests, and unpaid sales taxes achieved by purchasing goods from a non-sales tax state. According to Henry and Barr (1998), many tax payers fudge a little on their taxes and justify their decision by arguing that the tax laws are too complex, almost everybody does it, they need the money more than the government does, or that they paid their taxes on time and have always done so.

For another example of cheating at the low end of the continuum, consider your own choices. When you are selling something, such as your car, do you sell it for no more than its fair value, or do you try to get more for it? In our open market we accept the idea that it is fair for consumers to pay whatever price they are willing to pay. We do not object to the idea that the buyer can get cheated. If, as teachers and designers of instruction, we are willing to cheat on our taxes, in our cars, or elsewhere, then how can we hold students accountable when they cheat in our classes?

Of course the answer is simple: because we must. Students who cheat fail themselves and run the risk of failing society. One thing that the research reveals is that students often consider cheating as falling at the low end of the continuum (Olt, 2002; Whitley & Keith-Spiegel, 2002). But, the majority of faculty members and institutions deem cheating as falling nearer to the high end of the continuum, considering the potential for harm.

Be that as it may, it would be naïve for us not to consider the continuum of cheating when making decisions about this topic, especially when the stakes for our students can be very high. We need to understand our students' perspective on cheating and help them to understand the potentially far-reaching consequences. The issue of assessing student perspective on the subject and addressing it in the course will be discussed later in this chapter in the section on honor codes and honesty agreements.

Definitional Difference

Students and faculty members tend to disagree on their definition of cheating behaviors. Whitley and Keith-Spiegel (2002) explained, "Although there are many areas of agreement, research bears out that students generally view academic dishonesty more leniently than do faculty members" (p. 19). According to Whitley and Keith-Spiegel (2002),

> *students whose views of what constituted cheating on exams coincided with their university's definition of cheating were much less likely to report having cheated on exams than were students whose views did not coincide with the university's definition.* (p. 27)

One reason for disagreement between faculty and student definitions may be a lack of explanation from the institution or instructor. According to Olt (2002), "A recent study reveals that few instructors take up the topic of academic integrity and dishonesty with their students (¶ 14). Dirks (1998) explains, "15% of the syllabi collected had academic policies in them" (p. 18). Even with explanations, student understanding can continue to be problematic. The concept of plagiarism is a striking example. Whitley and Keith-Spiegel (2002), who have written one of the few books to undertake the topic of academic dishonesty, explained the challenge of student understanding of plagiarism: "Roig (1997) found that even when students understand the concept of plagiarism, they often cannot recognize it when presented with examples" (p. 28).

Additionally, there may be cultural differences between faculty members and students' definitions of cheating. For example, according to Whitley and Keith-

Spiegel (2002), "some cultures feel it is disrespectful to the author to alter the original words [of a source document]" (p. 21). In another example, according to Curry (1997), in Poland cheating is treated like a contest between instructors and students. The cultural norm, Curry said, is for students to help one another cheat: "Students copy one another's work during exams and get together in marathon minimizing sessions at copy machines to prepare cheat sheets—sometimes shrinking entire books to palm size. Not helping classmates on an exam is socially unacceptable" (p. 7). The reason for this viewpoint, according to Curry (1997), appears to be related to the nature of the exams:

> *Teachers emphasize specific knowledge over general understanding and critical thinking. In high school and college, cumulative final exams are the only grades students get. Most students see cheating as the only possible response to an educational system that values rote memorization more than personal creativity.* (p. 7)

Cost Versus Effect

Another critical consideration instructors need to undertake regarding online assessment is the cost of an intervention. Generally speaking, solutions that employ technology involve greater cost than do those that employ instructional design strategies. For example, it is possible to have students install software on their computers that monitors and reports their activities to the instructor. The software could report what programs ran during the time the student was taking a test. However, the cost of the software would need to be paid by the student or by the institution. Furthermore, technical support becomes challenging for the college as such an application is not often supported by the typical university help desk. This would demand additional training of help-desk personnel and development of support resources; that is if the help desk is even willing to support the product.

Assuming the software is installed and running correctly, the next concern is the effectiveness of the application. There is a dearth of literature on the topic. Faculty members are generally left to rely on the claims made by the product manufacturers. Students may be able to bypass or trick the system. Furthermore, it is difficult to determine the number of cheating attempts such a system might prevent. As such, determining the effectiveness of the system would be challenging. Without such data it may be hard to convince university budget officers to provide funding for the software, for helpdesk personnel training, or support resources.

On the other hand, an instructional design solution is likely to be less costly, and easier to implement, because it typically requires only the efforts of the course instructor. One example might be to offer multiple forms of assessment rather than just a midterm and a final exam. The cost for developing multiple assessments shows up in the form of labor hours for the instructor both in setting up the assessments and in grading them. The larger the class size, the greater the cost. The best approach is to find the combination of technological interventions and instructional design strategies that work for a given instructional situation within the resources available to the instructor and institution.

Ethical Considerations

Another topic that has received little attention in the literature is the question of ethics in monitoring student computer activity. Online course management systems (such as WebCT and Blackboard) offer monitoring capabilities. In some systems it is possible to view server logs that show where a student logged in from and what commands they gave. A consequence of the administrative capabilities is that it is even possible to read or alter private messages within these systems. (Many systems have an e-mail tool that is built into it and does not use the student's personal e-mail.) This kind of capability raises privacy questions which many institutions have yet to address.

Server log files and other features of an online course management system will allow instructors to look for cheating patterns or outright evidence of cheating, such as students sending copies of test questions to one another through the mail tool. However, this also enables instructors to see messages students send one another about topics unrelated to instruction. It may be argued that the students should not use the course management system for this purpose. However, the tools often give the appearance of privacy, making it likely that students will not understand the kind of access that instructors or server administrators have.

Some institutions inform students that such capabilities exist and defend their right to view all data collected on and by the server claiming that the server is university property. For example, one school we interviewed makes sure that students are advised and read their security policy prior to commencing an online course. The security policy clearly states that students can have no expectation to privacy. Students are advised to create Web-based e-mail for free and use that e-mail account for personal messages. For those institutions that have addressed this issue, what they have in common is a process for informing students of the policy and the extent of the students' right to privacy within the system.

For those who have yet to take up this issue, drawing a parallel to the face-to-face classroom might help institutions determine how to proceed. If a student passes a note in class, is that note the property of the university? How much right does the instructor have to see what a student has in his book bag once the student enters the classroom? Is reading a student's mail in an online course the same as getting into his book bag? We offer no position on this issue but encourage universities to ask these questions and endeavor to inform students about university policy and practice before a major incident lands them in a courtroom. So far we have not seen an incident reported in the literature, but we have also seen very few institutions with published policies about student privacy in online courses. We view this as an emerging issue.

Motivation for Students to Cheat

Through our understanding of the motivations to cheat we can begin to identify methods to reduce cheating. While little has been written about the cost of cheating interventions or the ethics of viewing student online activity, a good deal of research has been done on the motivation for cheating. In the interest of brevity, we will only briefly review the topic and encourage readers to explore it further through the references listed at the end of this chapter.

The primary motivation for cheating is simple, and not surprising, "Students cheat to raise their grades" (Finn & Frone, 2004, p. 115). Finn and Frone also cited a review of cheating research performed by Cizek in 1999 in which the correlation between achievement and cheating was discerned: "The research is consistent and unequivocal – cheating is inversely related to achievement; that is cheating occurs most often among students with low achievement" (p. 116). However, there are often underlying issues that motivate cheating which help reveal potential solutions and are, as such, worth exploring.

Finn and Frone (2004) offered several circumstances in which cheating is more likely to occur: "Cheating is more likely when students perceive external pressure to perform but fear they cannot succeed" (p. 115). Additionally, "cheating is more likely among students with low academic self-efficacy" (p. 115); and lastly, "cheating is more likely to occur when students feel alienated from school and dissociate from school rules and procedures" (p. 115). Additionally, their study found that students with "lower levels of self-efficacy and lower levels of school identification" (p. 116) were more likely to cheat. And, interestingly, "students who had high self-efficacy but lower performance reported the greatest amount of cheating" (p. 116). Their results indicated a finding that has been supported by others:

> *Emphasizing grades can increase cheating behavior. Elevated achievement standards, accompanied by increases in high-stakes*

testing, may have the unintended consequence of increasing students' motivation to cheat [sic]. [T]he risk is elevated when achievement stakes are high and there is personal consequence for failure. (p. 117)

In addition to considering the underlying motivation, it is informative to examine the explanations students give for their cheating. Whitley and Keith-Spiegel (2002), after reviewing the literature on the topic, offer a list of explanations. Students cheat when the following occurs:

- There is opportunity
- They have a perception of a permissive climate
- There is high pressure to perform (grade competition)
- They lack understanding of what constitutes cheating
- They have a positive attitude toward academic dishonesty
- They perceive a benefit will accrue to them if they are successful at cheating
- They perceive a low risk of getting caught
- They have a negative opinion of the instructor
- They lack a relationship with the instructor
- They have had prior success at cheating
- They have a moderate expectation for success without cheating
- They party more frequently
- They have a need for parental or peer approval (pp. 28-34)

The implication of the research is that if as instructors we can help students achieve a higher level of self-efficacy, avoid establishing highly evaluative classrooms, reduce the opportunity to cheat, educate students about our definition of cheating and what constitutes cheating behaviors, and establish relationships with students, then we may be able to combat cheating effectively without establishing an adversarial relationship.

The Role of Organizational Policy In Cheating

Organizational policy has a role in student and instructor motivation with regard to cheating. Policies and practices can contribute to reducing cheating by creating a climate that addresses the motivations behind student cheating. A climate of academic integrity is critical to achieving a reduction in cheating. An honor code is one aspect of the campus environment that can help to reduce cheating.

Another important role that organizations undertake is that of policing and enforcement. University support of instructors, communication of policies and procedures to students and faculty, and vigorous follow-through on reports of cheating are critical to reducing cheating. For example, a survey performed at Duke University in 2000 discovered an unwillingness on the part of faculty members to report cheating incidents: "Faculty chose not to bring cases of suspected dishonesty forward because they had bad experiences with the undergraduate judicial system in the past" (Ruderman, 2004, p. 8).

History

Although there has been considerable research on cheating in education, there is less information available on cheating from a historical perspective. Freeman and Ataov (1960) referred to research that dates as far back as 1928. However, the subject received little attention until the 1960s. Regardless of the availability of literature on the topic, we feel confident that academic dishonesty has most likely been practiced since the first classroom. Over time, the considerations related to cheating have changed little, even in distance education. Correspondence courses, instruction delivered on audio recordings, videotape, over television, and on the radio have all struggled with issues such as determining if the person receiving credit is indeed the person performing assessments, identifying plagiarism, preventing the use of cheating aids, and so forth.

However, online learning has added the possibility of synchronicity and many new opportunities for assessment that have greatly improved our ability to reduce the potential for cheating. More important, online learning has introduced the notion of learning communities to distance learning. One of the student motivations for cheating is a sense of disconnectedness from the instructor. With online learning communities that isolation can be removed, and in some ways students can enjoy an even greater individual relationship with the instructor. Online communication tools provide a means of developing community and they offer additional forms of assessment. The use of such tools will be discussed in the section on securing online assessments.

The amount written on the topic of cheating has grown dramatically in recent years, with many claims that the incidence of cheating has steadily increased. It is difficult to determine if the results are indicative of a true increase in cheating behaviors or merely an appearance of growth as a result of an increase in the attention being given to the subject. Furthermore, there is disagreement in the literature about whether an increase has truly occurred. Pino and Smith (2003) cited Spiller and Crown (1995): "There is little comparable longitudinal research on academic dishonesty and [Spiller and Crown] challenge the assumption that cheating has increased over the years" (p. 490). One early study that supports this assertion was published in 1960, by Syracuse University. In this study, Freeman and Atoav revealed a self-reported cheating rate in one classroom where, "33 out of 38 students (approximately 87%) admitted to having cheated at least one time" (p. 445). This result is similar to numbers reported recently. For example, Whitley and Keith-Spiegel (2002) cited a large survey of under-graduates performed in 1994 by McCabe and Trevino in which 78% of the students admitted to cheating.

In contrast to this evidence that cheating has not increased, the Center for Academic Integrity (2002-2003) reported the following:

> *Longitudinal comparisons show significant increases in serious test/examination cheating and unpermitted student collaboration. For example, the number of students self-reporting instances of unpermitted collaboration at nine medium to large state universities increased from 11% in a 1963 survey to 49% in 1993. This trend seems to be continuing: between 1990 and 1995, instances of unpermitted collaboration at 31 small to medium schools increased from 30% to 38%. (¶ 6)*

It is evident that cheating has been happening in higher education for many years but that the extent of cheating remains an issue of disagreement.

Recent Examples

In order to reduce cheating in online tests it is, first and foremost, important for instructors to stay informed about methods of cheating, old and new. Many studies have been done and continue to be completed on this topic and new information seems to emerge weekly. Cheat sheets are still a very common method in both the conventional and online classroom. Cheat sheets can take on an electronic form in the online classroom. Some of the methods of cheating in online tests include calling another person, looking up answers in a book, and getting a copy of the exam from someone who has already completed it. Here are a few more methods that have surfaced (Ford, 2000; Orlans, 1996; Richardson, 2002):

- Having someone else take the exam
- Storing notes in a cell phone, PDA, or programmable calculator (the electronic version of cheat sheets)
- Wearing and earphone tape recorder
- Looking up answers on the Internet during an exam
- Receiving pages on cell phones and using two-way pagers during an exam
- Purchasing papers from the Internet
- Copying text from Internet Web sites
- Copying text from online article databases
- Using an Internet service (e.g., YourWAP.com) through a data-enabled cell phone

The creativity of cheating schemes also improves with the technology to reduce cheating. For example, at one university, students in the same degree program taking online exams found a way to cheat on a test that had been structured to reduce cheating. The exam had been set up to deliver one question at a time with a limited time in which students could take the exam. The exams were also given in a proctored classroom. The students involved in the scheme agreed to have one student start the test early and then copy and paste questions into an e-mail message which was then sent to the rest of the students, giving them a chance to look up answers before beginning the exam themselves. With each exam a different student volunteered to start early. The cheating ring was caught only because of one student's conscience.

The Challenge of Cheating Research and Determining Effectiveness of Interventions

One other major concern about cheating is the difficulty of getting consistent and reliable research results, which implies a difficulty with evaluating the effectiveness of a cheating-reduction program. As mentioned previously, many studies depend on the survey method and on student self-reporting, which may not produce reliable information. There has been a wide range of results in the reports of the preponderance of cheating on college campuses. In a review of literature on college student cheating patterns, Stern and Havilicek (1986) reported a range of 50% to 91% of students who admitted to having cheated at some time while in college. However, Finn and Frone (2004) support the use of self-reporting: "Student self-report is the most common method for assessing cheating and has been shown to provide reasonably accurate estimates" (p. 116).

A few observational and nonobtrusive studies have been done in laboratories or on campuses. For example, in one study by Pullen, Ortloff, Casey, and Payne (2000), a team of professors searched for discarded cheat sheets and cataloged their locations in relationship to where courses were taught. Interesting enough, one of the programs whose classrooms were found to have a high percentage of discarded cheat sheets was also a program known for its highly evaluative teaching environments. However, because of the difficulty of studying actual cheating behavior by using unobtrusive research, very few such studies have been done (Karlins, Michaels, & Podlogar, 1988; Pullen et al., 2000).

Whatever system of cheating reduction is implemented in a classroom or on campus, it is clear that gathering data about the program's effectiveness will be challenging, especially when the purpose of the evaluation is to measure the degree to which a behavior does not exist. It may be impossible to determine how many students would have cheated if the reduction program were not in place. Considering the difficulty in confirming results and the amount of time and resources that might go into developing and evaluating such a program, we recommend being judicious in investing in a cheating reduction program. Not only can it be costly and time consuming, but it can create a negative learning environment. Draves (2002) concurs: "Do not create a trust issue between yourself and the vast majority of students who will participate in your course honestly and with full integrity" (p. 121).

Technological Solutions and Recommendations

There are two categories of strategies that can be employed to reduce cheating in online assessments: technological solutions and instructional design solutions. As mentioned previously, a combination of approaches is recommended. Any one solution may be defeated by students or simply ineffective. Furthermore, there are cost and benefit issues that must be considered when selecting an intervention. Generally, technological solutions are more costly than instructional design solutions. However, there are potentially powerful approaches available from technology that should be considered.

We may not be able absolutely to prevent students from calling a "lifeline" during an online test, but it is possible to make it difficult for students to share answers with one another. There are a wide variety of technologies that can be used to reduce the potential to cheat, ranging from the type of programs used to deliver a quiz to the computers used by the students and other equipment. Many applications used to deliver quizzes include security features. Additionally, there are add-on programs for computers that can improve security. Last, there are a number of additional technologies that can be used separately from the computer. Some of the options available in these categories will be discussed in this section.

Online course management systems (e.g., WebCT, Blackboard) or exam delivery systems (e.g., QuestionMark, LXR) can help in a many ways. The quiz delivery tools in these systems offer a variety of security and organizational features that can be used to make cheating difficult. Several of the systems include features in their quiz tools that allow the instructor to do the following:

- Randomize questions
- Randomize the sequence of answer options
- Create questions with multiple correct answers
- Create short-answer and essay questions
- Draw from a large database of questions
- Deliver different versions of quizzes to different students (e.g., by course section)
- Control the amount of time students have to complete an exam
- Control the time frame in which an exam is available
- Limit when and who can see each exam
- Limit from where students can log in to take an exam
- Require a password to enter an exam
- View student computer activity within the exam (e.g., time spent per question, sequence of clicks)
- Control the way the questions are displayed (e.g., one at a time, all at once, whether or not they can revisit questions)
- Control how many times students can take an exam
- Control what students see after they have completed the exam (i.e., their score, the questions themselves, what answer they selected, whether or not they got each question right, and the correct answers)
- Control when students can see exam results (e.g., immediately after completing the exam, only after a certain date)

There are many more features and numerous quiz delivery systems that could be discussed. We recommend that instructors contact campus technical support providers for these systems for information about the capabilities that are available in the systems used. If instructors have a course management system or a quiz delivery program, then it is advisable to learn as much as possible about its security and quiz design capabilities to take advantage of features such as those listed previously. While it is unrealistic to attempt to list in this chapter all of the features that exist, an example of how these features can be used may be helpful.

In an online psychology course, students are given one quiz per week to assess their progress and to encourage them to complete course readings. These quizzes need strong measures in order to reduce the potential to cheat and to protect the validity of the question bank. The quizzes, which are made up of 10 multiple-choice questions, are set to allow students only 15 minutes to answer all questions. The questions are delivered one at a time, and students are prevented from revisiting questions once answered.

The questions are presented in random order, as are the answer options for each question. This makes it unlikely that any two students will get the exact same quiz. When the students are done, the software gives them only a final score but does not allow them to see the questions or answers. It shows them whether they got the question right, and it provides feedback about the location in the course textbook where the question came from.

Additionally, the quizzes are only one component in the assessment scheme for the course. There are simulations the students must complete and submit. There are two midterm exams, and at the end of the course the students are given a comprehensive final exam in a proctored computer classroom. All of the other quizzes and exams are given online at the location of the student's choice.

The time limitation in this example makes it difficult for students to look up answers elsewhere and still finish on time. The randomization makes it very difficult for students to call one another and compare answers. Presenting one question as a time makes it difficult for the students to print all of the pages, because it takes time for each question to download and to print. The time invested in printing may prevent the student from finishing the exam. Furthermore, server activity logs make it possible to determine when and where a student was when taking the exam. The logs can be used to look for cheating patterns. The logs can also be used to verify aspects of a student's claims about technical problems if such a claim is made.

These strategies are typically used in conjunction with objective tests as in the example. But, online tests can also use short answer and essay questions to increase the difficulty of cheating. However, additional methods are needed to detect and reduce plagiarism with this type of question. Thankfully, the methods used to cheat on these types of questions are also powerful strategies for identifying plagiarism. When there are grounds for doubt of a specific response to a question, one simple approach is to copy a unique segment of a statement and use it as a search term on the Internet. Many times this quick technique will locate the original source. Another strategy is for schools to purchase a service that checks written submissions for plagiarism, one such provider is Turnitin.com. Of course, this adds cost to the delivery of instruction.

Another technology that can be very powerful is computer configurations on campus computers. Students can be required to use a campus computer that controls which programs can be opened. As such, the computers can be set to allow only the quiz to be opened and not any other applications. Some campuses have testing labs that include not only computers with this type of security, but also proctors who watch for cell phone use, books, talking to others, and so forth.

Biometric devices are an emerging technology that can be used to verify that the person taking the exam is the person who will receive credit for it. For example, some companies are offering fingerprint scanners built into computer keyboards or as separate devices (Levine, 1999). Although this helps with initial identification, there is still the potential for the student to rely on outside help while taking the test.

The telephone, while certainly not a new technology, can also be a powerful ally to instructors who are concerned about reducing cheating. McCabe and Drinan (1999) suggested a practice of making phone calls to students to discuss their answers on an exam. In fact, the mere suggestion prior to a test that such a phone call could be made might be all that is needed.

Webcams are another technique available for online exams. The instructor can require that students taking exams from off campus use a computer with a Webcam (which can cost as little as $20-$50 in U.S. currency). Then, the instructor can either watch the student or even give an oral exam. This does not mean a student cannot still use cheat sheets or have someone off camera providing assistance, however.

Cost and practicality are important considerations that need to be weighed when selecting technological interventions. We encourage instructors to take advantage of existing campus resources not only because it may be less expensive, but also because it may not be possible to get technical support otherwise. For example if an instructor wants to require that students use Webcams, then a question will arise about who will provide students with technical support in installing and operating them. In many cases the instructor ends up bombarded with student technical questions and the issue takes away time from course instruction. It has been our experience that telling students they must solve their own problems rarely resolves the issue to the instructor's or student's satisfaction.

Instructional Design Solutions and Recommendations

The second major category of solutions to reduce cheating comes from the discipline of instructional design. For the purposes of this discussion we use the term in its general sense to refer to the process of planning, developing, teaching and evaluating instruction. Many authorities recommend approaches to cheating interventions that fall in the arena of instructional design (Draves, 2002; Finn & Frone, 2004; McKeachie & Hoffer, 2002; Whitley & Keith-Spiegel, 2002). Since there is a tremendous amount of literature on the topic, we will use Hinman's three categories of cheating reduction strategies, as cited in Olt (2002): "There are three possible approaches to minimizing (online) cheating a plagiarism: first there is the virtues approach.... Second is the prevention approach.... Finally, there is the police approach" (¶ 5).

Strategies in the first category (virtues approach) include honor codes and communication with students about cheating. The second category, prevention, includes many of the strategies discussed previously in the section on technology solutions. However, instructional design also offers a rich variety of options for prevention strategies, some examples of which include developing student self-efficacy, adjusting grading strategies to avoid having highly evaluative classrooms, and improving communication with students. The final category, policing, includes detection and pursuit of cheaters.

Virtue Approaches

The use of institutional and classroom-based honor codes are the most widely discussed approaches in this category. Whitley and Keith-Spiegel (2002) explained institutional honor codes and their role in cheating reduction:

> *Colleges and universities with honor codes are those in which students pledge to abide by a code that specifies appropriate and inappropriate academic behavior and in which students are responsible for administering and enforcing the code.... In a survey of several thousand students enrolled at 31 institutions, McCabe and Trevino (1993) found that a smaller portion of students at these colleges and universities report engaging in academic dishonesty.* (p. 31)

They cautioned, however, not to view the honor code as a solution in itself: "The simple establishment of an honor code is not sufficient to reduce academic dishonesty; rather, the honor code reflects the presence of a normative climate that frowns on dishonesty" (Whitley & Keith-Spiegel, 2002, p. 31).

Honor codes, however, have been found to be highly effective as a method for reducing cheating. The Center for Academic Integrity (2002-2003.) reported the following:

> *Academic honor codes effectively reduce cheating. Surveys conducted in 1990, 1995, and 1999, involving over 12,000 students on 48 different campuses, demonstrate the impact of honor codes and student involvement in the control of academic dishonesty. Serious test cheating on campuses with honor codes is typically 1/3 to 1/2 lower than the level on campuses that do not have honor codes. The level of serious cheating on written assignments is 1/4 to 1/3 lower.* (¶ 3)

McKeachie and Hoffer (2002) suggested the use of a classroom honor system, or at least a class vote on it. They explained that an honor code can be implemented in his class if 100% of students vote for it: "Although a minority of classes vote for the honor system, a discussion of academic dishonesty is itself useful in helping students recognize why cheating is bad" (p. 99).

We have seen another form of classroom honor codes in which instructors place an honor question in each exam. Students select whether or not they affirm that they have used no disallowed methods for taking the test. In online tests this question usually takes the form of a multiple-choice question with two options, to agree or to disagree. The question acts as a reminder that cheating is not acceptable. In many instances the question includes a link to the definition of cheating and the explanation of disallowed behaviors. We have not seen any literature, however, to indicate the effectiveness of this approach. Nonetheless, because communication with students about cheating is recommended, we feel that this approach is likely to be beneficial.

Simply communicating about the university's policy on cheating and what constitutes cheating in your classroom is agreed upon by many researchers as an important strategy for cheating reduction. This can be particularly helpful in combating the definitional differences that students and faculty have. By informing students and discussing the issue, students are made aware of what behaviors are not allowed. Whitley and Keith-Spiegel (2002) recommended that the explanation address each kind of assessment the students will perform, explaining the requirements for quizzes and exams, group work, papers, and so forth.

Prevention Approaches

It is in this category that instructional design offers the greatest opportunities for cheating reduction, because solutions in this category address many of the motivations behind cheating. Finn and Frone (2004) cited school identification and self-efficacy as major motivational factors that influence cheating. Other significant motivational factors include highly evaluative classrooms, and student sense of isolation, which may be more pronounced in an online class.

School identification, according to Finn and Frone (2004) is "the extent to which students have a sense of belonging in school and value school and school-related outcomes" (p. 118). They explained the importance self-efficacy as well: "Students who are performing well and believe they have the intrinsic ability to continue performing well are less likely to cheat" (p. 118). Finn and Frone recommended that instructors endeavor to increase student identification with the school by encouraging class attendance, participation in class discussions, coming prepared for class, and engaging in extracurricular activities (p. 119). They also suggest taking steps to improve student self-efficacy by setting, "reasonable levels of acceptable performance and reward[ing] students at all achievement levels for hard work and learning" (p. 119).

Pino and Smith (2003), Finn and Frone (2004), and Hancock (2001), recommend adjusting assessment and grading strategies to avoid having highly evaluative classrooms. Hancock stated, "Classroom situations in which students perceive the need to compete with one another and in which professors exert significant influence over classroom procedures and student behavior negatively influence student performance on examinations" (p. 96). By removing competition and evaluative threat, not only will students do better on tests, but they are also less likely to cheat.

Pino and Smith (2003) found that, "If one… reject[s] the GPA perspective, he or she will be much more likely to resist the temptation to engage in academic dishonesty" (p. 492). Draves (2002) offered a strategy to achieve this purpose: "Give a pre-course assessment, short in duration, perhaps 10 questions" (p. 115). The benefit of the precourse assessment is that it discloses what the student will be held accountable for knowing, thereby increasing their level of self-efficacy.

Another way to minimize evaluative threat is to use multiple forms of assessment. McKeachie and Hoffer (2002) explained that "an obvious first answer [to cheating] is to reduce the pressure" (p. 98). To do so, McKeachie and Hoffer recommended "provid[ing] a number of opportunities for students to demonstrate achievement of course goals, make reasonable demands and write reasonable tests" (p. 7). Draves (2002) offered a list of assessment options, including: essays and papers, online discussion postings, timed online quizzes, proctored tests, individual projects, group projects, mentored practice, individual presentations, group presentations, reflective journals, analysis of case studies, debates, role plays, and games.

Draves (2002) explained the advantage of multiple assessments: "The variety and regularity of the testing reduces the chances of cheating, raises the cost of cheating, and most importantly, constitutes positive learning activities for the vast majority (of students)" (p. 7). Olt (2002) recommended utilizing several short assessments throughout the course, citing Cox's explanation of the strategy: "Multiple, individualized tasks are harder to counterfeit because of the necessary coordination and planning involved for the student to arrange for someone else to do the work in a timely and appropriately specific manner" (¶ 6).

An important way, according to Whitley and Keith-Spiegel (2002), to avoid a highly evaluative classroom is to use criterion referenced grading, where scores are based on student "individual student performance" (p. 53) of course outcomes instead of a curve (also known as normative referenced grading). For example, instructors can provide a list of specific outcomes students must demonstrate and the criteria by which they will be graded. Then, when students undertake tasks to perform the outcomes, the grading is based on each individual student's performance, rather than on a curve.

One of the challenges with this approach is the cultural norm in some academic programs among faculty that says if a "normal distribution" of grades is not given in a class, then something is wrong. We have heard many instructors say that they cannot use the criterion referenced approach to grading because it may cause too many students to get A's. It is this perspective that leads to evaluative threat and can result in an increase in cheating.

Yet another important instructional design strategy to reduce cheating is a high level of communications between instructors and students. Olt (2002) recommended "build[ing] into the course a high level of instructor/student interaction" (¶ 7). Communication will reduce the students' sense of isolation and anonymity. Online discussions are a widely used strategy for this purpose. Online discussions can be done in instructor-led format, wherein the instructor poses a question and asks each student to respond, in small-group format, where students in small groups discuss a question and then one group representative posts a collaboratively written response, or in one-on-one format, such as private mail or journals.

It is possible through online discussions to achieve a high degree of student interaction with one another as well as with the instructor (Palloff & Pratt, 1999). This is not only an effective learning strategy but also a useful cheating reduction strategy in that it can decrease the students' perception that an online course is impersonal. Furthermore, discussion messages offer another opportunity for assessment. Discussion and mail postings give the instructor a sense of the student's level of comprehension and ability to apply and synthesize new knowledge.

Also in the category of communication, feedback is very important. According to Hancock (2001), "Brophy (1987) found that feedback from teachers can direct students' future efforts by highlighting the topics that have been learned and those in need of improvement... [W]hen students know what is needed to improve their abilities, they perceive control over achievement outcomes, which often enhances their learning" (p. 284).

One final communication strategy, discussed in the section on technological solutions, bears mentioning again: the telephone call. Draves (2002) recommended giving a form of pop-quiz over the telephone. The quizzing can act as a form of graded assessment or simply as a means of checking that the student can answer a question impromptu that he or she answered on a test. Another use of the telephone is to establish meaningful communication with students, the purpose of which is to increase instructor–student interaction. This strategy may also be particularly valuable for students who have an auditory preference in their learning style or who are more feelings oriented in their personality style. A telephone may give them the type of contact that will take advantage of their strengths.

Last, it is critical that the communication strategy include specific information in the course syllabus. Whitley and Keith-Spiegel (2002) offered the following list of elements to be contained in the syllabus:

1.   A brief, general statement about the importance of academic integrity in higher education.

2.   A personal statement declaring your commitment to upholding academic honesty in your class.

3.   How you will deal with any incidents that you observe or that come to your attention.

4.   A brief list of any types of academic dishonesty in your school's policy (or reference to where the complete policy can be found).

5.   A brief list of any types of academic dishonesty that could occur in your particular course that could benefit from more detail (e.g., oral plagiarism in a class that requires an oral report).

6.   A brief list of campus resources that may help reduce the risk factors associated with cheating (e.g., writing clinic, counseling center, learning center or tutoring program).

7. An invitation to come directly to you to discuss anything that is unclear to confusing regarding the appropriate way to complete assignments.

8. An invitation to report incidents of academic dishonesty.

There is a great deal of literature on effective instructional design strategies for online learning. In many cases, an effective instructional design strategy is also an effective cheating reduction strategy. The reason for this is that instructional design encourages instructors to rethink the way they approach their courses and design according the best approaches to achieve the learning outcomes of the course. This approach usually leads to the very same recommendations that have been discussed throughout this section. We encourage faculty members to spend some time reading instructional design literature, attending courses in best practices, and taking advantage of teaching and learning resources on their campuses.

Policing Approaches

The final category of instructional design approaches to cheating includes those that use monitoring and discipline to detect and punish cheaters. Pino and Smith (2003) prescribe a cautious approach to these methods:

> *The deterrent effect against academic dishonesty is weak as well. If many students engage in academic dishonesty, but we don't see half of the student body on academic probation, then we must assume that the risk of detection and certainty of punishment is low.* (p. 493)

However, they add, "Vigorously disciplining students proven to engage in academic dishonesty and communicating this to students also adds a potential deterrent effect" (p. 493). The first step to take in policing cheating is to communicate when cheating has been detected and how it has been punished. Whitley and Keith-Spiegel (2002) explain, "Students who contemplate cheating tend to weigh the risk of detection and punishment against the benefits they expect to derive from dishonesty and will cheat if they perceive the expected benefits outweigh the potential costs (Michaels and Miethe, 1989)" (p. 85).

As part of the policing strategy it is critical that support is provided to faculty and students in the disciplinary process. Reporting and dealing with cheating incidents can create some of the most stressful experiences in the careers of faculty members and students. This means that ensuring instructors and students are informed about policies and procedures is essential. Whitley and Keith-

Spiegel (2002) recommend a training program for faculty and the use of various discussion forums for students, such as the classroom as well as the student newspaper and/or radio station.

Furthermore, it is critical that institutions stand behind their academic dishonesty policies. One of the major reasons listed by faculty members for not taking action on cheating incidents is fear that their action will not be supported (Whiley & Keith-Spiegel, 2002). An immediate add-on to informing students and faculty and providing support through the disciplinary process is to aggressively publicize cheating incidents and disciplinary action taken. Whitley and Keith-Spiegel (2002) explain, "People will take an academic integrity policy seriously if they know it is being enforced" (p. 140).

A widely used policing approach is the proctored test. Proctoring can be used with online assessment. There are several approaches to proctoring online tests. One approach is to bring students to a central location for testing. This could mean having a number of sites with proctors available for the students. Another approach is to employ proctors in locations near where the students live. A third approach was described in the section on technological solutions, which is to use a Web-cam and microphone or telephone.

Each approach has advantages and limitations. There is considerable cost associated with maintaining multiple testing centers. Additionally, the number and location of centers limits the distances from the university students can live. College and university programs need to carefully consider the importance of proctoring when planning an online program. The distance of the program will be limited by the availability of testing centers, if this approach is used.

Hiring proctors in the students' home locations offers a greater degree of freedom regarding the distance students can live from the university, but creates a need for administrative procedures and tracking. The proctors need to be approved, trained, and, many times, paid. The labor involved in establishing a network of such remote proctors may be prohibitive for some programs. Again, the issue needs to be carefully considered before an online program is established.

The third approach, using Webcams to proctor tests, was discussed in the section on technological solutions, and, as such, will not be reiterated here. What is important is that instructors take into consideration all of the options and select the approach that will work within each situation.

Reducing Cheating on Objective Tests

Although many of the recommendations discourage the use of objective tests, there are many situations in which the objective test is needed. For example, large-section classes with only one instructor and no teaching assistants make it difficult for instructors to use labor-intensive assessment activities such as papers, group projects, and so forth. Furthermore, online course delivery systems usually provide a quiz and exam tool that will automatically grade objective questions. As such, security on objective tests must be considered and solutions provided.

As explained in the section on technology strategies, there are a number of ways to design an online objective test to make it difficult for students to cheat. From an instructional design standpoint, one of the most formidable approaches instructors can use is to maintain an ever-expanding and developing question bank. First, this reduces the potential benefit student may attempt to gain by getting copies of the test or memorizing questions and passing them on to other students. Second, and more important, it keeps the course fresh. Writing new exam questions will cause the instructor to engage in reflective practice, continuously fine tuning the course.

We are not suggesting that instructors replace every question every time they teach the course, but rather, write alternate versions of a percentage of questions before each course offering. We have not encountered recommended strategies in the literature for doing this, but we encourage an approach that takes into account the average size of the class and frequency with which it is offered. The more students taking the course, the greater the likelihood that exam questions are being copied or memorized. The larger the course and the more often it's offered, the more often the question bank needs to be expanded.

One strategy that may be particularly useful is to write alternate versions of question and have the exam delivery software select from among the alternate versions and present them randomly. For example, an instructor in an anatomy class could write three versions of a question about the term *tissue* and have the exam randomly select one of the three for a given question. The first time the course is offered, only one version of the question may exist. But the second time it's offered, another version can be added, and so on.

It is important, however, to take care to design questions with similar levels of difficulty and discrimination values. This process takes time, because data needs to be collected to validate the questions. However, if objective tests are to be used online, the time investment is worthwhile.

Instructional Design Versus Technological Solutions

The overarching message of this section on instructional design was to consider nontechnological methods for reducing cheating before undertaking solutions that utilize technology. There are several advantages to this approach. First, nontechnical solutions can often be implemented immediately and they may cost little or nothing at all. Second, nontechnical solutions often do not require training. Third, non-technical solutions rarely involve resources outside the classroom, such as a campus computer services office. In other words, changing the course design rather than implementing a technological intervention, can be faster, cheaper, and less stressful for the instructor.

Future Trends and Conclusion

As technologies to prevent cheating advance, so do techniques to get around them. One trend that is likely to continue is that methods of cheating will evolve along with methods of prevention and detection. Regardless of this difficulty, it is necessary that methods of cheating reduction continue to develop and improve.

One area in which we expect tremendous advancement is the technological methods for reducing cheating. There are two avenues through which technological improvements will become available. First, as bandwidth increases, so will our ability to deliver synchronous instruction online. At this time, many students have slow Internet connections, making it difficult to use Webcams and other synchronous applications, such as desktop sharing. However, as more users gain access to wide bandwidth connections, it will become possible to use synchronous video applications to monitor students while taking test or performing other assessment activities, such as giving speeches, participating in group presentations, or performing a piece of music, and so forth.

The second major area in which technological improvements will come will be in equipment and software to prevent and detect cheating. The fingerprint scanner has already been developed, but its use has yet to gain favor in higher education. Other types of identity scanners may not be far off, such as a retinal scanner or voice identification software. Also, applications to monitor, limit, and report student activities on their computers while taking tests are likely to improve. As computer operating systems improve, so does our ability to control what students can do while on their computers. A future generation of online course management systems could include applications that restrict students from doing things such as opening other Web sites while taking an exam. This capability already exists with some systems but must be installed on the student's computer. In the future, it is likely that students will not have to install software in order for instructors to have this capability.

Another important future trend is the training of faculty in instructional design concepts. While many universities and college have teaching and learning centers, the topic of instructional design is low on the priority list in the typical faculty member job. Considering the tremendous costs involved with adoption of prevention technologies, and the merits of instructional design solutions, the need for faculty development in this area will urge colleges and universities to invest in it. This will not only result in better cheating reduction, but better instruction period.

With the fast pace of technology improvement and the increasing demand for online courses, the future directions for cheating reduction will be many and varied. We encourage instructors to subscribe to educational news feeds and journals on educational technology in their disciplines. Advances come quickly and sometime produce radical change. We recommend that instructors stay informed about the educational technology trends in their subjects.

Conclusion

Online learning has introduced a variety of new assessment methods as well as a variety of new cheating techniques. Furthermore, old cheating techniques still find usefulness in the online classroom. While cheating is likely to continue to occur in nearly every teaching environment, a combination of strategies utilizing technology and instructional design can inhibit cheating. The most important lessons to be learned from this chapter follow:

- Cheating happens in nearly every type of assessment
- Technology can be used to reduce cheating, but cost and technical support need to be considered before selecting a technological solution
- Many campuses have technology support personnel who can assist faculty with discovering what is possible on their campuses; avail yourself of their services
- Instructional design strategies offer a wide variety of options for reducing cheating and these strategies are often less costly and easier to implement
- Just as campuses offer technology support personnel they often also have instructional design personnel; avail yourself of their services as well
- Institutional policies and procedures can inhibit or encourage cheating, the issues needs to be on the institutional radar at all times

References

Center for Academic Integrity. (2002-2003). CAI research. Retrieved June 10, 2004, from *http://www.academicintegrity.org/cai_research.asp*

Corbett, B. (1999). The cheater's handbook: The naughty student's bible. New York: Regan Books.

Curry, A. (1997, July 10). Psst, got the answer? Many say 'yes'. *Christian Science Monitor, 89*(157), 7.

Dirks, M. (1998). How is assessment being done in distance learning? Retrieved May 4, 2005, from ERIC online database *http://www.eric.ed.gov* (ED 423273)

Draves, W. (2002). *Teaching online* (2nd ed.). River Falls, WY: LERN Books.

Finn, K., & Frone, M. (2004). Academic performance and cheating: Moderating role of school identification and self-efficacy. *Journal of Educational Research, 97*(3), 115-123.

Ford. (2000, May 22). High-tech cheating. *RCR, 19*(21), 12.

Freeman, L., & Ataov, T. (1960). Invalidity of indirect and direct measures of attitude toward cheating. *Journal of Personality, 28*(4), 443-447.

Hancock, D. R. (2001). Effects of test anxiety and evaluative threat on students' achievement and motivation. *Journal of Educational Research, 94*(5), 284-290.

Henry, E., & Barr, A. (1998, June). The official tax outlaw quiz. *Kiplinger's Personal Finance Magazine, 52*(6), 121-124.

Karlins, M., Michaels, C., & Podlogar, S. (1988). An empirical investigation of actual cheating in a large sample of undergraduates. *Research in Higher Education, 29*(4), 359-364.

Levine, D. (1999, June 7). Let your fingerprint be your password. *InternetWeek, 768*, 46-49.

McCabe, D., & Drinan, P. (1999, October 15). Toward a culture of academic integrity. *Chronicle of Higher Education, 46*(8), B7.

McKeachie, W., & Hoffer, B. (2002). *McKeachie's teaching tips: Strategies, research, and theory for college and university teachers.* Boston: Houghton Mifflin.

Olt, M. (2002, Fall). Ethics and distance education: Strategies for minimizing academic dishonesty in online assessment. *Online Journal of the Distance Learning Administration, 5*(3). Retrieved September 23, 2004, from *http://www.westga.edu/~distance/ojdla/fall53/olt53.html*

Orlans, H. (1996, September/October). How to cheat. *Change, 28*(5), 10.

Palloff, R., & Pratt, K. (1999). *Building learning communities in cyberspace: Effective strategies for the online classroom.* San Francisco: Jossey-Bass.

Pino, N., & Smith, W. (2003). College students and academic dishonesty. *College Student Journal, 37*(4), 490–501.

Pullen, R., Ortloff, V., Casey, S., & Payne, J. (2000). Analysis of academic misconduct using unobtrusive research: A study of discarded cheat sheets. *College Student Journal, 34*(4), 616-625.

Richardson, A. (2002). High-tech cheating: Where there's a will there's a gadget. *Black Issues in Higher Education, 19*(11), 32.

Ruderman, J. (2004). Faculty play a crucial academic integrity role. *Academic Leader, 20*(3), 8.

Stern, E. B., & Havlicek, L. (1986). Academic misconduct: Results of faculty and undergraduate student surveys. *Journal of Allied Health, 15*(2), 129-142.

Trotter, A. (2002, April 3). Plagiarism controversy engulfs Kansas school. *Education Week, 21*(29), 4-5.

Whitley, B., & Keith-Spiegel, P. (2002). *Academic dishonesty: An educator's guide.* London: Erlbaum.

Recommended Reading

Carnevale, D. (1999). How to proctor from a distance. *Chronicle of Higher Education, 46(*8), B7.

Driscoll, M. (1999). *Psychology of learning for instruction* (2nd ed.). Boston: Allyn & Bacon.

Fighting online plagiarism. (2001, July 27). *Chronicle of Higher Education, 47*(46), B17.

Gagné, R., Briggs, L., & Wager, W. (1992). *Principles of instructional design* (4th ed.). New York: Harcourt Brace.

Honor code keeps exam time flexible, calm. (2004, January). *Community College Week, 32*(1), 4.

Howlett, B. (2003, October 19). Integrity and a distance: Reducing cheating in online tests. *Proceedings of the 2003 NAWeb conference.*

Jerome, R., & Grout, P. (2002, June 17). Cheat wave. *People, 57*(23), 83.

Kansas college gives first "XF" grade to plagiarist. (2003, December), *Community College Week, 16*(9), 14.

Karlgaard, R. (2002, April). My Ken Lay. *Forbes, 169*(8), 35.

McCabe, D., & Bowers, W. (1994). Academic dishonesty among males in college: A thirty year perspective. *Journal of College Student Development, 35*(issue), 5-10.

McCabe, D., & Pavela, G. (2000, September/October). Some good news about academic integrity. *Change, 33*(5), 32-38.

McCarroll, C. (2001, August 28). Beating Web cheaters at their own game. *Christian Science Monitor, 93*(192), 16.

Sohn, E. (2001, May). The young and the virtueless. *U.S. News & World Report, 130*(20), 51.

Stolovitch, H., & Keeps, E. (1999). *Handbook of human performance technology* (2nd ed.). San Francisco: Jossey-Bass/Pfeiffer.

Chapter XV

Supporting and Facilitating Academic Integrity in Distance Education through Student Services

Brian F. Fox, Santa Fe Community College, USA

Abstract

This chapter briefly describes the growing concern over a lack of academic integrity in higher education and the traditional methods employed to detect and prevent it. Arguing that these possess inherent shortcomings, the author describes a systems approach that incorporates all aspects of student services: admissions, marketing, and orientation; instructional support; instructional technology; library services; and counseling and advocacy. For academic integrity policies and programs to truly be effective, they must be universal and preventative in scope and include all segments of student services and the student body itself. Regular assessment must be conducted and the topic incorporated into professional development. The primary goal for educational institutions should be to foster and support the development of academic integrity in their students.

Defining the Problem

The concern over a lack of academic integrity in education is certainly on the rise in recent years, with an increasing number of articles, papers, and presentations describing the results of surveys on academic integrity or the actions of colleges and universities against suspected cheaters. With regard to distance education (DE), regional accreditation groups are clearly requiring that institutions take steps to ensure the integrity of student work and the credibility of degrees and credits awarded (Commission on Colleges and Schools, 2000). At the same time, a growing number of organizations and institutions are actively pursing violations of copyright and intellectual property laws.

Although it is certainly impossible to determine the true extent of academic dishonesty, some statistics and examples provide both illustrations and indications. For example, in 2002, 47 students at Simon Frasier University turned in nearly identical economics papers (Hamlin & Ryan, 2003). The Center for Academic Integrity (CAI) at Duke University conducted a 1999 survey of 2,100 students on 21 campuses across the country, with about one-third admitting to serious test cheating, and half admitting to one or more instances of serious cheating on written assignments (Hamlin & Ryan, 2003). In a survey of 4,500 students at 25 high schools, over half admitted to having engaged in some level of plagiarism on written assignments using the Internet (Mayfield, 2001). Research by CAI members and others concluded that "student cheating is on the rise and that pressures and opportunities for dishonest behavior are increasing in many academic and professional contexts" (CAI, 1999, p. 4).

With regard to student populations, Dr. Diane Waryold, executive director for the CAI, stated that certain trends may be found in academic dishonesty: top students competing for spots in grad school; students with lower GPAs (survival); students who value grades over learning and honesty; females and males, though males tend to self-report more cheating; members of Greek organizations; business and engineering majors; younger students; and all cultural backgrounds. Waryold offers the following rule of thumb: 20% will never cheat, 20% will cheat whenever possible, and 60% are open to influence (Waryold, 2002). Although further research is certainly needed in terms of understanding the demographics of cheaters, the question remains as to how this data might be used in constructive ways.

Though the rise and development of the World Wide Web clearly cannot be blamed for a lack of academic integrity on the part of some students, it has certainly provided new opportunities for cheating, where the "age-old concerns about ethical practices in assessment ... take on new twists in the distance-learning environment" (Abbott, Siskivic, Nogues, & Williams, 2000). McMurtry

(2001) described several methods, including copying and pasting text directly from articles, sharing assignments through e-mail attachments, and simply purchasing and downloading papers through sites such as EssayWorld.com, Planet Papers, Evil House of Cheat, Other People's Papers, and School Sucks, all of which offer thousands of papers on a wide diversity of topics.

Hinman (2000) states that there are three possible approaches to minimizing online cheating and plagiarism: virtue (developing students who do not want to cheat), prevention (eliminating or reducing opportunities and reducing the pressure to cheat), and policing (catching and punishing cheaters). While some educators advocate an aggressive program of detection and punishment as the most effective method to deal with academic dishonesty, it may be argued that a primarily "downstream" approach lacks learner-centeredness and institution-wide coordination and will inevitably fail as a result. Therefore, in order to more effectively address these issues, this paper will argue for a systems approach to student services in order to better support and facilitate academic integrity.

Academic Integrity, Student Services, and the Systems Approach

Rumble (2000) argued that distance educators have generally been better at articulating what they mean by student services than traditional educators, and he adds that a systems approach to DE is embedded in the literature and that it is part of the culture of DE that includes student support. Lyons (1990) reminded us that student services exist to serve the institution's mission and objectives, and that this will determine to a large degree where resources are focused. As a result, it is absolutely critical that academic integrity be one of the guiding principles and objectives of the institution, embedded within all aspects of the college or university. Brindley (1995) further added that "any interventions which are made should be consistent with the unique context in which they are offered, reflecting institutional values and objectives," leading inevitably to the conclusion that there is no one right method to supporting and facilitating academic integrity for all institutions (Tait, 1995).

Bearing these guidelines in mind, the CAI (1999) stated that all institutions should have "clear academic integrity statements, policies, and procedures that are consistently implemented" (p. 10). Because all campus constituencies have a role in supporting and facilitating academic integrity (CAI, 1999), it is imperative that the institution's honor codes, conduct codes, and administrative policies and procedures dealing with academic integrity be developed through collaboration on the part of administrators, faculty, staff, and students, all of whom must "buy in" to them if success is to be achieved. Additionally, community input should be solicited to support the development and implementation of the program in order to increase its chances for success; for example, statements from local business leaders condemning academic dishonesty might lend support to institutional policies and statements (these might be gained through business advisory committees, marketing campaigns, etc.), while cooperation with local K-12 schools could serve to better prepare students for higher education.

Student Services Activities and Their Roles in Academic Integrity

The term *student services* has been defined in various ways throughout open and DE literature. This chapter will broadly define these functions, drawing primarily upon Brindley's (1995), Simpson's (2000), and Tait's (1995) definitions in order to outline the roles various systems might play in supporting and facilitating academic integrity.

Admissions, Marketing, and Orientation

One of the seven recommendations made by the CAI (1999) is to "inform and educate the entire community regarding academic integrity policies and procedures" (p. 10). To this end, institutions should take every opportunity to advertise their commitment to academic integrity. Materials should be developed in print, multimedia, and Web-based formats that would be included in information packets mailed to interested students and in advertising campaigns. Statements of support from community and business leaders and alumni might be included in these; for example, "Acme, Inc., strongly supports the high standards of academic integrity set by State U. Your graduates possess the values we are seeking in future employees."

Orientation programs have always been instrumental in terms of helping students effectively embark on their studies. DE students, removed from a great deal of campus culture and often nontraditionally aged and returning to formal education after many years, may benefit greatly from full orientation programs that prepare them for their new study activities. Topics might include organization, time-management, study skills, distance learning success, writing, independent study, stress management, and so forth (Granger & Benke, 1998). It is through the orientation process that campus academic integrity policies and procedures should be formally and thoroughly shared and discussed with all students. Because of DE students' physical separation from campus, accommodations should be made to support all of these objectives with videotape, multimedia CD-ROMs, telephone, and Web support methods being currently utilized by a growing number of institutions. Orientation processes should also ensure that students are properly assessed in terms of their preparedness for college-level work. For those students needing remediation, advisement should be provided to assist them through the requirements or to explain alternatives when appropriate.

Instructional Support (Tutoring-Teaching-Proctoring)

First, academic integrity must be modeled by all members of the faculty. The CAI (1999) stated, "Fair and accurate evaluation is essential in the educational process. For students, important components of fairness are predictability, clear expectations, and a consistent and just response to dishonesty" (p. 7). Faculty members also need to be aware that learners must first learn how to learn skills to be effective online learners, and that these skills need to be explicitly supported and taught. Drawing upon a range of theory, McLoughlin and Marshall (2000) maintained that effective online learning requires a variety of skills: articulation (connecting instructions and resources to course objectives and assessment measures), self-regulation (maintaining discipline in terms of class participation expectations and deadlines), a repertoire of learning strategies, and self-assessment and self-evaluation. The authors turn to sociocultural theory, which states that learning involves social interaction and dialogue, negotiation, and collaboration and that "scaffolded" or assisted learning (wherein instructors utilize a developmental approach, supporting individual students so that they gain increasing degrees of autonomy and competence in their learning) can increase cognitive growth and understanding. This can be demonstrated by instructors who require a progressive sequence of work from their students; as an example, for papers this might consist of a proposed topic, outline, research, draft, and then a final paper.

Due to the growing concern over Internet plagiarism, Web-based Internet detection services, both fee-based and non-fee-based, are being increasingly utilized (Hamlin & Ryan, 2003). These services operate in a variety of ways but frequently compare a student's electronic paper to a full-text database of thousands of documents and the Web in general, testing for possible matches of text, and then automatically feeding back a report with the findings. Turnitin.com© is a popular example of such a service, with over 20,000 registered users in 19 countries (McCarroll, 2001). While these plagiarism-detection services are attracting a growing number of proponents and customers, some concerns have been raised: Copyright considerations for student papers (Foster, 2002), cost in an era of tightening budgets, degradation of the relationship between faculty members and their students, incomplete databases providing false negatives, and false positives due to commonly utilized phrases (Foster, 2002).

McMurtry (2001) recommended eight suggestions for faculty to more effectively combat e-cheating:

1.   Take time to explain and discuss your college's academic integrity policy.

2.   Design writing assignments with specific goals and instructions.

3.   Know what is available online before assigning a paper.

4.   Give students enough time to do an assignment.

5.   Require oral presentations of student papers or have students submit a letter of transferal to you, explaining briefly their thesis statement, research process, and so forth.

6.   Have students submit essays electronically.

7.   When you suspect e-cheating, use a full-text search engine.

8.   Consider subscribing to a plagiarism search service.

With these last two recommendations, many educators view these services as being more valuable in terms of their deterrence than in their ability to detect plagiarism (Hafner, 2001); in order to be truly effective, therefore, students should be informed at the beginning of the semester that the instructor reserves the right to utilize these methods. Other educators add to these recommendations the use of pop quizzes and class participation requirements on discussion boards (Hamlin & Ryan, 2003), project-based assessments (Olt, 2002), signing academic integrity statements, proctored exams, and writing assignments that change each semester. The use of course journals, where students are required to summarize what they have learned and their thoughts and impressions as they progress through the course, makes plagiarism quite difficult, particularly if the journals are collected and reviewed frequently by the instructor. This assessment measure also clearly supports the goal of scaffolded learning. In addition, there is a growing movement toward the use of electronic portfolios; although there is to date little agreement over their makeup and format, many educators see them as a valuable assessment tool for both online and face-to-face classes (Ahn, 2004).

Another issue requiring attention is that of proctoring. Many distance and open educators argue for strict adherence to flexible assessment, striving for "anytime/anywhere"—which by its nature prohibits proctored examinations. Others, however, see DE as being on a spectrum with traditional education; while maximum flexibility is the goal, there are situations where proctored exams are required for pedagogical or licensure and certification reasons. When this is the case, institutions must ensure that proper facilities and staff are available, to include at a minimum supervised assessment centers with flexible hours and coordinators for off-site proctoring.

Instructional Technology

With the rapid implementation of Web-based technologies in DE, there is a growing concern over the verifiability of students' identities online. While this is certainly not a new concern of DE educators, new technologies are providing additional tools to institutions. It is now common for all course management systems (CMS), learning management systems (LMS), and college portals to require both a login identification as well as a password. With a widespread goal being ease of use, many institutions implement a single sign-on for all Web-based services and resources (registration, transcript review, online library, access to learning platforms, etc.), requiring the user to authenticate only once. Authentication is the process wherein a network user establishes a right to an identity, while authorization is the process of determining whether an identity—plus a set of attributes associated with that identity—is permitted to perform some action, such as accessing a resource (Mickool, 2004).

Although a variety of technologies have been or are being developed to address the issue of authentication (e.g., passwords, certificates, smart cards, biometric techniques), to date passwords are by far the most commonly used. This system easily lends itself to abuse; for example, student A might simply give his or her login identification and password to student B in order to allow student B access to an online classroom to take a test. In the traditional classroom, this is easily dealt with by simply recognizing one's students or by requiring a photo ID. Even if more advanced technologies are utilized, however, the potential for abuse will not disappear. For this reason, many online instructors prefer using assessment measures other than or in addition to tests or require proctoring (as described above). A single sign-on, however, might also serve a secondary role in preventing students' sharing of their login identifications and passwords if they understand that this gives others access to their financial aid information, grades, e-mail account, registration, and so forth. The key here is to adequately publicize this fact.

Library Services

Library services may support and facilitate academic integrity in a variety of ways. Self-paced online learning tutorials have been established at many institutions that allow students to learn research and writing skills, proper citation formats, and how to avoid plagiarism. Examples of such tutorials may be found at the University of Maryland University College's (UMUC) Virtual Academic Integrity Laboratory (UMUC, 2003). Additionally, support through chat, telephone, and e-mail may be provided to support students more flexibly as they perform their research. As proposed by some (Foster, 2002), plagiarism-detection services could be made available to students to allow them to review their own papers before submission to assist them in avoiding plagiarism.

Counseling and Advocacy

At every available opportunity, counselors should seek to support the institution's academic integrity policy through such activities as presentations to high schools and other groups, meetings with student government, and campus educational campaigns. Through dialogue with students, counselors should remain vigilant to signs of stress, frustration, and problems with grades, all of which can increase the probability of cheating. Counselors, be they full-time professional staff or faculty, should be proactive in advising moderation to their students with respect to course load and in consideration of the occasional need to withdraw from a course. Counselors may also serve as additional contacts for students wishing to report acts of academic dishonesty.

As Robinson (1995) reminded us, "Not all open and distance learners are adults, highly motivated or self-managing"(p. 223). Just as with any population of students, there will certainly be some DE students who violate the institution's academic integrity policy. In order to properly address such situations, a "clear, accessible, and equitable system to adjudicate suspected violations of policy" (CAI, 1999, p. 10) should be created. All students accused of academic dishonesty should be provided with counseling and possibly advocacy in such circumstances, as well as after a judgment is rendered. At a minimum, students should be fully informed of the charges against them, their rights and options, the official proceedings, and the consequences of being found guilty. Institutions might choose to go further than this, providing for optional representation for the accused as well as the accuser, thereby creating a true honor court.

Conclusion

Although most educators understandably find the topic of academic dishonesty unpleasant, Hamlin and Ryan (2003) reminded us that "unfortunately, cheating has always existed and will continue as long as there is temptation to do so." For academic integrity programs and policies to truly be effective, they must be preventive and universal in scope. All segments of student services must be involved in the development, implementation, and review of the policies. Regular assessment must be conducted, and the institution should remain ever vigilant to trends in higher education and technology that might impact academic integrity on campus (CAI, 1999). Just as students should be able to expect fair treatment in potential cases of academic dishonesty, faculty and staff must also have a right to expect the same (CAI, 1999). Additionally, institutions must include these topics in their professional development programs, particularly in faculty and staff orientation, and whenever possible students should be invited to share their views and concerns.

Although it is certainly the case that the effective use of technology may help to prevent or expose academic dishonesty, it should never be viewed as a cure-all. Academic integrity or dishonesty is based upon the actions of human beings, and it is here that our attentions and efforts should be largely focused. Every technology can be circumvented and every rule broken if students are dedicated to these goals and believe that they can do so with little or no chance for detection and punishment. It should also be understood that any institution that sets high standards and actively monitors academic integrity will inevitably discover it, and that this is not a sign of failure. The trick for educational institutions is to encourage and foster the development of academic integrity in their students so that they make no attempt to cheat in the first place.

References

Abbott, L., Siskovic, H., Nogues, V., & Williams, J.G. (2000). *Student assessment in multimedia instruction: Considerations for the instructional designer* (ERIC Document Reproduction Service No. ED 444 516). Retrieved June 11, 2002, from *http://newfirstsearch.oclc.org*

Ahn, J. (2004, April). Electronic portfolios: Blending technology, accountability & assessment. Retrieved May 11, 2004, from *http://www.thejournal.com/magazine/vault/A4757C.cfm*

Brindley, J. E. (1995). Learner services: Theory and practice. Retrieved August 28, 2003, from *http://www.uni-oldenburg.de/zef/cde/support/readings/brind95.pdf*

Center for Academic Integrity. (1999). *The fundamental values of academic integrity* [Brochure].

Commission on Colleges and Schools. (2000). Distance education: Definitions and principles—A policy statement. Retrieved November 3, 2003, from *http://www.sacscoc.org/pdf/distance.pdf*

Foster, A. (2002, May 17). Plagiarism-detection tool creates legal quandary. Retrieved February 27, 2003, from *http://chronicle.com/free/v48/i36/36a03701.htm*

Granger, D., & Benke, M. (1998). Supporting learners at a distance from inquiry through completion. In C. C. Gibson (Ed.), *Distance learners in higher education* (pp. 127-137). Madison, WI: Atwood Publishing. Retrieved September 20, 2003, from *http://www.uni-oldenburg.de/zef/cde/support/readings/grang98.pdf*

Hafner, K. (2001, June 28). Lessons in Internet plagiarism. Retrieved February 27, 2003, from *http://www.nytimes.com/2001/06/28/technology/28CHEA.html?0628i*

Hamlin, L., & Ryan, W. (2003). Probing for plagiarism in the virtual classroom. Retrieved May 1, 2003, from *http://www.syllabus.com/article.asp?id=7627*

Hinman, L. M. (2000). Academic integrity and the World Wide Web. Retrieved February 28, 2003 from *http://ethics.acusd.edu/presentations/cai2000/index_files/frame.htm*

Lyons, J. W. (1990). Examining the validity of basic assumptions and beliefs. In M. J. Barr, M. L. Upcraft, & Associates (Eds.), *New futures for student affairs* (pp. 22-40). San Francisco: Jossey-Bass.

Mayfield, K. (2001). Cheating's never been easier. *Wired*. Retrieved February 27, 2003, from *http://www.wired.com/news/school/0,1383,45803,00.html*

McCarroll, C. (2001, August 28). Beating Web cheaters at their own game. *Christian Science Monitor*. Retrieved February 27, 2003, from *http://www.csmonitor.com/2001/0828/p16sl-lekt.html*

McLoughlin, C., & Marshall, L. (2000, February 2-4). Scaffolding: A model for learner support in an online teaching environment. *Proceedings of the 9ᵗʰ Annual Teaching Learning Forum*, Perth, Australia. Retrieved September 18, 2003. from *http://www.uni-oldenburg.de/zef/cde/support/readings/loughlin2.htm*

McMurtry, K. (2001). E-cheating: Combating a 21st century challenge. Retrieved February 27, 2003, from *http://www.thejournal.com/magazine/vault/A3724.cfm*

Mickool, R. (2004, April 1). The challenge of single sign-on. Retrieved April 10, 2004 from *http://www.syllabus.com/article.asp?id=9194*

Olt, M. (2002). Ethics and distance education: Strategies for minimizing academic dishonesty in online assessment. Retrieved February 27, 2002, from *http://www.westga.edu/%7Edistance/ojdla/fall53/olt53.html*

Robinson, B. (1995). Research and pragmatism in learner support. In F. Lockwood (Ed.), *Open and distance learning today* (pp. 221-231). London: Routledge.

Rumble, G. (2000). Student support in distance education in the 21st century: Learning from service management. *Distance Education, 21*(2), 216-235.

Simpson, O. (2000). *Supporting students in open and distance learning.* London: Kogan Page.

Tait, A. (1995). Student support in open and distance learning. In F. Lockwood (Ed.), *Open and distance learning today* (pp. 232-241). London: Routledge.

UMUC. (2003). Virtual academic integrity laboratory. Retrieved November 4, 2003, from *http://www-apps.umuc.edu/forums/pageshow.php?forumid=3&s=bf1d29e4e3b1752749478eb5cfb7836d*

Waryold, D. (2002, November 14). *Establishing a climate of academic integrity on campus.* Workshop conducted at Santa Fe Community College in Gainesville, FL.

Chapter XVI

User Authentication and Academic Integrity in Online Assessment

Boris Vilic, Duquesne University, USA

Marie A. Cini, City University, USA

Abstract

This chapter reviews the issues surrounding user authentication and academic integrity in online assessment and offers a number of academic and technological solutions for dealing with student identification and plagiarism. It argues that even though violations of academic integrity are seemingly ubiquitous across all forms of educational delivery, the relative recency of online education has led to growing concerns among faculty and administrators. Although technological solutions for addressing the issues of user authentication and academic integrity are increasing in number, the chapter emphasizes the need for effective instruction and authentic assessment as the strongest means of deterring and reducing the number of academic integrity violations.

Introduction

Faculty and administrators alike—especially those new to online learning—often raise the issue of authentication of students or the process by which educators determine the identity of students. They raise this concern because of a series of related questions that most educators do not have answers to, including, How do we know that students taking a class (or earning a degree) are who they claim to be? Could it be that someone else—an imposter—is pretending to be that particular student? and Are institutions of higher education granting distance learning degrees to students who have hired others to do the work?

Less drastic than hiring others to do their work, online learners are often suspected of higher levels of plagiarism. Though often discussed anecdotally, we have no data to answer the concern that online students may be more prone to cheating than their face-to-face counterparts. Lacking good research, how do we know, without watching over them, that online students' writing samples are indeed theirs and not merely downloaded from one of the Internet sites selling academic papers? To complicate matters even more, cultural issues can arise with international students, who may have differing views of what it means to plagiarize. Educators who have little knowledge of a particular culture may be unable to differentiate a student who intentionally plagiarizes from one who is following a cultural norm (e.g., Chinese students consider using an expert's exact words as a sign of respect, not stealing; Xueqin, 2002).

These related problems of user authentication and academic integrity in online education are not unique to distance learners. Most institutions that offer face-to-face higher education do not ask their applicants to present a photo ID when applying; thus, they are unable to determine if the student who attends class is the same student who is officially registered for the course. Likewise, when graduate students submit papers in face-to-face classes, instructors rarely check to make sure that the student generated the work and not the student's spouse, for example. Moreover, no matter the type of course modality (face-to-face or online), the proliferation of the Internet has made it easy for students to simply claim the works of others as their own.

The fact that the two teaching formats—face-to-face and online—generally lack viable user authentication or plagiarism-detecting mechanisms may not be a satisfactory answer to most educators, as concerns over academic integrity violations increase. Therefore, this chapter introduces a viable set of strategies that an instructor can employ to help minimize and deter violations of academic integrity in the online environment. Our intent is to offer best practices in dealing with violations of academic integrity in online assessment and to address the pedagogical and technological implications of these practices.

Review of Effective Academic Practices

In an effort to deter plagiarism, many institutions of higher education have found that several academic practices can help them achieve that goal. The practices include proctored exams (i.e., administering exams in a secure environment in which students have to present proper identification to take the exam), systematic course design (i.e., creating a course structure that builds upon a student's own background and prior academic experience), portfolios (i.e., a collection of student work in a program or a course), and effective academic integrity policies.

Proctored exams. Prior to the advent of online learning, many universities offered paper-and-pencil correspondence courses as their method of offering education to distant learners. In this form of distance learning, students study lessons on their own at home and submit written homework and papers to instructors through regular mail. Instructors review and grade the work and then return it to the student through the mail. This form of distance learning is still a viable delivery mode in some institutions, but it is being rapidly replaced with online approaches to learning. However, these early adopters of distance modalities have provided today's online learning institutions with one solution to ensuring that exams are completed by the student who registered for the course: proctored exams at a distance.

Students enrolled in correspondence courses may be required to report to a preapproved site wherein a proctor will oversee the conditions under which the student takes the exam. Generally, the student will contact the instructor who will ensure that the exam is securely delivered to the preapproved site and proctor. The proctor keeps the exam under secure conditions until the student reports to take the exam. For example, the public library in a small town may, as a community service, serve as the approved site for local correspondence students from a variety of institutions to take exams. The exams are delivered to the library and staff members keep them securely stored until a student arrives to take the exam. The proctor verifies that the student who reports to take the exam is the student who is registered for the course (by checking photo identification) and then monitors the conditions under which the student takes the exam. When the student has completed the exam, the proctor returns the exam to the university for grading.

Some institutions that offer online courses have borrowed this methodology to ensure that, at the very least, examinations are authenticated as the registered student's original work. Faculty may be concerned that online discussions and papers submitted through e-mail may not necessarily be the student's work; thus, proctored exams are one possible remedy for this concern. While the idea of using proctors in this way may be appealing to some faculty and administrators, it is not without problems. First, if an online program has a substantial number of students who are in need of proctors each semester, a staff to oversee the entire process must be put in place; this can represent substantial overhead to the university. Second, the process of sending and receiving secure exams can be complex and fraught with errors. One misplaced exam can result in the need to create an entirely new exam, again at a cost to the institution. However, at institutions where faculty are concerned about authenticating distant learners' knowledge, the proctored exam may be a viable alternative.

Systematic course design. The great majority of institutions that offer online courses are concerned about ensuring student identity and reducing instances of plagiarism. However, proctored exams, as explained earlier, can result in expensive infrastructure and can be susceptible to implementation accidents. In addition, educators are increasingly focusing on authentic assessment rather than solely on the assessment of memorized information, as is common on most exams. Thus, a growing number of educators use systematic course design as an approach to decrease the number of instances of plagiarism occurring in online courses.

An increasing number of faculty members develop courses based on learning outcomes as a preferred way of designing educational experiences. Designing courses with learning outcomes outlined first allows for more authentic assessment activities to be designed into the course. Particularly for students in professionally based programs such as business, teaching, or counseling, the preferred forms of assessment are authentic in nature, including business plans, case studies, lesson plans, or patient intake summaries. These forms of assessment provide students with a way to integrate their newly acquired knowledge with actual or simulated experiences in their professional field. This type of assessment requires students to develop their ideas more fully and to link these ideas to current activities in ways that exams cannot. Thus, the chance of students plagiarizing decreases dramatically when they are asked to use their community, job, or family as the context for application of new learning. It is far more difficult to plagiarize when asked to integrate new knowledge with actual experience.

How might an instructor learn about online students' actual experience? Beginning on the first day of class, an online faculty member can structure course requirements to build upon students' introductions of themselves. For example, a student who notes that she works in the telecommunications industry will be asked to apply the course concepts to her industry and office environment. The student will analyze her work experience by utilizing theories read in the textbook and discussed in class (in discussion forums). Likewise, online discussion questions should prompt the student to integrate new material with personal work experiences. Throughout the semester, the online faculty member becomes knowledgeable about a student's work experiences. Faculty members who frame online discussions around application of learning to life experience also become familiar with their students' professional experiences. Final papers that require integration of course material with these experiences make it more likely that students will have completed their own work. It is nearly impossible to purchase a previously written term paper when the topic is specific to the student's experience.

Another approach that faculty members can use to decrease the opportunities for plagiarism is to design multistage assignments. For example, an instructor can require students to submit a linear progression of increasingly complex and developed outlines of the paper. An instructor might ask for a simple outline, then an annotated bibliography, then a first draft before the final version is submitted. Students who produce these parts of their final project are less likely to have plagiarized the work.

Systematic course design is primarily employed by instructors to deter plagiarism, but this technique does not deal with the issue of user authentication. That is, a student could pay another individual to produce each phase of the paper. A good online instructor should be involved and engaged enough with online students to notice any discrepancy in the students' online discussions and final written work. The astute instructor will notice a difference in the quality of work, thus prompting an investigation of the suspect written products.

Portfolios. An increasing number of universities are using the portfolio as a more authentic form of assessment for students' overall educational experience. The portfolio is a collection of students' best work over some period of time, highlighting learning outcomes that students have mastered and can demonstrate. For example, accounting students might collect samples of exceptional proposals, P and L statements, financial analyses, and other examples of competency in accountancy. Students may organize these in a binder or join the increasing number of students who are utilizing e-portfolios, or online versions of the more traditional paper-and-pencil portfolio.

More than a Web page, the e-portfolio allows students to store and organize an up-to-the-minute collection of artifacts of their best work. The primary benefits of the portfolio are the ability to assess the quality of an educational program by demonstrating student outcomes (e.g., improved critical thinking, improved writing), and the opportunity for students to demonstrate their best work to interested others. By providing instructors, future employers, and anyone else who is interested with a password, students can grant access to their portfolio only to those people they choose. Many course management systems (e.g., Blackboard) now include e-portfolio capabilities in their product releases. The use of e-portfolios is spreading quickly, and instructors must be prepared to help students use them effectively.

How can a portfolio decrease the chances that a student will plagiarize another's work? First, instructors may request to view a student's overall portfolio to reveal any discrepancies in the quality of work over many courses and disciplines. Thus, if a student submits a paper that is written at a level far above the typical work in his or her portfolio, an instructor might investigate the origins of the paper more closely. Second, an instructor may request that the student complete a portfolio for the course itself. This smaller version of the undergraduate portfolio requires students to compile their own work drawn from many sources. The more relevant the portfolio is to the student's life experiences, the less likely plagiarism can or will occur. Again, if students are required to document their work in starting a small business, for example, it is unlikely that they could buy such a plan from another source or have someone else write the material.

Effective academic integrity policies. Plagiarism and cheating are hot topics on campuses across the country, so much so that Duke University has established The Center for Academic Integrity (*www.academicintegrity.com*). The site assists colleges and universities in assessing the extent of plagiarism and cheating on their own campuses and provides resources to reduce the level of cheating. Of particular interest is the list of honor codes from a large number of colleges and universities across the country. For example, the Honor Pledge at the University of Maryland "is a statement undergraduate and graduate students should be asked to write by hand and sign on examinations, papers, or other academic assignments not specifically exempted by the instructor" (University of Maryland,). The Pledge reads,

> *I pledge on my honor that I have not given or received any unauthorized assistance on this assignment/examination.* (University of Maryland)

Many universities are requesting that students write a similar statement on every assignment they submit, be it in a face-to-face or in an online course. In fact, universities that have used honor pledges find that instances of plagiarism decrease (McCabe & Pavela, 2000*)*. The reason for this is because "public commitments, even seemingly minor ones, direct future action" (Cialdini, 2001, p.76). That is, individuals who publicly commit to certain behaviors are more likely to follow through.

Academic Integrity and Plagiarism Considerations

In addition to the aforementioned academic practices used to minimize or deter plagiarism, several considerations should be explored that provide a broader context for issues of academic integrity. Some of these factors can, by their very nature, help to minimize plagiarism, such as national board exams or copyright protection under the law. Other factors can actually contribute to a higher incidence of plagiarism (e.g., cultural differences).

National board exams. Student expectations for national board testing, Tulloch and Thompson (1999) assessed, can help deter instances of plagiarism, "as students know that they will be held accountable in future, highly secured testing situations" (p. 2). These expectations can also help decrease the number of student "imposters," because national board tests typically require proper (and multiple) identification documents. Thus, for example, students in a Pharm.D. program who need to pass board exams to qualify for professional practice are less likely to violate academic integrity. Tulloch and Thompson also found that the likelihood of plagiarism is also diminished in courses that serve as prerequisite courses for other courses. Students are less likely to cheat in those situations because they feel they will lack requisite knowledge to continue their studies in the courses that follow.

Cultural differences. Though Western educators grapple with a seeming epidemic of plagiarism, some cultures view the topic quite differently. These differences in cultural views of plagiarism may be heightened in the online environment because of the possibility of students from many different cultures enrolling in the same online course. As online programs break down the boundaries of geography, cultural differences among students can bring new challenges for educators. Online courses allow students from a variety of different cultures and time zones to learn together, thus setting up potential clashes of cultural norms.

For example, in China, copying from others' work has been an acceptable practice for many years (Xueqin, 2002). Chinese students reportedly find it honorable to copy works of national experts but find it insulting (to national pride) to copy works of Western experts (Xueqin, 2002). This "culture of copying" (Xueqin, 2002) may also be found in other nations; for example, public outrage in India forced officials to retract legislation banning plagiarism (Desruisseaux, 1999). Faculty members and administrators alike need to be aware of these cultural differences and take them into consideration when drafting academic integrity policies or resolving corresponding violations.

Legal considerations. U.S. copyright laws provide protection against plagiarism and as such can be used to help deter plagiarism. Plagiarism, under copyright law, is considered a misdemeanor and can be "punishable by fines anywhere between $100 and $50,000 – and up to one year in jail" (Plagiarism FAQs, 2003). However, faculty members should keep in mind that the copyright law does not protect all published work. Examples of noncopyrighted materials include work published prior to 1923, work published by the government, and work that is readily available (such as the standard calendars or weight charts), to name just a few. Furthermore, the U.S. copyright laws provide protection only in the United States. Though most countries do offer some form of copyright protection, there are no international copyright laws that offer universal protection. This is of special importance to U.S.-based distance learning faculty, because their students may not be located in the U.S., and the plagiarized work those students claim as their own may not have been published in the U.S. (or in the English language, for that matter).

Potential Technological Solutions

Technology plays an increasingly important role in the deterrence of plagiarism. Many technology companies that cater to higher education (e.g., creators of course management systems, productivity software suites, hardware vendors) are responding to the academic community's concerns about academic integrity by including proposed solutions in their software and hardware platform releases. Thus, a proliferation of assessment technologies, plagiarism-detection software products, and even computer security features have been recently offered to address academic integrity concerns.

Assessment technologies. As faculty members voice concerns that online, nonproctored assessments may contribute to an increased number of violations of academic integrity, manufacturers of assessment technologies continue to improve their software products to address those concerns. Although technologies that would prevent student imposters from taking a nonproctored exam (or even prevent collusion by limiting the number of people present while taking the exam) still do not exist, a myriad of features that are being incorporated into assessment technologies make it increasingly difficult for students to cheat during online assessments.

Security features of online assessment technologies include the following:

- **Time limitation:** Students are given only a certain amount of time to complete the exam. Faculty members can specify that the exam has to be taken on a given day (or days) and during a specified time frame. For example, an exam may be available only between 6:00 p.m. and 9:00 p.m., and students may only take up to 30 minutes to answer all questions on the exam. Setting this time limit fairly low can help minimize the "open book" atmosphere of online exams.

- **IP address protection:** Faculty members can specify the Internet Protocol (IP) addresses that are allowed to access the exam, thus limiting access to the exam to only specific computers. IP addresses serve as a unique identifier of a computer (although, e.g., with a dial-up Internet connection, it is likely that a different IP address is assigned to the same computer every time that computer connects to the Internet). This feature can be very useful if students are asked to complete the exam in a proctored environment.

- **Proctor password:** Faculty members can specify a password for the exam that is then given to the exam proctor. Without knowing the password, students are unable to access the exam.

- **Question randomization:** Faculty members provide question sets from which the software randomly selects questions for each student. This results in no two exams being identical.

- *Adaptive testing:* Some assessment technologies will allow faculty members to create a dichotomy within the exam that is based on a student's knowledge of the subject matter. For example, depending on whether a student answers a specific question correctly or incorrectly, the program will select subsequent questions. This method of adaptive testing is used in some standardized entrance exams, such as the Graduate Record Examination.

- **Restriction of computer use:** Some assessment technologies allow faculty to disable certain computer features, such as printing the exam (and therefore being able to distribute it to other students), copying and pasting into or from the exam, surfing the Internet or accessing other computer programs, and using special keys (such as CTRL-ALT-DELETE or ALT-F4 to exit the exam). These features allow faculty to improve the security of proctored exams and make it more difficult for students to violate academic integrity in the nonproctored environment.

- **Exam statistics:** After the student completes the exam, a faculty member can see how long it took a student to complete each question. However, reaching a conclusion that a particular student answered most questions in less time than it takes a good speed reader to actually read the text of those questions may not be sufficient to prove a violation of academic integrity; but it may help the faculty member realize that the exam (or answers thereto) are circulating among students and that is time to create a new exam.

Unfortunately, there does not seem to be an assessment technology that would include all of the aforementioned features into its capabilities and yet serve as a course delivery tool. For example, both Blackboard (*http://www.blackboard.com*) and WebCT (*http://www.webct.com*), as market leaders in the area of course management systems, incorporate only some of those security features into their assessment technologies. QuestionMark's Perception (*http://www.questionmark.com*) includes most of the aforementioned features and can be incorporated for use with Blackboard, WebCT, and eCollege, but it does present an added cost— in terms of software, hardware, and training and support for institutions.

Plagiarism-detection software and Internet resources. Many tools exist that can help faculty detect instances of plagiarism. One type of these tools compares strings of text (e.g., any 30 consecutive characters) found in student papers against the works of others (e.g., online journal databases or paper mills). Another type of plagiarism-detection tools uses the cloze procedure to remove every fifth word from a student's work; students are then asked to "fill-in-the-blanks" to prove the work is indeed theirs. Some software programs allow faculty to maintain collections of papers submitted for a particular class and compare newly submitted papers against those papers that were submitted in the past. Finally, Internet search engines can be an effective tool in detecting plagiarism.

Examples of tools that compare student work against previously published works include the following:

- Turnitin.com (*http://turnitin.com*) is arguably one of the most popular tools for plagiarism detection. It allows faculty to check student work for plagiarism by examining the Internet sources, academic journals and published books, and student work previously submitted to Turnitin.com. In addition to its plagiarism-detection service, Turnitin.com also features Peer Review (allowing students to comment on one another's papers), Grade Book (allowing faculty members to maintain an electronic class grade book), Digital Portfolio (allowing students to maintain an archive of their work), and GradeMark (allowing faculty to use a modified MS Word's track-changes tool to provide students with feedback on written assignments).

- MyDropBox.com (*http://www.mydropbox.com*) utilizes the plagiarism-detection methods similar to those of Turnitin.com by comparing submitted papers against Internet sources, academic journals and password protected sites, as well as its customers' databases of papers. Previously known as Plagiserve, the company "suffered from allegations that it is actually a front for Eastern European essay mills—selling copies of the essays that faculty submit for checking" (Humes, Stiffler, & Malsed, 2003). These allegations surfaced after the discovery that the principal of Plagiserve was connected to several paper mills (Young, 2002) for fear that papers submitted to Plagiserve for plagiarism detection would be sold through one of these paper mills.

- EVE2 (http://www.canexus.com/eve/index.shtml) stands for Essay Verification Engine, and it too compares student work with sources available on the Internet. When instances of plagiarism are suspected, EVE2 presents the faculty member with links to the Web site where similarities in text were found.

- MOSS (http://www.cs.berkeley.edu/~aiken/moss.html) or Measure of Software Similarity is used to compare computer programs written by students in several programming languages to those programs developed by others. It is therefore only used to detect plagiarism in computer programming classes.

- OrCheck (http://cise.sbu.ac.uk/orcheck/) is an originality checking desktop software that compares student papers against Google search engine results.

An example of a tool that uses the cloze procedure to detect plagiarism includes the following:

- Glatt Plagiarism Service (http://www.plagiarism.com/) eliminates every fifth word from a paper submitted and asks the student who submitted the paper to reinsert the eliminated words. This tool is particularly effective when the source of plagiarism is not known or when a faculty member fears that a bilingual or international student simply translated someone else's paper (because most other tools only compare documents written in the same language).

Examples of tools that compare student work against a faculty's own collection of past papers include the following:

- WCopyFind (http://plagiarism.phys.virginia.edu/Wsoftware.html) was developed by Lou Bloomfield, Professor of Physics at the University of Virginia. The tool identifies incidents of plagiarism by comparing student papers to a faculty's own collection of past student assignments or a faculty-specified Web site (as opposed to the entire Internet).

- WordCheck Keyword DB (*http://www.wordchecksystems.com/*), much like WCopyFind, will compare a student's paper to a collection of other papers located on the faculty member's own computer.

Examples of Internet search engines that compare student work with materials found on the Internet include the following:

- Google (*http://www.google.com*) can be used to check if certain phrases a faculty member suspects were plagiarized can be found on the Internet. Google's search results, due to the powerful technology used for searching the Internet, are also used by many other Internet search engines.
- Amazon.com (*http://www.amazon.com*) can be used to check if certain "suspicious" phrases can be found in book excerpts of this digital bookstore. If Amazon.com cannot find the phrase, it will parse the phrase to a traditional Internet search engine and will thus search the entire Internet as well.

An important consideration is that plagiarism-detection tools are just that—tools. They help find evidence of plagiarism, but it is still up to the individual faculty member to prove that plagiarism did in fact occur. For example, *The New York Times* reported that plagiarism-detection tools, when used to "inspect" several biographies of famous people, found "175 instances of plagiarism in the [Abraham] Lincoln biography, 200 instances of plagiarism in the [William Faulker] biography and 240 instances in the [Martin Luther] King biography" (Eakin, 2002, p.1). However, Eakin notes that despite technological tools identifying instances of plagiarism, no evidence was found to support the allegations of plagiarism against the author of the biographies, but rather only overreliance on documentation used.

Likewise, some faculty members question the effectiveness of the cloze procedure method in proving plagiarism because some paper mills direct plagiarists to revise their papers by using the cloze method to ensure that the plagiarized work reflects a student's own writing style. Furthermore, when relying solely on the cloze procedure, faculty members are not presented with links to original sources, as they are with other types of plagiarism-detection tools, making it difficult to prove plagiarism beyond a reasonable doubt.

Another important consideration when selecting (or using) plagiarism-detection tools is the Family Educational Rights and Privacy Act (FERPA). FERPA, as a federal law, protects the privacy of student educational records. Thus, faculty members, if using their own database of previously collected student work to detect plagiarism, may not be able to disclose the contents of their own collection when building a case against plagiarism without the express consent of the original author. Without being able to present the original work to the student found guilty of plagiarism, there may be little (if anything) left to actually prove plagiarism.

The costs associated with the use of these tools vary. Some tools are available for free, some tools charge a fee per each paper submitted, and some tools charge a flat fee for unlimited submissions. Many fee-based tools offer discounts when whole departments or institutions license their product. The fees range from 1 dollar per submission to thousands of dollars for unlimited submissions (and as such may be cost-prohibitive to individual faculty members, departments, or even institutions).

Computer security in user authentication. With the advent of the information age, concerns about user authentication have become more prevalent–and not only in higher education. The military, government, and corporate worlds all share a common concern for protecting the information that can be accessed digitally. Fears of "imposters" who access (and manipulate) information pretending to be someone else have been fostered by many real life examples of thieves, terrorists, and scam artists.

To protect information assets and ensure that the user of the information is legitimate, computer systems administrators typically impose security measures that fall into the following three disparate categories (Haag, Cummings, & McCubbrey, 2004):

- **What You Know:** Access to information is restricted by passwords or PINs and as such represents the weakest form of security. Survey results reported by Cushnie and Jones (2003) found that 90% of respondents were easily coaxed into giving away their computer passwords in exchange for a free pen.

- **What You Have:** Access to information is restricted by tangible items in one's possession—such as a key, magnetic stripe card (e.g., credit card), smart card, and the like. Unfortunately, these items can be lost, stolen, or counterfeited.

- **What You Are:** Access to information is restricted by using biometrics. Biometrics can be defined as using one's physical or behavioral properties for user authentication (Gallagher, 2001). Physical biometrics utilize fingerprint, hand print, iris of the eye, face, or voice analyses to identify a user. Behavioral biometrics analyzes a person's typing pattern (i.e., the pattern of pressing keys when typing on a keyboard) to determine the legitimacy of users. This is the most secure type of information protection.

In higher education, the use of biometrics to identify students has been rare, if at all. Most institutions do not even check photo IDs when admitting face-to-face students, let alone when admitting online students to their programs. However, if the concerns over violations of academic integrity continue to grow, that may change. Institutions could begin utilizing biometrics to attest that their online (and face-to-face) students are indeed who they claim to be.

This authentication of students can be achieved by either using physical or behavioral biometrics. However, it is important to note that no course management system (such as WebCT, Blackboard, or e-College) currently supports the integration of biometrics into their software platforms. When it comes to *physical biometrics* in online assessment, a student could be asked to provide a sample of their voice. This voice recording could then be attached to the student's transcript (assuming that paper-only transcripts will eventually cease to exist) and consequently be available to employers of those students. Faculty could (perhaps even randomly) call their students to either discuss projects or papers submitted or conduct oral examinations, using the "voice print" to verify the student's identity. By utilizing *behavioral biometrics*, a student's typing pattern could be tracked for consistency across online courses. This could then be compared to the typing pattern in a proctored setting to verify that the student really did the work.

Although the aforementioned possibilities are still not on the horizon of many institutions of higher education, the U.S. government and corporations have already made strides in utilizing biometrics to protect their information. For example, after an inmate managed to simply walk out of a Lancaster County, Pennsylvania, prison pretending to be someone else (who shared a close resemblance to the inmate), the prison started to use iris scans to identify all inmates (Isaacs & Cutts, 2002). Sprint, a telecommunications giant, provides its customers with the ability to bill calls made from any phone to their home phone number by using voice recognition. Customers are asked to speak a 10-digit number, which the computer system then compares to a previously saved template (Newham, 1996). "California, Georgia, and Colorado are [also] in varying stages of adding biometrics ... to driver's licenses" (Krebsbach, 2003, p.1) by utilizing fingerprint, face recognition technologies, or both. Likewise, Hewlett-Packard is including fingerprint scanning capabilities in the next release of its iPAQ handheld computer (Williams, 2002) as a way of deterring thefts and securing information on those devices. Finally, Microsoft also announced that it plans to improve security features of its Windows XP operating system by integrating "tamper-resistant biometric ID card[s]" (Mace, 2003, p.2) that use a combination of facial images, iris, or fingerprint scans to secure access to information.

As with any information technology, costs of these systems are not negligible. The systems range in price from approximately $100 for desktop fingerprint-recognition solutions, to more than $100,000 for voice recognition systems. In addition to these costs, institutions adopting biometric solutions also face training, support, and personnel costs associated with the implementation and maintenance of such solutions. With such high price tags, these systems may not even be on the horizon of most educational institutions.

Ethical and Legal Implications of User Authentication and Academic Integrity

Although online learning is a relatively new phenomenon in higher education, the ethical and legal implications surrounding user authentication and academic integrity are not new. The same issues surrounding online assessment have surfaced in face-to-face classes and older versions of distance learning (such as correspondence courses).

Ethical implications. When dealing with user authentication and academic integrity, one cannot help but ask questions such as "Do systems for user authentication and plagiarism detection impose more work on faculty?" "Are we punishing ethical students by imposing upon them systems (and obstacles) designed for the unethical students?" "Is it our role—as educators—to worry about user authentication and plagiarism?"

It is obvious that these questions lack simple answers. In an effort to address these questions, Raynolds, in "Ethics in Information Technology" (2003), suggested utilizing the following seven-step approach to ensure ethical decision making:

1.   **Get the facts:** Whether the facts pertain to an incidence of plagiarism or the role of a faculty member, it is important to consider all the facts. Violations of academic integrity, as prevalent as they are, have generated enough interest and press that situations facing both faculty and students are more than likely not unique, and consequently, one can find information about similar situations in newspaper articles, academic journals, and the like, that can guide decision making.

2.   **Identify stakeholders and their positions:** Stakeholders include students (those who plagiarize and those who do not), faculty members (whose workload may increase by tracking plagiarism), alumni (whose degrees may be devalued by plagiarists), community members (whose values may be hindered), and anyone else who is affected by violations of academic integrity.

3. **Consider the consequences:** The consequences and their impact on stakeholders need to be evaluated. For example, if a student fails a class due to a violation of academic integrity, the student's career may be affected; students who work in high security areas of the government may lose their clearance status if they are found guilty of plagiarism. Also, policies that require faculty to use plagiarism-detection software may add substantially to faculty workload.

4. **Weigh guidelines and principles:** Consider not only the institutional guidelines and principles, but also those guidelines and principles published by other institutions. (As previously mentioned, refer to sites such as Duke University's Center for Academic Integrity for examples of effective academic integrity policies.)

5. **Develop and evaluate options:** Identify several alternative solutions for the issue at hand. For example, if a student unintentionally plagiarizes, might the consequences be less severe than for a student who buys and submits a paper from an online paper mill? Can faculty be asked to submit only a subset of student papers to a plagiarism-detection service in order to reduce workload?

6. **Review your decision:** Decide which alternative identified in the previous step should be implemented. Does one solution best meet the needs of most stakeholders involved and or does it most closely align with the academic honesty philosophy of the institution?

7. **Evaluate the results of your decision:** Through follow up, evaluate if your solution has indeed achieved its intended outcome. Are fewer students plagiarizing work? Are faculty members spending too much time "policing" plagiarism?

Though simple, Raynolds' (2003) model can provide a framework for a structured decision-making process and help minimize oversight. This model, however, only provides a structure for reaching a resolution; it does not provide the answer. Therefore, the ethical dilemmas presented by the online assessment ought to be discussed by the academic community of individual departments, institutions, or in its entirety.

Legal implications. Aside from ethical considerations, institutions that engage in user authentication or utilize plagiarism-detection tools should also be aware of the legal implications surrounding the two issues. Under FERPA, students enjoy protection of privacy of their educational records. These educational records include "records, files, documents and other materials which contain information directly related to a student; and are maintained by the educational agency or institution or by a person acting for such agency or institution"

(FERPA, p. 1). Excepted from the definition of protected educational records is information contained in the institution's directory (e.g., student's name, campus e-mail address, enrollment status) that generally cannot violate a student's privacy or be harmful to the student if disclosed.

In user authentication, institutions must exercise care when collecting and maintaining any information about the students' physical or behavioral properties. For example, an institution cannot collect photographs or voice recordings of students and make those records accessible to the public without the express permission from those students. Again, unless it is a part of an institution's directory (and student photographs cannot be included in the directory because they are considered confidential), those pieces of information about a particular student cannot be disclosed without permission from the student.

Likewise, when it comes to plagiarism-detection tools, faculty members have to respect the privacy of a student's educational record if they decide to create collections of previous assignments and papers. For example, if a faculty member utilizes a plagiarism-detection tool to check a student's work against a database of previously collected assignments from other sections of the class, then the privacy of the original author has to be respected, and the original work cannot be disclosed without the express permission of the author.

As new tools and technologies become available, it is often not clear if they violate FERPA regulations. If in doubt, faculty should seek advice from the institution's FERPA officer (a role typically assigned to the registrar) to avoid FERPA violations and infringing upon a student's right to privacy.

Conclusion

Online learning, and consequently online assessment, pose seemingly new challenges for educators. However, it is important to keep in mind that the challenges associated with academic integrity are probably equivalent (though not necessarily the same), regardless of the teaching format (online or face to face). In the online environment, faculty members are concerned that students have instantaneous access to either purchased or "borrowed" materials and that the true identity of their students is difficult (if not impossible) to verify. In the face-to-face environment, however, we witness similar problems: cheating is considered to be an epidemic.

With those issues in mind, this chapter outlined some key considerations that can help deter (e.g., national board exams or prerequisite courses) or reinforce (e.g., cultural differences) violations of academic integrity. These issues not only provide a framework for violations of (or adherence to) academic integrity, but also help provide faculty members with the broader context within which plagiarism is more likely to occur. For example, as online assessment continues to break down the boundaries of geography, faculty members need to be aware of cultural differences that may help increase the number of academic integrity violations and adjust their course and assessment methodologies accordingly.

This chapter has also reviewed academic and technological solutions (and ethical and legal considerations attached thereto) for educators who are in the process of designing or delivering online courses, programs, or assessments. These solutions represent a broad range of alternatives aimed to provide choices that align with the culture of the department or institution, address specific concerns among the faculty, and utilize available levels of human and financial resources.

The number of technological solutions for plagiarism detection and user authentication will continue to proliferate in the areas of online learning, teaching, and assessment. However, these advanced technological solutions are likely to remain vulnerable to the creative mind of the cheater for a long time to come; even the most technologically protected and secured environments can be breached. Although technological solutions may intrigue faculty members, it is important to note that the effective instructor will be the one who helps diminish plagiarism through excellent teaching and assessment. The instructor who engages students with interesting, relevant, and learner-centered assignments and corresponding assessments and who takes the time to get to know students will continue to be the most effective tool in the fight against plagiarism.

References

Cialdini, R. (2001, February). The science of persuasion. *Scientific American,* 76-82.

Cushnie, L., & Jones, R. (2003, August 2). When careless talk costs a fortune. *The Guardian,* 2.

Desruisseaux, P. (1999, April 30). Cheating is reaching epidemic proportions worldwide, researchers say. *The Chronicle of Higher Education*, A45.

Eakin, E. (2002, January 26). Stop, historians! Don't copy that passage! Computers are watching. *The New York Times,* B6.

Family Educational Rights and Privacy Act, 20 U.S.C. § 1232g; 34 CFR Part 99. Retrieved June 20, 2004, from *http://www.fontanalib.org/ Family%20Educational%20Rights%20and%20Privacy%20Act.htm*

Gallagher, J. (2001, June). Biometrics: As plain as the nose on your face. *Insurance & Technology,* 39-42.

Haag, S., Cummings, M., & McCubbrey, D. (2004). *Management information systems for the information age* (4th ed.). New York: McGraw-Hill/Irwin.

Humes, C., Stiffler, J., & Malsed, M. Examining anti-plagiarism software: Choosing the right tool. Retrieved June 22, 2004, from *http://www.educause.edu/ir/library/pdf/EDU03168.pdf*

iParadigms. (2003). Plagiarism FAQs. Retrieved June 22, 2004, from *http://www.plagiarism.org/research_site/e_faqs.html*

Isaacs, L., & Cutts, B. (2002, March). Body language: Using biometric technology. *The American City & County*, 22-28.

Krebsbach, K. (2003, November). The rise of biometrics. *Bank Technology News,* p. 54.

Mace, S. (2004, February 24). Gates previews security projects. Retrieved May 15, 2004, from *http://www.pcworld.com/news/article/0,aid,114 916,00.asp*

McCabe, D., & Pavela, G. (2000, September/October). Some good news about academic integrity. *Change Magazine,* 32-39.

Newham, E. (1996, April). Knowing me knowing you. *Communications International,* 55-58.

QuestionMark. Press Release. Retrieved May 8, 2004, from *http://www.questionmark.com/us/news/pressreleases/ecollege _october_2003.htm*

Raynolds, G. (2003). *Ethics in information technology* (1st ed.). Boston: Course Technology.

Tulloch, J. B., & Thompson, S. (1999). The AGENDA. Retrieved May 1, 2004, from *http://www.pbs.org/als/agenda/ articles/testing.html*

University of Maryland. Honor pledge. Retrieved June 22, 2004, from *http://www.jpo.umd.edu/aca/honorpledge.html*

Williams, M. (2002, October 9). HP's new PDAs offer secutiy at your fingertips. Retrieved May 15, 2004, from *http://www.pcworld.com/news/article/ 0,aid,105764,00.asp*

Xueqin, J. (2002, May 17). Chinese academics consider a "culture of copying." *The Chronicle of Higher Education,* A45-A46.

Young, J.R. (2002, March 12). Anti-plagiarism experts raise questions about services with links to sites selling papers. *The Chronicle of Higher Education.*

About the Authors

Mary Hricko is an associate professor of libraries and information services at Kent State University (KSU), USA. She serves as the library director of the KSU Geauga campus library. She has published and presented numerous articles and papers on academic support services in distance education, information literary, and Web accessibility.

Scott L. Howell is the assistant to the dean for the division of continuing education at Brigham Young University (BYU), USA. He assisted BYU in launching its online learning and assessment initiative (1999-2003) as the director of a new center of instructional design (CID). Dr. Howell is widely published and respected for his work in distance education and online assessment. He received his PhD in instructional science, specializing in assessment and measurement, his MS in community education, and his BS in business management.

*   *   *

Jeanette M. Bartley is the associate vice president for continuing education, open and distance learning at the University of Technology, Jamaica (UTech). Her portfolio includes primary responsibility for facilitating non-traditional educational opportunities and virtual access to UTech, the leading poly-technical educational institution in the Anglophone Caribbean. With more than 22 years of collective experience working with Caribbean and North American institutions, Dr. Bartley offer her diverse expertise as educator, corporate trainer, consultant and project manager in facilitating distance learning, organizational development and human resource development. She has presented at many conferences and developed several in-house training manuals. Her professional affiliations have included membership in the World Council for Curriculum & Instruction, Jamaican National Education Advisory Committee for UNESCO, Jamaican Association for Open and Distance Learning, among others. Dr. Bartley's academic and professional involvement in the field of education has always been linked to the role of pioneer. Throughout her career history, there has been a pattern of breaking new ground and creating new paradigms. She considers it her mission and responsibility as a change agent. Her personal goals and values are therefore directed toward facilitating the improvement of human conditions and quality of life through education and training.

Betty Bergstrom, PhD, vice president of testing services, leads the content development, technical implementation, and measurement services teams at Promissor, USA. Her staff supports client programs for job analysis, virtual and in-person question development, building and updating computer-based tests (CBTs), statistical analysis of questions and tests. Dr. Bergstrom's teams also aid clients in the implementation and use of Promissor software systems, including item banking and CBT delivery. Dr. Bergstrom has research expertise in the areas of item response theory (IRT), equating, standard setting, computerized testing, and adaptive testing. Dr. Bergstrom earned her MS and PhD in measurement, evaluation and statistical analysis from the University of Chicago.

Cheryl Bielema, instructional development specialist, works with the Center for Teaching and Learning, Office of Academic Affairs, at the University of Missouri (UM)-St. Louis, USA. She conducts faculty workshops and orientation programs and leads evaluation studies of technology integration. Dr. Bielema teaches part time in the online adult education program in the UM-St. Louis College of Education. She earned advanced degrees in adult education and human resource development at the University of Illinois, Champaign-Urbana, where she studied factors affecting adoption of computer-mediated communication by faculty and students.

Bryan D. Bradley is a faculty development coordinator in the faculty center at Brigham Young University, Provo, Utah, USA, and is a specialist in issues concerning the assessment of student learning. Dr. Bradley has worked for over 20 years in the education, industry, and government services arena as a skills-training and performance-measurement consultant. His work and research interests include assessing learner performance at the higher levels of cognitive behavior. He holds a doctor of education degree in instructional technology and distance education from Nova Southeastern University, Fort Lauderdale, Florida.

Jake Burdick is the curriculum development director for University of Phoenix's School of Advanced Studies, USA, as well as a faculty member for that institution's online campus. He holds bachelor's and master's degrees from Northern Arizona University, where he focused on composition theory, English education, and creative writing. Currently, Jake is working with University of Phoenix in creating new assessment and development approaches to doctoral education, for which he was selected to present at the 2004 Trends in Higher Education conference in Phoenix, Arizona.

Marie A. Cini serves as the associate dean for the School of Business and Management at City University in Bellevue, Washington, USA. She is also the facilitator for the online task force at the university. Dr. Cini has extensive experience in distance education and in adult and continuing education. She has published articles on the pedagogy of online leadership education, on faculty perceptions of their experiences teaching in traditional and in accelerated formats, and on student satisfaction and learning in distance formats. Her work has been published in *The Journal of Public Management and Social Policy, The Journal of Leadership Studies,* and *The Journal of Excellent in Teaching.* She earned her PhD in social psychology from the University of Pittsburgh in 1994, with emphases in group processes and research methodology. Her awards include the Duquesne University Innovative Excellence in Teaching, Learning, and Technology award.

Brian F. Fox is an assistant professor in the business programs department at Santa Fe Community College in Gainesville, Florida, USA, where he teaches a variety of face-to-face and online courses in office and Internet technologies. His studies include an undergraduate degree from the University of Florida and a master of distance education degree from the University of Maryland University College. He has several presentations and publications to his credit.

John Fremer is a founder of Caveon Test Security, USA, a company formed in 2003 to help test program sponsors, testing agencies, states, school districts, and others to improve security practices in all phases of test development, administration, reporting, and score use. He has 40 years of experience in the field of test publishing and test program development and revision, including management level positions at Educational Testing Service and The Psychological Corporation/Harcourt. In his 35-year career at Educational Testing Service, Fremer led the ETS team that worked with the College Board to develop the current version of the SAT. Fremer also served as director of exercise development for the National Assessment of Educational Progress, and was director of test development for School, Professional, and Higher Education Programs. During 2000-2003, Fremer designed and delivered measurement training programs to international audiences for the ETS Global Institute. Fremer is a past president of the National Council on Measurement in Education (NCME) and a former editor of the NCME journal *Educational Measurement: Issues and Practice.* Fremer also served as president of the Association of Test Publishers (ATP) and the Association for Assessment in Counseling (AAC). He was co-chair of the Joint Committee on Testing Practices (JCTP) and of the JCTP work group that developed the testing-industry-wide *Code of Fair Testing Practices in Education*; one of the most frequently cited documents in the field of educational measurement. Fremer is a co-editor of *Computer-Based Testing: Building the Foundations for Future Assessments* (2002, Erlbaum.) and author of "Why use tests and assessments?" in the 2004 book, *Measuring Up: Assessment Issues for Teachers, Counselors, and Administrators.* John has a B.A. from Brooklyn College, City University of New York, where he graduated Phi Beta Kappa and Magna Cum Laude, and a PhD from Teachers College, Columbia University, where he studied with Robert L. Thorndike and Walter MacGinitie.

Jim Fryer, CPCU, director, Regulatory Professional Services, has over 17 years of experience in the insurance industry, provides strategic planning support for Promissor's, USA, regulatory professional services department. Dr. Fryer is also responsible for all aspects of Promissor's continuing education program, including provider, course, and instructor approvals, as well as the credit banking group. Before joining Promissor, Dr. Fryer was director of information technology education and continuing education at the American Institute for Chartered Property Casualty Underwriters (CPCU) and currently serves on the board of directors for the Insurance Regulatory Examiners Society (IRES) Foundation. Dr. Fryer earned his EdD in educational administration from Temple University.

Margaret Gunderson has worked within the field of distributed or distance education for over 17 years. Currently, she is the associate director of educational technologies at the University of Missouri-Columbia, USA. Her areas of expertise include instructional design, online course development, the effective use of educational technologies for teaching and learning, and evaluation of online education. In addition to her administrative role, Dr. Gunderson is also an adjunct faculty member for the Department of Educational Leadership and Policy Analysis in the College of Education at the University of Missouri-Columbia. Her educational background includes a doctorate in educational technology (with emphasis in instructional design), an MEd degree in counseling and personnel services, and a BS in education.

Eric G. Hansen is a development scientist in the Center for Assessment Innovations and Technology Transfer, in the Research and Development Division of the Educational Testing Service (ETS) in Princeton, New Jersey, USA. He has an undergraduate degree from Harvard and received a doctorate in Instructional Psychology from Brigham Young University. He has served as principal investigator on a variety of ETS and U.S. government-funded projects involving accessibility and technology in educational testing and instruction. Hansen was co-editor of the World Wide Web Consortium's *User Agent Accessibility Guidelines (version 1.0)* and contributes to the accessibility activities of the IMS Global Learning Consortium.

Beverly Hewett is an assistant professor of nursing at Idaho State University, USA. She is also the Learning Resource Center (LRC) coordinator for the department. As the LRC coordinator, she is responsible for, among other things, a 15 cpu computer lab for proctoring tests and for general use of students in the College of Health Professions. She has an MS in nursing from Idaho State University and is currently in the adult education PhD program at the University of Idaho.

Clark J. Hickman is the associate dean for continuing education and assistant professor in the division of educational psychology, research and evaluation in the College of Education at the University of Missouri-St. Louis, USA. He has worked in the field of continuing education as a conference coordinator, director, and associate dean for 27 years. He earned a masters in adult education in 1984 and a doctorate in educational psychology and research methods in 1993. Throughout his career, Dr. Hickman has attended and presented research at regional and national conferences on the topic of distance education. Currently, he is engaged in evaluating distance education courses and their implications for the changing roles of continuing education.

Bernadette Howlett is an instructional designer and research instructor with the Physician Assistant Program at Idaho State University. She has an MS in instructional and human performance technology from Boise State University and she is currently in the adult learning PhD program with the University of Idaho. Howlett has worked as an online course systems administrator and instructional designer in higher education since 1996, starting with one of the earliest online degree programs at Marylhurst University in Marylhurst, Oregon. She has delivered numerous conference sessions on securing online assessments and has published peer-reviewed papers related to online instructional design systems.

Robert R. Hunt is vice president for product development, general counsel, and co-founder of Caveon Test Security, USA, a company established in 2003 to provide innovative security products and consultation to the testing industry. Prior to Caveon, Mr. Hunt directed the design and development of performance-based Microsoft certification exams and programs while at Certiport, Inc.; held a variety of administrative positions in Utah institutions of higher education; and practiced law in both private and public settings. Mr. Hunt holds a law degree and a PhD in higher education leadership and policy from the University of Utah. Mr. Hunt and has written widely on higher education law and policy as well as the legal aspects of testing and assessment, and is a charter member and general counsel of the Performance Testing Council.

John Kleeman personally developed the first version of Questionmark testing and assessment software in 1988, when he founded the Questionmark company, USA. In addition to his current role as company chairman, he has been involved in the creation of e-learning standards. He led the BSI panel that created the British Standard, BS 7988: Code of Practice for the use of IT for the delivery of assessments. He has a first class degree in mathematics and computer science from Trinity College Cambridge and is a chartered engineer.

Katrina A. Meyer is currently associate professor of higher and adult education at the University of Memphis, USA, specializing in online learning and higher education. She is the author of *Quality of Distance Education: Focus on On-Line Learning*, a 2002 publication of the ASHE-ERIC Higher Education Report Series. Her articles on online learning have appeared in the *Journal of Asynchronous Learning Networks*, the *Online Journal of Distance Learning Administration*, *Planning for Higher Education,* and *TC Record*. For over three years she was director of distance learning and technology for the University and Community College System of Nevada. Prior to this, she served 8 years as sssociate director of academic affairs for the Higher Education Coordinating Board in the state of Washington and was responsible for technology planning and online learning issues.

Robert J. Mislevy is professor of measurement, statistics, and evaluation at the University of Maryland, College Park, USA, and before that was a distinguished research scientist at ETS. His research applies developments in statistical methodology and cognitive research to practical problems in educational measurement. His work includes a multiple-imputation approach for integrating sampling and psychometric models in the National Assessment of Educational Progress and, with Linda Steinberg and Russell Almond, an evidence centered assessment design framework. Dr. Mislevy received AERA's Raymond B. Cattell Early Career Award and the National Council of Measurement's Award for Career Contributions, and has been president of the Psychometric Society.

Chris Morgan and **Meg O'Reilly** are lecturers and educational designers at the Teaching & Learning Centre at Southern Cross University, Lismore, Australia. They have co-authored two books on assessment—*Assessing Open & Distance Learners* and, more recently, *The Student Assessment Handbook*.

Jamie Mulkey, EdD, is the senior director for Caveon Test Security Services, USA, where she is a security consultant to high stakes testing programs. Previously, Jamie was manager of exam design and management for Hewlett-Packard's Certified Professional program. With more than 17 years of professional experience, she has worked in a variety of certification and training development positions. Jamie has held a number of leadership positions in the testing industry including: chair of the board of directors for the Association of Test Publishers (ATP), ANSI Personnel Certification Accreditation Council, jCert Initiative, and a three-year appointment to the editorial board of *Certification Magazine*. Jamie currently sits on the board of directors for the Performance Testing Council (PTC) and is a research fellow for the American Society for Training and Development (ASTD). Jamie holds a doctorate in educational psychology and technology from the University of Southern California.

Joel Norris, manager, measurement services, has worked in the adult learning market as a business development project coordinator, product manager, and most recently as manager of measurement services at Promissor, USA. Mr. Norris has 8 years of experience in the assessment industry and has worked in both the academic and for-profit sides of the field. Mr. Norris gained his theoretical grounding in assessment at Washington State University, where he was appointed assistant director of writing programs after taking his degree. During that time, he oversaw a variety of qualitative assessment programs, including several highly innovative portfolio-based assessments of student writing and reading performance. Mr. Norris earned his MA in rhetoric and composition from Washington State University.

Joan Phaup, a former newspaper and radio journalist, handles corporate communications for Questionmark, USA. She edits the monthly Questionmark newsletter, which is read by educators and trainers in the United States and abroad.

Rachel F. Quenemoen is the senior research fellow for the National Center on Educational Outcomes (NCEO) at the University of Minnesota, USA. Ms. Quenemoen specializes in technical assistance, training, and networking for NCEO. She has worked for 25 years as an educational sociologist on educational change processes and reform efforts and on building consensus and capacity among practitioners and policy makers. As NCEO's technical assistance team leader, she works with multiple partners and collaborators to plan and carry out joint efforts to build the capacity of local, state, and national educators and other stakeholders. Her current research interests and publications include reporting of assessment results for all students, options for accountability indexes that include all students' performance, principles of inclusive assessment and accountability systems, and continuous improvement of assessment and accountability systems.

Richard Schuttler is a dean of the School of Advanced Studies at University of Phoenix, USA, where he oversees the Doctor of Management in organizational leadership and doctor of business administration degree programs. He is an *Organizational Troubleshooter,* with 20 years of diversified, domestic and international management and leadership improvement expertise within academia, federal and state governments, and Fortune 1,000 environments. He has mentored executives, faculty, and students from around the world in a variety of professional settings. He has an extensive and proven background applying the Malcolm Baldrige National Quality Awards criteria. He is the co-author of *Working in Groups: Communication Principles and Strategies.*

Eric Shepherd is president of Questionmark Corporation, USA, which has provided testing and assessment software to schools, universities, colleges, businesses and government organizations since 1992. He leads workshops on secure assessment technology around the world and on the role that assessments play in the learning process. He advises colleges, universities, corporations, and government agencies on implementation strategies for testing software.He assisted the IMS Global Learning Consortium and the Association of Test Publishers in developing recommendations for delivering and exchanging assessment content.

Sandra J. Thompson, PhD, is a research associate for the National Center on Educational Outcomes (NCEO) at the University of Minnesota, USA. Dr. Thompson has coordinated multiple research activities, including an online survey of state directors of special education on the inclusion of students with disabilities in state accountability systems. She has assisted several states in the design of inclusive assessment systems, developing tools for determining inclusive assessment practices in the areas of accommodations, alternate assessment, and universal design. Prior to joining NCEO, Dr. Thompson spent nearly 10 years with Minnesota's department of education, as a special education administrator focusing on educational experiences and outcomes for students with disabilities. She also spent 10 years as a special education teacher, working with students with developmental disabilities.

Martha L. Thurlow is the director of the National Center on Educational Outcomes at the University of Minnesota, USA. In this position, she addresses the implications of contemporary U.S. policy and practice for students with disabilities and English language learners, including national and statewide assessment policies and practices, standards-setting efforts, and graduation requirements. Dr. Thurlow has conducted research for the past 30 years in a variety of areas, including assessment and decision making, learning disabilities, early childhood education, dropout prevention, effective classroom instruction, and integration of students with disabilities in general education settings. Dr. Thurlow has published extensively on all of these topics, authoring numerous books and book chapters, and publishing more than 200 articles and reports.

Boris Vilic is currently the director of technology for the School of Leadership and Professional Advancement at Duquesne University, USA. In this capacity, he was responsible for the implementation of several award-winning online degree programs, as well as the creation of an array of online student services. Mr. Vilic is also the team leader for computer technology faculty in the school's undergraduate curriculum and was the recipient of the school's 2002 Distinguished Faculty Award. His areas of interest include online teaching and learning and utilizing computer technology to improve the efficiency of administrative processes in higher education. His areas of research include customer satisfaction, innovative technologies, and online teaching, and his work has been published in *Syllabus* and *Ed Journal*.

Index

A

academic
   dishonesty 331
   integrity 331, 342
      policies 343
accessibility 69, 215
accommodations 103
adaptive testing 349
adult professionals 46
affective assessment 175
alternative assessment 19
American Association for Higher Education
   (AAHE) 7
application program interfaces (APIs) 69
assessment 7, 67, 86, 133, 182, 201,
   263, 342, 348
assistive technology 108
Association of Real Estate License Law
   Officials 61
Association of Registered Diagnostic
   Medical Sonog 53
asynchronous learning networks (ALN) 23
authentic assessment 14, 19
authentication 284, 342

B

behavioral biometrics 353
best practices 118
biometrics 293, 353
Blackboard 345
Bloom's Taxonomy 169
braille 222

C

CATs 23
Center for Academic Integrity 346
CEO Forum on Education and Technology
   34
certification 52
cheating 301, 323, 346
classroom assessment techniques (CATs)
   23
cloze procedure 350
collaborative
   assessment 174
   learning 87
computer
   adaptive testing 11
   security 348
   use 349

computer-based tests 281
computerized adaptive tests (CATs) 58
conceptual framework 215
construct validity 187
constructive assessment 14
constructivist 87
content
    analysis 121
    validity 187
contextualized assessment 14
continuing education (CE) 61
copyright 284, 348
correspondence courses 343
cost 306
Council on Certification of Nurse
    Anesthetists 58
course
    evaluation 135, 266
    management systems 345
course-evaluation 133
courseware 51
criterion validity 187
culture 347
curriculum 168

D

data 67
    warehouses 11
Data Forensics™ 293
design alternatives 215
diagnostic 52
    assessments 49
digital generation 2
disabilities 103
distance education (DE) 134, 331
distance-learning environment 331

E

e-learning 46
eCollege 350
effectiveness 306
essay mills 350
ethics 307, 355
evaluation 8
evaluative classrooms 309

evidence centered assessment design
    (ECD) 214, 218
exam statistics 349
extended testing time 214
extensible markup language (XML) 71

F

Family Educational Rights and Privacy Act
(FERPA) 352
fingerprint 353
font enlargement 214
formative 13, 52
    assessments 49
frameworks 121

H

high-stakes tests 56
historical perspective 310
honor pledge 346

I

imposter 342
incidence report 283
individuals with disabilities 214
instructional design 324
IP address 349

L

LAN 57
language learners 103, 214
law 201
learner-centered 16
licensure 52
linguistic
    qualifiers 121
    simplification 223

M

magnetic stripe card 353
motivation 309

N

national board exams 347
National Research Council 14

No Child Left Behind 3
nonproctored 53

O

online
      assessment 4, 11, 20, 48, 300
      course 133
      dictionaries 214
      discussions 16, 120
      instructional systems 3
      learning 5, 87
      methods 280
      proctored testing 54
      test 156, 264

P

paper mills 350
password 345
pedagogical 13, 86, 302
peer
      assessment 14
      review 350
performance-based assessment 14
physical biometrics 353
plagiarism-detection tools 350
policing 310, 322
portfolio 14, 343
prerequisite courses 347
prevention approaches 319
proctor 57, 343
proctored exams 343
promissor 60

Q

question randomization 349
QuestionMark 350
quiz design 315

R

reduction 312
reflective journaling 14
reliability 190
rubric 121, 177

S

screen magnification 214
secure
      site 51
      tests 280, 283
security 193, 280
      safeguards 55
self-assessment 14
self-efficacy 309
skills-based certifications 183
stakeholders 355
standards 67
strategies 315
student services 332
summative 13
      assessments 49
supervision 283
support 63
synchronous assessment 12
systematic course design 343

T

teacher-centered behaviorist model 13
technological
      innovations 264
      solution 324
technology 312
      -mediated
            environment 16
            learning 7
            models 12
test
      blueprint 189
      cheating 280
      delivery anomalies 283
      security plan 290
testing-environment infrastructure 283
text-to-speech 214
trained test administrators 283
transformative assessment model 37
transformative model 37
Turnitin.com 350

U

U.S. Department of Commerce 47
universally designed assessments 103

V

validation of learning 17
validity 185, 215
video camera 57
virtual security 291
virtue approaches 317
visual alteration 222

W

Web 6
    Patrolling™ 294
    -based assessments 51
    -centric courses 3
    -enhanced courses 3
WebCT 350